MW01130284

The Life and Death of Stars

In this well-illustrated text, Kenneth R. Lang explains the life cycle of stars, from the dense molecular clouds that are stellar nurseries to the enigmatic nebulae that some stars leave behind in their violent ends. Free of mathematical equations and technical jargon, Lang's lively and accessible text provides physical insights into how stars such as our Sun are born, what fuels them and keeps them bright, how they evolve, and the processes by which they eventually die. The book demonstrates the sheer scope and variety of stellar phenomena in the context of the universe as a whole. Boxed focus elements enhance and amplify the discussion for readers who want more depth. Featuring more than 150 figures, including color plates, *The Life and Death of Stars* is a modern and up-to-date account of stars written for a broad audience, from armchair astronomers and popular-science readers to students and teachers of science.

KENNETH R. LANG is Professor of Astronomy at Tufts University. He is the author of many popular astronomy books including *The Cambridge Guide to the Solar System*, 2nd ed. (2011); *Sun, Earth, and Sky*, 2nd ed. (2006); and *Wanderers in Space* (1994). An expert in radio astronomy and astrophysics, his research examines how magnetic energy generates explosions on the Sun.

The Life and Death of Stars

KENNETH R. LANG
Tufts University, Massachusetts

32 Avenue of the Americas, New York NY 10013-2473, USA

Cambridge University Press is part of the University of Cambridge.

It furthers the University's mission by disseminating knowledge in the pursuit of education, learning, and research at the highest international levels of excellence

www.cambridge.org
Information on this title: www.cambridge.org/9781107016385

First published 2013
Reprinted 2013

Printed in the United States of America

A catalog record for this publication is available from the British Library.

Library of Congress Cataloging in Publication data
Lang, Kenneth R.
The life and death of stars / Kenneth R. Lang.
p. cm.
Includes bibliographical references and index.
ISBN 978-1-107-01638-5 (hardback)
1. Stars – Formation. 2. Stars – Evolution. 3. Stellar dynamics. I. Title.
QB806.L36 2013
523.8′8–dc23 2012033415

ISBN 978-1-107-01638-5 Hardback

Contents

Focus Elements

Tables

Preface

How did the Sun and other stars come into being, what keeps them hot and makes them shine, how do they change with time, and what will be their ultimate fate? These are questions of interest to people of all ages; this book, *The Life and Death of Stars*, provides a lively and comfortably accessible account of them.

It begins with a discussion of radiation, which carries a message from the stars and tells us just about everything we know about them. The text continues with a description of gravity, which rules the universe, and the motion that holds everything up. We then take a voyage inside the atom to discover the subatomic particles that govern how energy is liberated inside stars, including the related topic of radioactive transformation of the elements. Heat, temperature, and pressure also are vital to our understanding of the interiors of the stars and their birth, growth, and decay.

These fundamental physical concepts provide the foundation for what follows, which is the approach that George Gamow used more than a half-century ago in his classic account of *The Birth and Death of the Sun*. This book made a tremendous impression and inspired an entire generation, but many of its conclusions are completely out of date. Although consistent with what was known at the time, subsequent improvements in our knowledge have shown that Gamow was misled about the dominant nuclear reactions in the Sun, the course of stellar evolution, and the origin of the elements. However, he had a marvelous physical insight and applied fundamental physics to our understanding of the Sun, without an equation in sight.

This book, *The Life and Death of Stars*, is written in a light and friendly style that can be appreciated by all readers, without being unnecessarily weighed down by specialized material, scientific jargon, or mathematical equations. Throughout this book, the basic concepts are translated into a common language with apt, down-to-earth metaphors and analogies, making them accessible to general readers and adding to the material. The text also is humanized with historical anecdotes about significant contributors to our celestial science.

Separate focus elements enhance and amplify the discussion with interesting details. Vital facts and physical information are presented in numerous tables. The focus elements and tables will be read or used by an especially curious person or serious student; however, they do not interfere with the general flow of the text and can be bypassed by readers who want to follow the main ideas.

There also are excellent line drawings, prepared by Kacha Bradonjich, and stunning images from the ground and space that help cement our newfound knowledge. They help crystallize a new concept with a visual excitement that adds another dimension to our understanding.

The book provides a comprehensive account of the enormous recent advances in our detailed understanding of the Sun using instruments aboard spacecraft. Sound waves have been used to peer deep within the Sun, and invisible x-rays have been employed to investigate its million-degree outer atmosphere. The mismatch between the observed and expected amounts of the ghostlike neutrinos has been resolved using massive underground detectors. These results all serve to confirm and embellish our understanding of how energy is generated by nuclear reactions at the center of the Sun and transported to its glowing disk that warms our ground, lights our days, and sustains life on the Earth. *The Life and Death of Stars* also describes how explosions on the Sun and powerful gusts in its supersonic winds interact with our planet, threatening humans and satellites in nearby space.

In the past few decades, our knowledge of all the other stars also has expanded enormously. The book portrays the tremendous range in how bright, luminous, hot, big, and massive the stars are. It also describes the nuclear reactions that keep different stars hot and luminous and how this is related to their growth and transformation. We place the Sun within this story of stellar lives and demonstrate how the life and death of former stars, which lived and died before the Sun was born, resulted in the creation of elements required for the very existence of the Earth and people living on it.

Star birth and death are continuing before our very eyes. We can see how new stars arise from interstellar material and detect planets around those nearby. Stellar destinies are just as fascinating, for dying stars do not simply disappear. They are reborn in another form, as white dwarfs, neutron stars, or black holes.

This brings *The Life and Death of Stars* to the larger questions of what lies beyond the stars and how the first stars began. Here, the book provides a concise account of the observable universe, which was propelled into expansion by "the big bang." We are still immersed within its background radiation, which is now being scrutinized with instruments aboard spacecraft. The text then wonders how it all began and explores the ultimate destiny of the stars, when they all will cease to shine.

This book tells a story of discovery and the wonderful, exciting diversity of the stellar universe. It is an amazing collective portrait of birth, transformation, decay, and rebirth. *The Life and Death of Stars* also provides for readers the background needed for a greater understanding and appreciation of those inevitable, currently unknown, celestial discoveries that will unfold during their lifetime.

The author also writes more advanced texts that include mathematical equations and references to original research papers and comprehensive up-to-date reviews. For this complementary approach, the reader is referred to *Essential Astrophysics* (New York: Springer, 2013).

Special gratitude is extended to my friend and neighbor Paul Strauss for his encouragement and careful reading of the page proof.

Kenneth R. Lang
Tufts University
November 16, 2012

1

Light of the Sun

1.1 Ultimate Power

The Sun is the source of all of our power. Its radiation energizes our planet, warms the ground and sea, lights our days, strengthens our bodies, and sustains life on the Earth. Green leaves of growing plants absorb sunlight, giving them the energy to decompose atmospheric carbon dioxide into carbon and gaseous oxygen. The oxygen is liberated back into the atmosphere, where we breathe it, and the carbon is deposited in plants. When we burn the wood of a tree, the carbon reunites with atmospheric oxygen. Long-dead, compressed plants provide the petroleum, coal, and natural gas that energize the lights of our house and power the vehicles we drive while also increasing the amount of carbon dioxide in the atmosphere.

The Sun's warmth is the source of our weather and the arbiter of our climate, producing the winds and cycling the water from sea to clouds and rain. Hydroelectric power plants are energized by water running back to the sea. Wind power also is driven by the Sun. Uneven solar heating of different parts of the Earth produces the winds, which blow from hot to cold regions.

The Earth glides through space at exactly the right distance from the Sun for life to thrive on our planet's surface, whereas other planets in the solar system freeze or fry: We sit in the "comfort zone." Any closer and the oceans would boil away, as on Venus; farther out, the ground would be a frozen wasteland, resembling Mars, which is now in a global ice age. We receive exactly the right amount of energy from the Sun to keep most of our water in liquid form, which is a requirement for life as we know it. Turn off the Sun's powerhouse and in only a few months, we would all be under ice.

The rising Sun has been celebrated since ancient times, in many religions, and by artists and poets. Its radiance dispels the darkness, brings us joy, and warms our souls. A smile comes to our face when the Sun returns, in a moment captured in the lyrics of a popular Beatles song:

Here comes the Sun
Here comes the Sun and I say
It's all right[1]

1.2 The Closest Star

How far away is the Sun? The mean distance separating the Earth and the Sun is known as the *astronomical unit* (AU) and it provides the crucial unit of planetary distance. Yet, for a very long time, no one knew exactly how big it was. We now know that this distance is 150 million kilometers, but it took a long time to find that out.

By the end of the seventeenth century, astronomers and other scientists had a good understanding of how the planets move around the Sun, but they could produce a scale model of the solar system that only provided relative distances of the planets from the Sun. The true distances and speeds of motion of the planets remained unknown.

It is no wonder then that obtaining a precise value for the Sun–Earth distance played an important role in the astronomy of the eighteenth and nineteenth centuries. The quest for accuracy in its measurement involved hundreds of trips to remote countries, tens of thousands of observations and photographs, and the lifetime work of several astronomers. They first determined the separations of the Earth and a nearby planet, such as Venus or Mars, and then used this planetary distance to infer the separation of the Earth and the Sun.

The distance of a nearby planet can be estimated by measuring the angular separation in the apparent direction of the planet when observed simultaneously from two widely separated locations on the Earth. This angle is known as the *parallax*, from the Greek *parallaxis* meaning the "value of an angle." If both the parallax and the separation between the two observers are known, then the distance of a planet can be determined by triangulation. This is based on the geometric fact that if we know the length of one side of a triangle and the angles of the two corners, then all of the other dimensions can be calculated.

The parallax technique to estimate a planet's distance is similar to the way our eyes infer how far away objects are. To see the effect, hold up a finger in front of your nose. First look at your finger with one eye open and the other eye closed; then alternate the open and closed eyes. Any nearby background object at one side of your finger seems to move to the other side, making a parallax shift. When this is repeated for a more distant object, the shift is less. In other words, the more distant an object, the smaller the parallax shift, and vice versa.

Because angular measurements were involved, the AU was naturally specified by an angle called the *solar parallax*, which is defined as half of the angular separation of the Sun as viewed from opposite sides of the Earth. More than a century of estimates for the solar parallax are shown in Figure 1.1 and discussed in Focus 1.1.

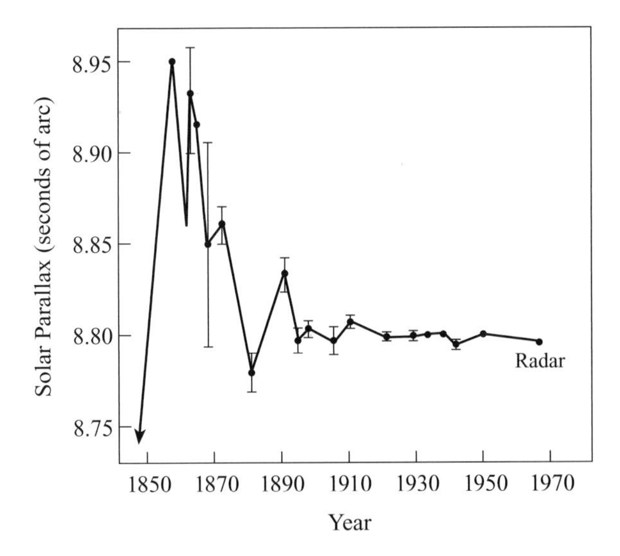

Figure 1.1. Distance to the Sun. Values of the solar parallax obtained from measurements of the parallaxes of Venus, Mars, and the asteroid Eros between 1850 and 1970. The solar parallax, designated by $\pi_{\odot}$, is half the angular displacement of the Sun viewed from opposite sides of the Earth. The error bars denote the probable errors in the determination; the points for 1941, 1950, and 1965 all have errors smaller than the plotted points. In the 1960s, the newly developed radar (i.e., radio detection and ranging) technology enabled determination of the Sun's distance with an accuracy of about 1,000 meters. The radar value of the solar parallax is 8.79405 seconds of arc.

Focus 1.1 Distance of the Sun

The distance separating the Earth and the Sun, known as the astronomical unit (AU), is determined by first estimating the distance between the Earth and a nearby planet. This planetary distance then can be used to specify the AU. The distance of Venus from the Sun, for example, is equal to one half of the difference between the Earth and Venus when it is closest and farthest away, on the other side of the Sun. When the Venus–Sun distance is known, we can infer the distance of any other planet from the Sun using Kepler's third law (see Section 2.1), which relates the orbital periods and orbital distances of the planets.

For more than a century, the distances of Venus and Mars were determined by triangulation from different points on the Earth. It involved measurements of the parallax, or angular difference in the apparent direction of the planet, as observed from widely separated locations.

In 1672, Giovanni Domenico Cassini, an Italian astronomer and the first director of the Paris Observatory, obtained an early triangulation of Mars, combining his observations from Paris with those taken from Cayenne, French Guiana. The planet was then in opposition, at its closest approach to the Earth. From the two sets of observations, obtained 7,200 kilometers apart, it was possible to estimate the distance to Mars and to infer a value of 139 million kilometers for the AU.

(continued)

Focus 1.1 *(continued)*

Astronomers in the eighteenth and nineteenth centuries attempted to improve the measurement accuracy of the Sun's distance during the rare occasions when Venus crossed the face of the Sun in 1761, 1769, 1874, and 1882. The method also involved comparison of observations from widely separated locations to determine the distance by triangulation. Subsequent measurements of the distance to the nearby asteroid, named 433 Eros, during its closest approaches to the Earth, resulted in an estimated 150 million kilometers for the AU.

Significant improvements in the precision of planetary distances came in the late 1960s by bouncing pulsed radio waves off of Venus and timing the echo. The roundtrip travel time – about 276 seconds when Venus is closest to the Earth – was measured using accurate atomic clocks. A precise distance to Venus then was obtained by multiplying half of the roundtrip time by the speed of light. The distance of Venus from the Sun is equal to one half of the difference between the Earth and Venus when it is closest and farthest away, on the other side of the Sun. The value of the AU, inferred from the radar determination of the distance of Venus, is 149,597,870 kilometers, with an accuracy of about 1 kilometer; for the accuracy required in most astronomical calculations, 1 AU is equal to 149.6 million kilometers.

Nowadays, the Earth–Sun light travel time – or the time for light to travel across 1 AU – is given as a primary astronomical constant and has the exact value of 499.004782 seconds. The AU is equal to the product of this travel time and the speed of light, which moves at precisely 299,792.458 kilometers per second.

The time for light to travel across the 1-AU distance is used now as a primary astronomical constant, and it is approximately 499 seconds(s), which corresponds to an AU of about 149.6 million kilometers – approximately 10,000 times the diameter of the Earth. Once an accurate value for the Sun's distance is obtained, the Earth's mean orbital velocity can be determined by assuming – to a first approximation – a circular orbit and dividing the Earth's orbital circumference by its orbital period of one year. The Earth's orbital velocity is 29,800 meters per second, abbreviated as 29,800 m s^{-1}. This is equivalent to approximately 107,000 kilometers per hour, much faster than a vehicle moves on a highway.

By way of comparison, it takes 4.24 years, or approximately 134 million seconds, for light to travel from the next nearest star other than the Sun. This star is called Proxima Centauri and it is located at a distance of 268,000 AU.

Therefore, the Sun is about a quarter-million times nearer to the Earth than the next nearest star. Because of this closeness, the Sun is approximately 100 billion times brighter than any other star. This brilliance and proximity permit detailed investigations that are not possible for any other star. As a result, studies of the Sun provide the foundation and benchmark for an understanding of other stars.

Because of its proximity, we are linked more closely to the life-sustaining Sun than to any other star. Although our unaided eyes can see about 6,000 stars in the night sky – and telescopes reveal hundreds of billions of them – our own daytime star, the Sun, is special;

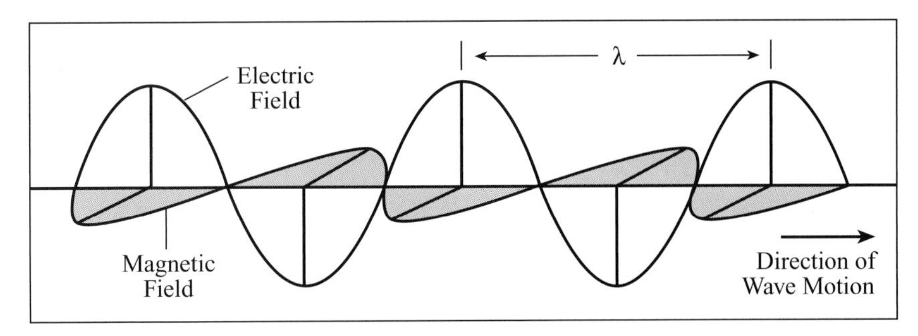

Figure 1.2. Electromagnetic waves. All forms of radiation consist of electrical and magnetic fields that oscillate at right angles to each other and to the direction of travel. They move through empty space at the speed of light. The separation between adjacent wave crests is called the *wavelength of the radiation* and usually is designated by the lowercase Greek letter lambda, λ.

nothing else in the universe is so critically important to us. As the English poet Francis William Bourdillon wrote:

> The night has a thousand eyes,
> And the day but one;
> Yet the light of the bright world dies,
> With the dying Sun.
>
> The mind has a thousand eyes,
> And the heart but one;
> Yet the light of a whole life dies
> When love is done.[2]

The Sun draws us to the other stars, and to understand how the Sun or any other star operates, we must examine its radiation, which spreads out and carries energy in all directions.

1.3 Waves of Light

Our perception of the universe is governed almost solely by the electromagnetic radiation received from cosmic objects. This radiation, which is the primary way that energy is transported in the universe, is known as *electromagnetic radiation* because it propagates by the interplay of oscillating electrical and magnetic waves.

In common with any wave, electromagnetic radiation has a wavelength, denoted by the lowercase Greek letter λ; different types of electromagnetic radiation differ in their wavelength, although they propagate at the same speed. Like waves on water, radiation waves have crests and troughs; but, unlike water waves, electromagnetic waves can propagate in empty space. A wavelength is the distance between successive crests or successive troughs (Fig. 1.2) and usually is measured in units of a meter, abbreviated m.

We are all familiar with the colorful display of a rainbow, which is sunlight bent into separate wavelengths by droplets of water. A crystal chandelier or compact disk similarly display the spectrum of visible light, arranging the colors by their different wavelengths. From short to long waves, the colors in the spectrum of visible light correspond to violet,

Table 1.1. *Approximate wavelengths of the colors*[a]

Color	Wavelength (nm)
Violet	420
Blue	470
Green	530
Yellow	580
Orange	610
Red	660

[a] The wavelengths are specified in nanometers, in which 1 nanometer, abbreviated nm, is equal to 1/billionth of a meter, or 1 nm $= 10^{-9}$ meters. Another wavelength unit used in optically visible light astronomy is the *Ångström*, abbreviated Å, where 1 Å $= 0.1$ nm $= 10^{-10}$ meters.

blue, green, yellow, orange, and red (Table 1.1). Light from the Sun or a light bulb often is called *white light* because it contains all of the colors, whereas black denotes the absence of color, when we see no light.

Sometimes radiation is described by its frequency, which indicates how fast the radiation *oscillates*, or moves up and down. The frequency of a wave is the number of wave crests passing a stationary observer every second.

In 1865, the Scottish scientist James Clerk Maxwell derived the equations that describe electromagnetic waves. He demonstrated that all waves move with the speed of light, which is designated by the symbol c. Two years later, the German physicist Heinrich Hertz extended Maxwell's results by building equipment to produce and detect certain invisible electromagnetic waves, which are now known as *radio waves*. In his honor, the unit of frequency is named the *Hertz*, abbreviated Hz; 1 Hz is equivalent to 1 cycle per second. Radio stations, for example, are denoted by their call letters and the frequency of their broadcasts in MHz, in which 1 MHz equals 1 million Hz.

The product of wavelength and frequency is equal to the speed of light; therefore, radiation at a shorter wavelength has a higher frequency and a longer wavelength has a lower frequency. All electromagnetic waves, regardless of wavelength or frequency, travel through empty space at the same constant speed – that is, the speed of light, at precisely 299,792,458 m s^{-1}. This speed is independent of reference in space and time, and it is the maximum speed allowable anywhere (Focus 1.2).

Focus 1.2 Light, the Fastest Thing Around

It was once thought that light moves instantaneously through space, but we now know that it travels at a very fast but finite speed, which was first inferred from observations of Jupiter's moon Io in the seventeenth century. The king of France had directed Giovanni Domenico Cassini, director of the Paris Observatory, to use such observations to improve knowledge of terrestrial longitude and maps of France. The Danish astronomer Ole Rømer, who also worked at the observatory, and Cassini noticed a varying time between eclipses of Io by Jupiter. Although Io's orbital period was approximately 42 hours, the duration of the orbit seemed to decrease when Jupiter was closer to the Earth and increase when it moved away, with a total time difference of approximately 22 minutes. Both Cassini and Roemer concluded that it was not Io's orbit around Jupiter that varied but rather the time it took Jupiter's light to reach the Earth when our planet was in different parts of its orbit around the Sun and closer to or farther from Jupiter.

Neither astronomer gave a value for the speed of light, which would have been equal to the diameter of the Earth's orbit divided by the time difference – or approximately 227,000,000 m s^{-1}.

More refined laboratory measurements during subsequent centuries indicated that light is always moving at a constant speed with the precise velocity of $c = 299{,}792{,}458$ m s^{-1}. Light emitted by any star moves through empty space for all time, never stopping or slowing down and never coming to rest. Moreover, nothing outruns light; it is the fastest thing around.

1.4 Invisible Rays

There is much more to the observable universe than meets the eye. In addition to visible light, there is invisible radiation. The invisible domains include infrared and radio waves – with wavelengths longer than that of red light – and ultraviolet rays, x-rays, and gamma rays, whose wavelengths are shorter than violet light. They all are electromagnetic waves and part of the same family, moving in empty space at the speed of light, but we cannot see them.

A display of the intensity of radiation as a function of wavelength is known as a *spectrum* – Latin for "appearance" or "apparition" – and the electromagnetic spectrum describes the types and wavelengths of electromagnetic radiation (Fig. 1.3). From short to long wavelengths, the electromagnetic spectrum includes gamma rays, x-rays, ultraviolet radiation, visible light, infrared radiation, and radio waves. Our eyes can detect a narrow range of wavelengths, which include the visible colors, and these wavelengths comprise only one small segment of the much broader electromagnetic spectrum. This band of light also is termed *visible radiation* to distinguish it from *invisible radiation*, which cannot be seen by the human eye. The most intense radiation of the Sun – and many other stars – is emitted at these wavelengths, and our atmosphere permits it to reach the ground.

In 1800, the German-born English astronomer William Herschel discovered infrared radiation when he put a beam of sunlight through a prism to spread it into its spectral components. He noticed that an unseen portion of sunlight warmed a thermometer placed beyond the red edge of the visible spectrum. The thermometer recorded higher temperatures in the invisible infrared sunlight than in normal visible sunlight. Herschel called them "calorific rays" because of the heat they generated; the term *infrared* did not appear until the late nineteenth century.

Humans "glow in the dark," emitting infrared radiation, but cannot see the heat; it is outside our range of vision. The military uses special infrared detectors to sense and locate heat generated by the enemy in total darkness. Soldiers can locate the enemy at night by

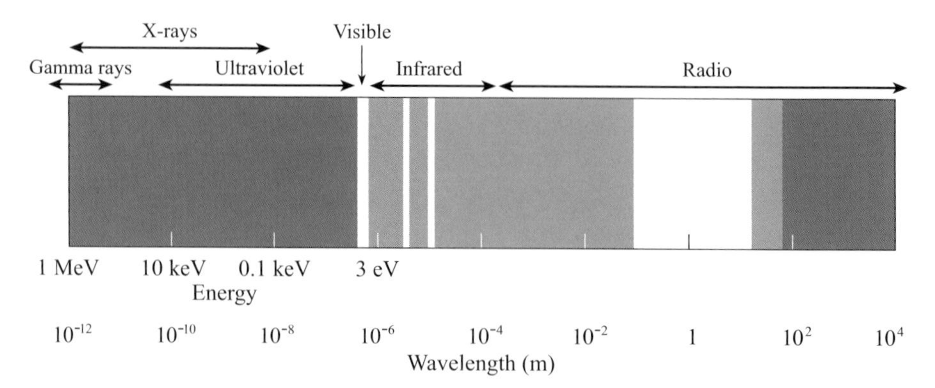

Figure 1.3. Electromagnetic spectrum. Radiation from cosmic objects can be emitted at wavelengths from less than 10^{-12} m to greater than 10^4 m, where m denotes meters. The visible spectrum that we see with our eyes is a very small portion of the entire range of wavelengths. Lighter shading indicates a greater transparency of the Earth's atmosphere to cosmic radiation. It only penetrates the Earth's atmosphere at visible and radio wavelengths, respectively represented by the narrow and broad white areas. Electromagnetic radiation at short gamma-ray, x-ray, and ultraviolet wavelengths, represented by the dark areas, is absorbed in our atmosphere; therefore, the universe is now observed in these spectral regions from above the atmosphere in Earth-orbiting satellites.

using night-vision goggles with infrared sensors that detect a person's heat, and spy satellites use infrared telescopes to detect heat radiated by rocket exhaust and large concentrations of troops and vehicles.

Most ultraviolet and infrared radiation coming from cosmic objects does not pass through our atmosphere, and all cosmic x-rays and gamma rays are absorbed there, so these rays never reach the ground. To look at the universe with these invisible wavelengths, we must loft telescopes above the obscuring atmosphere. This was done first by using balloons and sounding rockets, followed by Earth-orbiting satellites with telescopes that view the cosmos at invisible ultraviolet, infrared, x-ray, and gamma-ray wavelengths.

Radio waves are the only type of invisible radiation that is not absorbed in the Earth's atmosphere, but they are too long to enter the eye and not sufficiently energetic to affect vision. Radiation at radio wavelengths even can pass through rain clouds; therefore, the radio universe can be observed on cloudy days and in stormy weather just as a car radio or cell phone works even when it is raining or snowing.

1.5 The Radiation Energy of the Sun

The Unit of Energy

The *joule* is the unit of energy in the International System of units, abbreviated as SI from the French *Système International d'unités*. The SI unit of energy is named for the English physicist James Prescott Joule. When an SI unit is used in written English, it begins with a lowercase letter (e.g., joule) but it is abbreviated with an uppercase first letter (e.g., J).

One joule is the work required to produce 1 watt of power for 1 second; therefore, a power of 1 J s^{-1} is equivalent to 1 watt.

One joule is twice the kinetic energy of a mass of 1 kilogram, abbreviated 1 kg, moving at a speed of 1 meter per second, or 1 m s^{-1}. This amount of energy is a small number relative to the mass and speed of cosmic objects. The Sun, for example, has a mass of about 2,000 billion billion billion kilograms, or 2×10^{30} kg, and moves through space at a speed of approximately 220,000 m s^{-1}.

Even an ordinary table lamp with a 100-watt light bulb uses only 100 J s^{-1}, whereas the Sun liberates significantly more power: some 382.8 million billion billion J s^{-1}, written as 3.828×10^{26} J s^{-1}.

The Solar Constant

As solar radiation spreads out into space, it is dispersed into an ever-increasing volume; the distant Earth therefore collects only a small fraction of the total energy radiated by the Sun. The *solar constant* specifies the amount of the Sun's radiation that arrives at our planet. It is denoted by the symbol $f_{\odot}$ and is defined precisely as the total amount of radiant solar energy per unit time per unit area reaching the top of the Earth's atmosphere at the Earth's mean distance from the Sun. (Any physical parameter of the Sun is denoted by a subscript $\odot$, a circle with a dot at the center.) Artificial satellites have been used to accurately measure the solar constant, obtaining a value of $f_{\odot} = 1{,}361$ J s^{-1} m^{-2}.

The Sun's Luminosity

The intensity of radiation per unit area falls off as the inverse square of the distance from the source of radiation. This effect is observed when watching the increased brightness of a vehicle's headlights as it approaches and its distance decreases, or when watching the dimming taillights as the vehicle moves away to greater distance.

We can use the solar constant and the Earth–Sun distance to determine the total amount of energy radiated by the Sun every second. At the Earth's mean distance of 1 AU from the Sun, the solar radiation per unit area is diminished by $4\pi\,(\mathrm{AU})^2$, which is the surface area of a sphere at this distance. We therefore infer the Sun's luminosity, denoted $L_{\odot}$, by multiplying the solar constant with this area to obtain $L_{\odot} = 4\,\pi\,f_{\odot}\,(\mathrm{AU})^2 = 3.828 \times 10^{26}$ J s^{-1}, where 1 AU $= 1.496 \times 10^{11}$ m.

1.6 The Size and Temperature of the Sun's Visible Disk

Any incandescent body shines because it is hot. The wire filament in an incandescent light bulb, for example, is heated to a white-hot temperature of about 3,000 K to produce its luminous glow. As it turns out, the visible solar disk is just about twice that hot, and its much greater luminosity is due to its vastly larger size.

The solar radius, denoted $R_{\odot}$, can be determined from observations of the Sun's angular size and distance (Fig. 1.4), and these measurements indicate that the solar radius $R_{\odot} = 6.955 \times 10^{8}$ m, which is 109 times the radius of the Earth. With a radius of nearly 0.7 billion meters, or 700 million kilometers, the Sun could contain a million planets like ours.

Once we know the radius and luminosity of the Sun, we can determine the effective temperature of the Sun's visible disk. The luminosity of any hot, incandescent gas increases with the square of the radius and the fourth power of the temperature, according to a famous

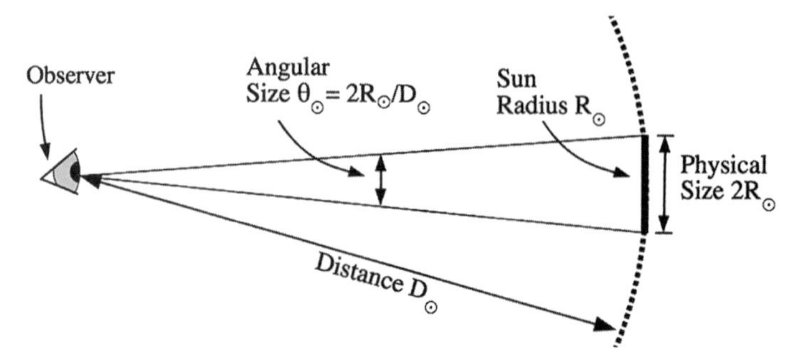

Figure 1.4. The Sun's angular size and radius. The solar radius can be determined from the Sun's angular size and distance. As long as this angle is small, the physical size is only a small arc of a large circle, denoted by the dashed line, and the angular size is the ratio of the physical size to the distance. Astronomers specify this angle as a partial arc of a full circle of 360 degrees; for the Sun, it is about 32 minutes of arc, in which there are 60 minutes of arc in 1 degree. This angle has been enlarged to display it in this illustration. In mathematics, the *radian* is the standard unit of angular measure. It describes the angle subtended by a circular arc as the length of the arc divided by the radius of the arc. When the arc length is equal to the arc radius, the angle is 1 radian. We can convert between the two methods of describing angles by noting that the circumference of a circle is 2π times its radius; therefore, 1 radian is equal to 360 degrees/(2π), or 57.2958 degrees. For the Sun, the angular size $\theta_\odot = 2R_\odot/D_\odot$ radians, where $R_\odot$ denotes the Sun's radius and the mean distance of the Sun, $D_\odot$, is 1 AU. The observed angular size of the Sun corresponds to a radius of 695.5 million meters.

expression known as the *Stefan–Boltzmann law*. It means that a bigger and/or hotter star shines with greater energy per unit time, which is analogous to saying a bigger or hotter fire burns brighter. The relevant numbers for the Sun's absolute luminosity and radius indicate that the visible solar disk has a temperature very close to 6,000 K and precisely 5,780 K.

No solid material can exist at this temperature, not even the wire filament of a light bulb; the material will melt and completely evaporate when made this hot. This is what happened at the visible solar disk, where all elements are present in gaseous form.

Incidentally, astronomers use the kelvin temperature scale that starts from absolute zero, which is the temperature at which atoms and molecules cease to move. The unit for this scale is written kelvin, without an uppercase K, or is simply denoted by an uppercase K. Water freezes at 273 K and boils at 373 K.

Any hot gas with the radius and disk temperature of the Sun will emit the Sun's luminosity. This emission, like that of any hot gas, is known as *thermal radiation*.

1.7 Thermal Radiation

An ideal thermal radiator is known as a *blackbody*. By definition, a blackbody absorbs all of the radiation that falls on it and reflects none – hence, the term *black*. A black shirt similarly will absorb most of the visible sunlight falling on it and reflect no colors.

Thermal radiation is emitted by a gas in thermal equilibrium and arises by virtue of an object's heat, or temperature. A single temperature characterizes thermal radiation.

Any hot gas that is in *thermal equilibrium*, with a temperature above absolute zero, will attempt to radiate away its energy. The emission from such a thermal radiator is found at all

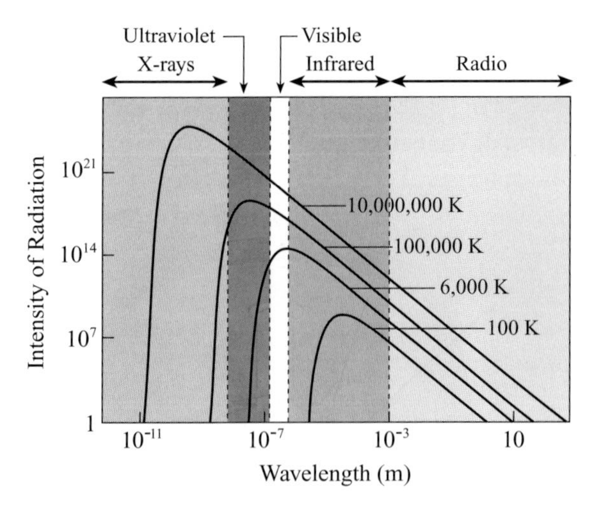

Figure 1.5. Blackbody radiation. The spectral plot of blackbody radiation intensity as a function of wavelength depends on the temperature of the gas emitting the radiation. The German physicist Max Planck (1858–1947) derived the formula that describes the shape and peak of this spectrum in 1900, proposed that the radiation energy was quantized, and provided a foundation for quantum theory. At higher temperatures, the wavelength of peak emission shifts to shorter wavelengths, and the thermal radiation intensity becomes greater at all wavelengths. At a temperature of 6,000 degrees on the kelvin scale, or 6,000 K, the thermal radiation peaks in the visible, or V, band of wavelengths. A hot gas with a temperature of 100,000 K emits most of its thermal radiation at ultraviolet wavelengths, whereas the emission peaks in x-rays when the temperature is 1 million to 10 million K.

wavelengths, or frequencies, but with a varying intensity that depends on the temperature (Fig. 1.5). As the temperature increases, more energy is radiated at all wavelengths. Moreover, the wavelength of maximum radiation shifts toward the shorter wavelengths when the temperature rises.

Because the emission of thermal radiation is present at all wavelengths, astronomers say that it emits a continuum spectrum. A display of its radiation intensity as a function of wavelength, known as the *spectrum*, shows no gaps, breaks, or sudden increases or decreases. It is an unbroken continuum ascending to peak intensity and then descending again as the wavelength increases.

No real object emits a perfect thermal or blackbody spectrum but, to a rough approximation, the Sun shines with such a spectrum. It closely matches the radiation spectrum of a blackbody at a temperature of 5,780 K.

In 1900, the German physicist Max Planck derived the formula for the radiation spectrum of a perfect absorber, or blackbody, thereby introducing the idea that it radiates energy in fundamental indivisible units, which he called *quanta*, whose energy is proportional to the frequency of the radiation. The constant of proportionality between the frequency and energy of the radiation now is known as the *Planck constant*.

The continuum spectrum of a blackbody is markedly asymmetric. It falls off rapidly with decreasing wavelength on the short wavelength side of the maximum and decreases gradually with increasing wavelength at long wavelengths. The maximum intensity occurs at a wavelength that is inversely proportional to the temperature. Hotter thermal radiators emit their most intense radiation at shorter wavelengths; and cooler thermal radiators are most luminous at longer wavelengths. In other words, as the temperature of a gas increases, most of its thermal radiation is emitted at increasingly shorter wavelengths.

We can sum, or integrate, the contributions at every wavelength to obtain the total luminosity of a thermal radiator. When this is done, we obtain the important Stefan–Boltzmann law in which the luminosity increases with the square of the radius and the fourth power of the effective temperature. At a given disk temperature, bigger stars are more luminous and, for a particular size, hotter stars have greater luminosity. This luminosity is intrinsic to a star, establishing its power and energy output per unit time.

1.8 The Energy of Light

When radiation moves in space from one place to another, it behaves like trains of waves. However, when radiation is absorbed or emitted by atoms, it behaves not as a wave but rather as a package of energy, or like a particle, and is given the name *photon*. Photons have no electric charge and travel at the speed of light. A photon is a discrete quantity of energy associated with electromagnetic radiation.

Photons are created whenever a material object emits electromagnetic radiation, and they are consumed when matter absorbs radiation. Moreover, each atom, ion, or molecule can absorb and radiate energy only at a specific set of photon energies. (An *ion* is an atom that has lost one or more electrons; a *molecule* consists of two or more atoms.)

The ability of radiation to interact with matter is determined by the energy of its photons. This photon energy depends on the wavelength or frequency of the radiation. Waves with shorter wavelengths, or higher frequencies, correspond to photons with higher energy. The product of the Planck constant and the frequency results in the photon energy.

Radio waves have a small photon energy compared with the photons of visible light. The low energies of radio photons cannot easily excite the atoms of our atmosphere; therefore, these photons easily pass through the air. Visible radiation also can slip through the Earth's atmosphere with little interference. Its photons are too energetic to resonate with molecular vibrations and they are too feeble to excite atoms.

Ultraviolet photons are sufficiently energetic to remove electrons from atoms and many of the molecular constituents of the Earth's atmosphere, particularly in the ozone layer. Most of these energetic photons cannot reach the ground; if they could, they would cause considerable damage to our skin and eyes.

Astronomers often describe energetic, short-wavelength radiation, such as x-rays or gamma rays, in terms of its energy rather than its wavelength or frequency. At the atomic level, the natural unit of energy is the *electron volt*, or eV. One electron volt is the energy that an electron gains when it passes across the terminals of a 1-volt battery. A photon of visible light has an energy of about 2 eV. Much higher energies are associated with nuclear processes; they often are specified in units of millions of electron volts, or MeV. A somewhat lower unit of energy is 1,000 electron volts, called *kilo-electron volts* and abbreviated keV; these often are used to describe x-ray radiation.

1.9 Observing the Radiation

Different types of telescopes are used to gather the various forms of electromagnetic radiation at different wavelengths, including the gamma ray, x-ray, ultraviolet, visible, infrared, and radio regions of the electromagnetic spectrum. They are used to observe both the visible (like stars) and the invisible universe, including the cold interstellar spaces or objects that are much hotter than stars.

A bigger telescope gathers more radiation than a smaller telescope, permitting the detection of fainter objects and providing a brighter image of any cosmic object for analysis. Big telescopes also provide greater *angular resolution*, which is the ability to see the separation between objects that are close together. Better resolution permits observation of finer detail on the object emitting the radiation.

The lenses and mirrors used to focus and collect visible radiation are described by the science of optics; therefore, the study of visible light from cosmic objects is called *optical astronomy*. There are two types of optical telescopes, the refractor and the reflector, which respectively use a lens and a mirror to gather and focus optically visible light (Fig. 1.6).

Cosmic radiation that arrives at the Earth travels in rays that are parallel to one another. A lens bends the incoming rays by refraction, focusing them to a point where they meet, which is called the *focal point*. A curved mirror reflects the incoming rays, sending them to the focal point. If we place a detector at the focal point, in the plane parallel to the lens, we can record an image of whatever the telescope is observing. The distance from the lens to the focal point is called the *focal length*, which determines the overall size of the image.

Professional astronomers place electronic detectors at the focal point of telescopes. These detectors generate digitized signals that are analyzed, manipulated, and recorded in a computer. A Charge-Coupled Device (CCD) might be used to efficiently detect the radiation and form an image of it; nowadays, CCDs are used in this way in everything from digital cameras to the Hubble Space Telescope. A diffraction grating alternatively might be used to separate incoming radiation into its component wavelengths, dispersing it into fine wavelength intervals to form a spectrum that might then be recorded by a CCD.

The most important property of a telescope is the diameter of its reflecting mirror or focusing lens. For optical telescopes, this is known as the *diameter of the aperture*, which may be somewhat smaller than the physical size of the main mirror. The larger the diameter of a lens or mirror, the finer the detail that can be resolved and the more light the telescope captures. The amount of radiation that can be collected is proportional to the area of the mirror or lens and the square of its diameter.

Optical telescopes are named by the diameter of their mirror. The 2.5-m (100-inch) Hooker Telescope at the Mount Wilson Observatory in California and the nearby 5-m (200-inch) Hale Telescope at the Palomar Observatory, also in California, are of great historical importance. Recent important, large optical telescopes include the four Very Large Telescopes (VLTs), each with an 8.2-m (323-inch) effective aperture, located at the Paranal Observatory in Chile; the two 10-m (400-inch) Keck telescopes at the Mauna Kea Observatory in Hawaii; the 10.4-m (410-inch) Gran Telescopio Canarias on the Canary Islands; and the Large Binocular Telescope

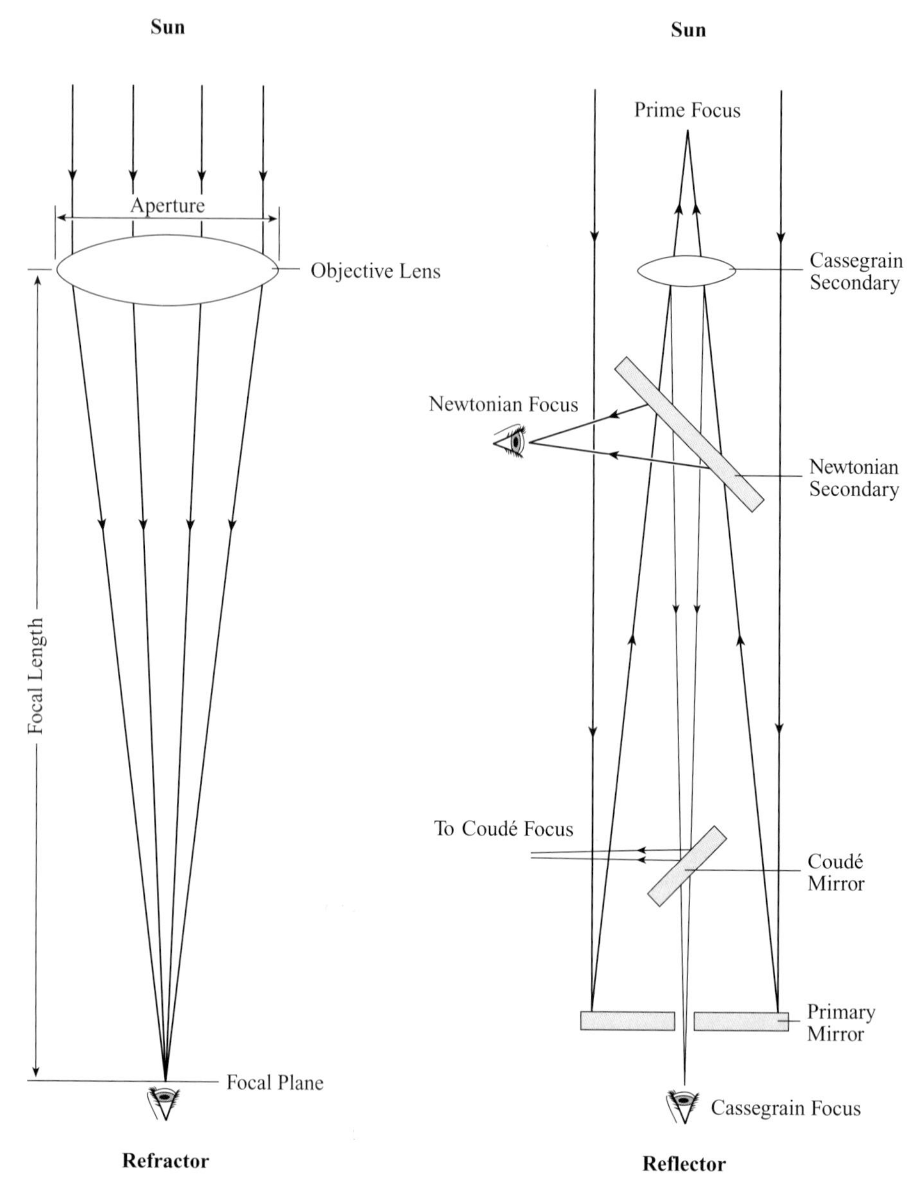

Figure 1.6. Telescopes. Light waves that fall on the Earth from a distant object are parallel to one another, and they are focused to a point by the lens or mirror of a telescope. Early telescopes were refractors (*left*). The curved surfaces of the convex objective lens bend the incoming parallel light rays by refraction and bring them to a focus at the center of the focal plane, where the light rays meet and an image is created. A second smaller lens, called the *eyepiece*, was used to magnify the image in the early refractors; later versions placed photographic or electronic detectors at the focal plane. In 1670, the English physicist Isaac Newton (1643–1727) constructed the first reflecting telescope (*right*), which used a large concave, or parabolic, primary mirror to collect and focus light. A small, flat, secondary mirror, inclined at an angle of 45 degrees to the telescope axis, reflected the light sideways, at a place now known as the *Newtonian focus*. Other light-deflecting mirror arrangements can be used to obtain any desired focal length, which varies with the curvature and position of small convex mirrors.

located on Mount Graham in Arizona – it consists of two 8.4-m (330-inch) mirrors on a binocular mount.

Angular resolution is inversely proportional to telescope size and directly proportional to the observing wavelength. The resolution specifies the ability of a telescope to discriminate between closely spaced parts of the object under view and usually is specified in *seconds of arc*, denoted by the symbol ″. The smaller the angular resolution, the finer is the detail that can be seen.

The diameter of a lens or mirror determines a telescope's angular resolution at a given wavelength. The resolving power of an optical telescope, which operates at the wavelengths that we can detect with our eye, is about $0.13/D_T''$ if the diameter D_T is in m. By way of comparison, the typical angular resolution of the unaided human eye is about 60″, so the eye acts like a lens with a diameter of about 0.005 m; however, some people have sharper vision.

The angular resolution also depends on the wavelength of observation, which means that vastly bigger telescopes are needed at long radio wavelengths. Because radio waves are millions of times longer than those of light, a radio telescope needs to be at least 1 million times bigger than an optical telescope to obtain the same resolving power. For this reason, the first radio telescopes provided a myopic, out-of-focus view. At a radio wavelength of 0.1 m, an angular resolution of 1″ would require a telescope with a diameter of 20 km.

Turbulence in the atmosphere limits the resolution of any telescope operating at visible wavelengths to about 1 second of arc; therefore, the angular resolution cannot be improved by building an optical telescope larger than about 0.13 m in diameter. Similar atmospheric variations cause the stars to "twinkle" at night. This atmospheric limitation to angular resolution at visible wavelengths is called *seeing*. The best seeing, of 0.2″ in unusual conditions, is found at only a few sites in the world, and optical observatories are located in most of them. Better visible images with even finer detail can be obtained from the unique vantage point of outer space, using satellite-borne telescopes unencumbered by the limits of the atmosphere.

A bigger lens or mirror also collects more light than a smaller one, thereby permitting the detection of fainter sources. The human eye, for example, is severely limited by its inability to gather light. It can store the images it sees for no more than a few tenths of a second. The telescope overcame this limitation by collecting light from a large area and storing it on a photographic plate or electronic chip.

Optical astronomy began slightly more than four centuries ago, in 1609, when Galileo Galilei turned the newly invented spyglass, or telescope, toward the night sky and discovered four previously unknown moons that circle Jupiter, resolved small craters on the Moon, and detected numerous stars in the Milky Way that cannot be seen by the unaided eye. His rudimentary telescope was a refractor with a lens the diameter of which was only 0.04 m, or about the size of a hand. The angular resolution of his telescope was about 2.8″, and an angular resolution more than about 10 times better than this cannot be achieved with any optical or visible-light telescope on the planet. Nevertheless, bigger telescopes collect more light, enabling detection of fainter objects or stronger signals from bright objects even when there is no improvement in resolving power.

The advantage of radio signals is that the atmosphere does not distort them or limit the angular resolution. Nowadays, relatively small radio telescopes – separated by large distances

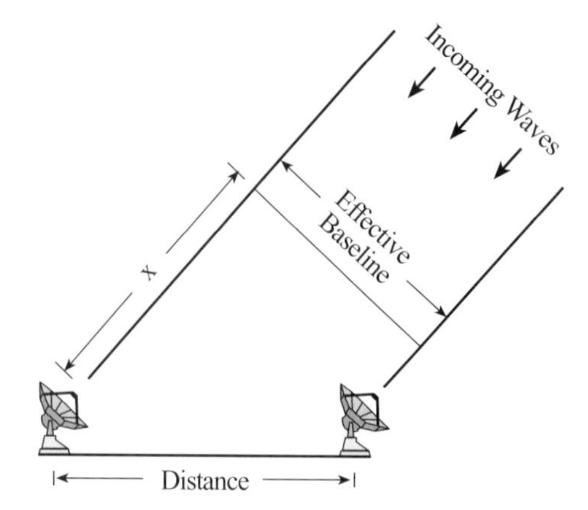

Figure 1.7. Interferometer. When incoming radiation approaches the Earth at an angle, the crests will arrive at two separated telescopes at slightly different times. This delay in arrival time is the distance X divided by the speed of light. If X is an exact multiple of the wavelength, then the waves detected at the two telescopes will be in phase and add up when combined. If not, they will be out of phase and interfere. The angular resolution of such an *interferometer*, or interference meter, is equal to the wavelength divided by the effective baseline. When the object being observed is directly overhead, the effective baseline is equal to the distance between the two telescopes.

or baselines – are combined and coordinated electronically, achieving radio images that are as sharp as optical ones (Fig. 1.7). Because it is spread out, an array of small telescopes has the property that is crucial for high resolving power – namely, great size relative to wavelength. The technique is known as *interferometry* because it analyzes how the waves detected at the telescopes interfere when they are added together. The simplest example is a pair of telescopes with a computer to reconstruct the waves from the combined data.

Even better angular resolution is obtained using radio signals recorded at widely separated radio telescopes when they are observing the same cosmic object at the same time. The recorded signals can be combined to effectively turn the Earth into a giant radio telescope with transcontinental interferometer baselines and the sharpest vision of any telescope on the Earth or in space.

The largest optical telescopes also are multitelescope interferometers that have longer baselines than the diameters of individual telescopes. Examples include the two Keck telescopes and the four VLTs, which have auxiliary telescopes to improve the baseline.

Radio telescopes do not provide the only window on the unseen cosmos. There are the invisible gamma ray, x-ray, ultraviolet, and infrared telescopes. This radiation is absorbed in our atmosphere and must be collected by telescopes in satellites that orbit the Earth above its atmosphere. All of these space telescopes measure the intensity of the incoming signal and convert these measurements into radio transmissions that are sent to radio telescopes and receivers on the ground. The National Aeronautics and Space Administration (NASA), for example, has launched four Space Telescopes: the Great Observatories named Hubble, Compton, Chandra, and Spitzer, which respectively operate at visible, gamma ray, x-ray, and infrared wavelengths. The American, European, and Japanese space agencies have sent a host of satellites into space with telescopes designed to observe specific cosmic phenomena (e.g., activity on the Sun) using telescopes on the Solar and Heliospheric Observatory (SOHO),

Hinode, the Solar Terrestrial Relations Observatory (STEREO), and the Solar Dynamics Observatory (SDO). The distances and motions of stars have been observed from HIPPARCOS, and the cosmic microwave background radiation has been delineated from the Cosmic Background Explorer (COBE), the Wilkinson Microwave Anisotropy Probe (WMAP), and PLANCK.

Astronomy from space has several advantages over ground-based observations. The weather in space is always perfectly clear and the atmosphere does not blur images obtained from telescopes in space. Furthermore, a large telescope is not needed to observe the short ultraviolet and x-ray wavelengths with high angular resolution. The mirror must be only 0.002 m across to achieve an angular resolution of $1''$ at extreme ultraviolet wavelengths and only 0.00002 m in diameter for the same resolution at x-ray wavelengths. Moreover, in space, the sky is truly dark and observations do not need to be limited to the night.

2

Gravity and Motion

2.1 Wanderers in the Sky

Look up at the Sun as it moves slowly across the bright blue sky, or watch the Moon's nightly voyage. On dark, moonless nights, you also might notice a planet such as Mars or Jupiter traveling against the stars.

Ancient astronomers thought that the Moon, Sun, and planets all moved in circles, forever wheeling around the central, unmoving Earth. After all, a wheel moves easily across the ground because it is round; because the circumference of a circle has no beginning or end, the cosmic motions could continue forever.

The Moon does indeed revolve about the Earth but our home planet is no longer considered to be the center, focus, and fulcrum of all things. The Earth and other planets revolve about the Sun, which is an important lesson in perspective. Motion is relative, perceived only in relation to something else, by comparison with another object that is either at rest or moving in a different way. It was once thought that the Earth was still and that the Sun revolved around it, but no, it is the other way around.

The earliest theories of planetary motion around the Sun had one fatal flaw. Like the mistaken Earth-centered interpretation, the Sun-centered one also initially assumed that the planets move in circular orbits. This explanation could not be reconciled with careful observations of the changing positions of the planets in the sky, meticulously carried out by the Danish astronomer, Tycho Brahe. As discovered by Johannes Kepler, Brahe's assistant and eventual successor, the architecture of the solar system had to be described by noncircular shapes.

In 1609, after eight years of computations, Kepler found that the observed planetary orbits could be described by ellipses with the Sun at one focus. This ultimately became known as *Kepler's first law of planetary motion*. Although the planetary orbits are nearly circular, they are slightly elliptical in shape. A planet moves from nearest the Sun, at a point called the *perihelion*, to the farthest point of its orbit, known as the *aphelion*.

Kepler also described how a planet moves at different speeds as it travels along its elliptical orbit. He was able to state the relationship in a precise mathematical form, which can be explained with the help of Figure 2.1. Imagine a line drawn from the Sun to a planet. As

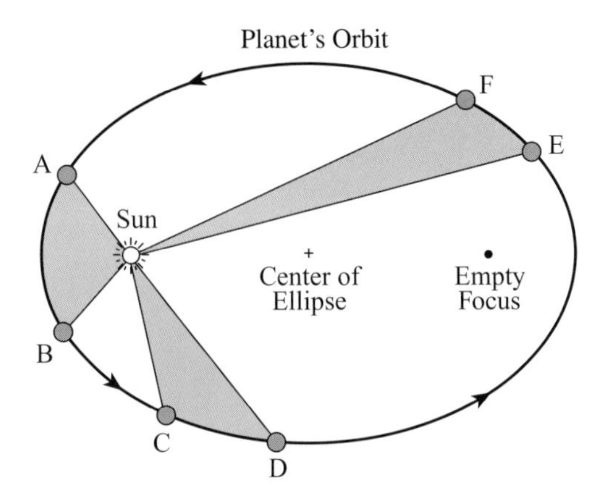

Figure 2.1. Kepler's first and second laws. The German astronomer Johannes Kepler (1571–1630) published his first two laws of planetary orbital motion in 1609. His first law states that the orbit of a planet about the Sun is an ellipse with the Sun at one focus. The other focus of the ellipse is empty. According to Kepler's second law, the line joining a planet to the Sun sweeps out equal areas in equal times. This also is known as the law of equal areas, and it is represented by the equality of the three shaded areas ABS, CDS, and EFS. It takes as long to travel from A to B as from C to D and from E to F. A planet moves most rapidly when it is nearest the Sun, at perihelion; a planet's slowest motion occurs when it is farthest from the Sun, at aphelion.

the planet swings about its elliptical path, the line (which increases and decreases in length) sweeps out a surface at a constant rate. This is known as *Kepler's second law* and also as the *law of equal areas*. During the three equal time intervals shown in Figure 2.1, the planet moves through different arcs because its orbital speed changes, but the areas swept out are identical.

So, a planet moves faster when it is closer to the Sun, and the modern explanation for this involves one of the fundamental concepts of physics, known as *the conservation of angular momentum*. The conservation law states that as long as no outside force is acting on a planet, its angular momentum cannot change. It explains why a planet keeps whirling around the Sun and why its speed is fastest at perihelion. For a given mass, the angular momentum is equal to the product of speed and distance; therefore, when the distance from the Sun decreases, at perihelion, the speed increases to compensate and keep the angular momentum unchanged. At aphelion, the distance increases so the speed must decrease, and the planet continues moving along without anything pushing or pulling it.

Strictly speaking, momentum involves velocity, which has an amount, magnitude, and direction. Speed is the magnitude of the velocity. In astronomy, the velocity often is given just by its observed magnitude in a given direction, the speed; therefore, the orbital velocity is given as its speed along the orbit.

An external force can change the momentum, which otherwise is conserved and unchanging. Here on the Earth, gravity exerts an outside force, pushing an automobile downhill and making it go faster; therefore, momentum is gained when going downhill, which is applicable to some people's life.

Kepler's third law took another 10 years of work to discover. It describes a musical pattern sought and found by Kepler for "the glory of God and the salvation of souls."[3] Each planet

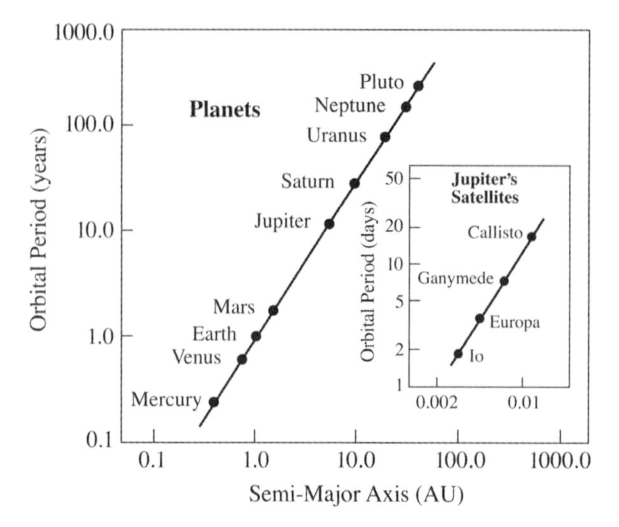

Figure 2.2. Kepler's third law. The orbital periods of the planets in years are plotted against the semimajor axes of their elliptical orbits in astronomical units, abbreviated AU, using a logarithmic scale. The straight line that connects the points has a slope of 3/2, thereby verifying Kepler's third law that states that the square of the orbital periods increases with the cubes of the planetary distances. The German astronomer Johannes Kepler (1571–1630) published this third law in 1619. This type of relation applies to any set of bodies in elliptical orbits, including Jupiter's four largest satellites shown in the inset, with a vertical axis in units of days and a horizontal axis that gives the distance from Jupiter in AU units.

produces its own unique note as it moves around the Sun, with an orbital period that increases with a planet's distance from the Sun. Kepler's harmonic relationship states that the squares of the planetary periods are in proportion to the cubes of their average distances from the Sun. This expression is illustrated in Figure 2.2 for the major planets and the brighter moons of Jupiter. It also implies that a more distant planet moves with a slower speed.

These were amazing discoveries, but no one knew what holds up the Moon and planets in space, confining them within their ceaseless, repetitive orbits. The explanation awaited the discovery of gravity, which rules the universe.

2.2 Gravitational Attraction

What moves the planets along within their well-defined orbital trajectories? Kepler supposed that some invisible magnetic force emanated from the rotating Sun, pushing the planets through space and controlling their motion. The farther the planet is from the Sun, the weaker is the solar force and the slower a planet's motion – as described by Kepler's harmonic relationship, his third law.

The great English scientist Isaac Newton proposed another unseen agent, the invisible gravitational force of the Sun, to grasp the planets and hold them in place. Newton showed that the pull of gravity is universal, with an unlimited range and capacity to act on all matter, thereby holding the Moon and comets in their trajectories, as well as the planets. Gravitation is the driving and organizing force of the universe. It binds stars and galaxies together and is responsible for their formation. Universal gravitation draws everything together, connecting every cosmic object to all of the others.

Long ago, it was thought that people living on the opposite side of a round Earth would fall off into empty space below, but gravity saved the situation. It keeps our feet on the ground at all places on the spherical Earth. Gravity has pinned us here, so we rotate with the spinning Earth and stay on it. The atmosphere and oceans similarly are held close to the planet by its relentless gravitational pull. Gravity is always there, trying to pull us down.

It is gravity that explains why and how things fall. Every object on the Earth falls in just one direction, down toward the ground. We might suppose, as Aristotle once did, that a heavy object will fall faster than a lighter one, in direct proportion to its weight, but that is not the case. All objects on the Earth fall at the same rate, regardless of their weight – a theory that was proposed in 1638 by Galileo Galilei, and apparently 17 centuries before that by the Roman poet Lucretius.

Galileo also proposed that any undisturbed body will fall with uniform acceleration, and he showed that the distance traveled by an object falling from rest is proportional to the square of the elapsed time. He thereby provided a scientific foundation for Newton's subsequent theory of universal gravitation.

According to tradition, Newton was sitting under an apple tree when an apple fell next to him on the grass. This reminded him that the power of gravity, the pull of which influences the motion of falling bodies, seems undiminished even at the top of the highest mountains. He therefore argued that the Earth's gravitational force extends to our Moon, and he showed that this force, diminished by distance to the Moon, can pull the Moon into its orbit. It was as if the Earth's Moon is perpetually falling toward the planet while always maintaining the same mean distance from it. Without the Earth's gravitational pull, the Moon would not orbit our planet but instead would travel out into space, never returning to the Earth. The Sun's gravity similarly deflects the moving planets into their curved paths, so they forever revolve around the Sun.

The English genius was a solitary loner, a self-isolated intellect, somewhat obsessed, famously distracted, and frequently depressed. Newton did not like interacting with people. He declined most invitations, avoided personal contact, never traveled outside of England, and, legend says, he died a virgin at the age of 84. He also spent much of his life immersed in experiments in alchemy and theological or mystical speculations, hoping to understand the origin of the elements and the eternal mysteries of health and mortality by examining mystic clues left by God.

Newton showed that motions everywhere, whether above in the cosmos or below on the ground, are described by the same concepts and that all material objects are subject to gravitation. Therefore, everything in the observable universe moves in predictable and verifiable ways. The basic ideas are that a moving body will continue to move in a straight line unless acted on by an outside force, and that every object attracts every other object as the result of universal gravitation. Such insights resulted in Sir Newton becoming the first person in England to be knighted for his scientific work.

It was his friend, the English astronomer Edmond Halley, who persuaded the secretive Newton to write his greatest work, the *Philosophiae naturalis principia mathematica* (*Mathematical Principles of Natural Philosophy*), commonly known as the *Principia*. It was presented to the Royal Society of London, which withdrew from publishing it due to insufficient funds; Halley, a wealthy man, paid for the publication the following year, in 1687.

The enormous reach of gravity can be traced to two causes. First, gravitational force decreases relatively slowly with distance, which gives it a much greater range than other natural forces, such as the strong force that holds the nucleus of an atom together. Second, gravitation has no positive and negative charge, like electricity, or opposite polarities, like magnets. This means that there is no gravitational repulsion between masses. In contrast, the attractive forces among unlike electrical charges in an atom cancel one another, shielding it from the electrical forces of any other atom.

Any two masses attract one another with a gravitational force that varies in proportion to the product of the masses and the inverse square of the separation between their centers. The constant of proportionality – the Newtonian constant of gravitation G – was not measured until 71 years after Newton's death and then indirectly by Henry Cavendish. Cavendish's aim was to determine the mass density, or mass per unit volume, of the Earth and, because our planet's radius was known, the mass density could be used to determine the mass of the Earth (Focus 2.1). The gravitational force between two masses is equal to the constant G times the product of the two masses divided by the square of their separation; that force is given appropriately in units of *newton*.

Focus 2.1 Weighing the World

The gifted English scientist Henry Cavendish, child of Lady Anne Grey and Lord Charles Cavendish, was just as secretive and solitary as Newton, working at home and avoiding visitors or most other human contact. When someone from a distant country arrived at the front door, Cavendish was known to disappear out the back door.

An ingenious torsion-balance apparatus had been designed and constructed for the measurement of the Earth's mass density by the Reverend John Michell, who died without completing the work. His experimental device eventually passed on to Cavendish, who rebuilt it, closely following Michell's concept. The apparatus consisted of a 6-foot wooden rod suspended from a wire with two small lead spheres attached to each end. A separate suspension system was used to hold larger lead balls in place, and the gravitational attraction between the small and larger balls was determined from the twisting of the wire supporting the rod.

To prevent air currents, temperature changes, and other extraneous forces from interfering with the measurements, the entire apparatus was placed inside a closed wooden shed and observed from outside using a telescope aimed through peepholes in the shed walls. After nearly a year of meticulous observations, Cavendish announced in 1798 that the Earth has a mass density of 5,488 kg m^{-3} (when corrected for a small mathematical error in his paper), which meant that the mass of the Earth is about 6×10^{24} kg, using the approximate radius of the Earth. (Distances part way around the surface of the Earth had been found by the surveying technique of *triangulation* and combined to determine the Earth's circumference and a radius of about 6,400 km.)

In Cavendish's time, mass and weight were assumed to be equal and, as he stated in his correspondence, he succeeded in weighing the world. It weighed in at a little more than 6 billion trillion metric tons. (A metric ton is 1,000 kg or 2,205 pounds, and the precise mass of the Earth is 5.974×10^{24} kg.)

Mass is an intrinsic aspect of an object, different from its weight that alters with distance from the main source of gravity. An astronaut, for example, weighs less when leaving the Earth but retains the same mass.

Any object has a gravitational potential stored within it due to its efforts at overcoming relentless gravity. Two separated objects, for example, have worked against the gravitational attraction that pulls them together, achieving a reserve of energy and a potential for future action. Gravity is always there, pulling on things, and it can take some effort to overcome it.

When we lift a book to our desk or drive our vehicle to the top of a hill, a gravitational potential energy has been acquired, which can be released when we drop the book or drive downhill. Energy cannot be created or destroyed, only transformed. (This conservation of energy is another fundamental law of physics.) Therefore, the energy that went into overcoming the pull of gravity is stored in an object, and this stored potential energy can be converted into the kinetic energy of motion when the book is dropped or the vehicle goes downhill. We notice the increased kinetic energy when the book hits the floor or the vehicle speeds up without putting our foot on the gas pedal.

This gravitational potential energy is due to an object's position and is associated with the gravitational force. It depends on the height of the object, its mass, and the strength of the gravitational field it is in. The stored potential resembles an internal strength and power that might be obtained from hard work – a reserve of energy available when needed and released.

2.3 Tidal Forces

The Ocean Waters Rise and Fall

While walking along the beach, we might notice that the waves are rising increasingly farther up the shore, spreading across the parched sand, steadily advancing and enlarging the bounds of the sea. The tide is flooding the beach. The water races over flat places, stopping at nothing; but, several hours later, it will turn around and retreat. We say that the tide is rising and falling, while the sea runs in and out; Newton showed that the Moon's attraction is the main cause for these tides.

Because the Moon's gravitational force decreases with increasing distance, the Moon pulls hardest on the ocean facing it and least on the opposite ocean, whereas the Earth between is pulled with an intermediate force. In this way, the Moon's gravity draws out the ocean into the shape of an egg and creates two high tides. As the Earth's rotation carries the continents past the two tidal humps, we experience the rise and fall of water, the ebb and flow of the tides, twice every day (Fig. 2.3).

The Moon creates most of the ocean waves that softly caress us when we swim or sometimes crash violently against the shore, but the Sun also contributes to the size and rhythm of the waves. Although more massive than the Moon, our Sun also is much farther away; as a result, the tide-producing force of the Moon is about 2.2 times that of the Sun. Near both full and new Moons, the tide-raising forces of the Sun and the Moon are in the same direction, producing the spring tide (Fig. 2.4). They reinforce one another's tides and produce high tides that can be a few times higher than normal. A quarter-month later, the two tidal forces are in opposition and interfere with one another, and the range of these neap tides is then lower than any others.

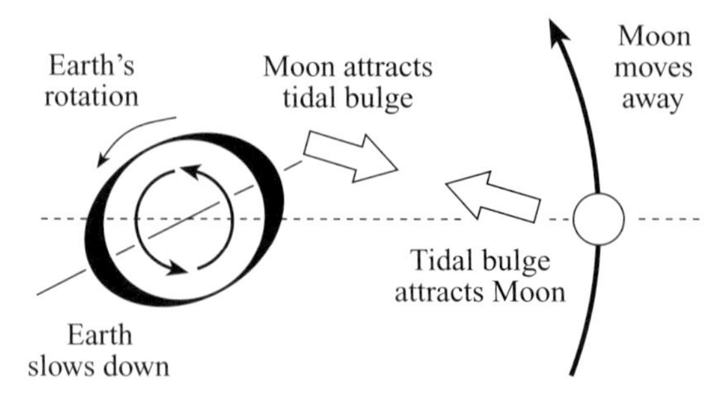

Figure 2.3. Cause of the Earth's ocean tides. The Moon's gravitational attraction causes two tidal bulges in the Earth's ocean water, one on the closest side to the Moon and one on the farthest side. The Earth's rotation twists the closest bulge ahead of the Earth–Moon line (*dashed line*), which produces a lag between the time the Moon is directly overhead and the time of highest tide. The Moon pulls on the nearest tidal bulge, slowing down the Earth's rotation. At the same time, the tidal bulge nearest the Moon produces a force that tends to pull the Moon ahead in its orbit, causing the Moon to spiral slowly outward.

Why We See Only One Face of the Moon

A planet's gravitational force pulls any natural satellite, or moon, it might have into a slightly elongated shape along an axis pointing toward the planet. That is, a planet's gravitation produces two tidal bulges in the solid body of the satellite: one on the closest side to the planet and one on the satellite's farthest side. If the satellite's rotation twists the

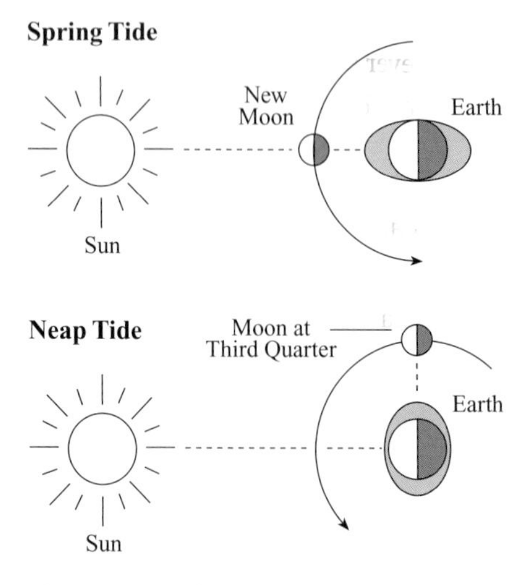

Figure 2.4. Earth's spring and neap ocean tides. The height of the tides and the phase of the Moon depend on the relative positions of the Earth, Moon, and Sun. When the tide-raising forces of the Sun and the Moon are in the same direction, they reinforce one another, making the highest high tides and the lowest low tides. These spring tides (*top*) occur at either new or full Moon. The range of tides is least when the Moon is at first or third quarter and the tide-raising forces of the Sun and the Moon are at right angles to one another. The tidal forces are then in opposition, producing the lowest high tides and the highest low tides, or the neap tides (*bottom*). In this diagram, the height of the tides is greatly exaggerated in comparison to the size of the Earth.

closest bulge ahead of the planet–satellite line, the planet pulls back on it. As a result, one hemisphere of the satellite always faces the planet, and the satellite takes as long to rotate as it does to orbit the planet. We say that the satellite has been tidally locked into synchronous rotation with the planet.

The Moon is in synchronous rotation about the Earth, so the Moon's rotation period is the same as the time it takes for the Moon to orbit the Earth, which is 27.32 Earth days. As a result, we always see the Moon's near side and never its far side, unless a passing spacecraft sends images home. Most of the major moons, or large natural satellites, in the solar system have synchronous rotation.

If the mass of two orbiting bodies is comparable and their physical separation is relatively small, they both may be tidally locked to one another. This is the case for Pluto and its nearby large moon, Charon. Mutual tidal locking also occurs for close binary stars.

Tidal forces also explain why the days are getting longer, why the Moon is moving away from the Earth, and why planets have rings (Focus 2.2).

Focus 2.2 Longer Days, the Retreating Moon, and Planetary Rings

The Earth Slows Down and the Moon Moves Away

As the Earth rotates, the bulge raised on its oceans by the Moon's gravity is always a little ahead of the Moon rather than directly under it. The Moon pulls back on the bulge and, in the process, slows down the planet.

When the ocean tides flood and ebb, they create eddies in the water, producing friction on the ocean floor, which heats the water ever so slightly and dissipates energy at the expense of the Earth's rotation. The tides therefore act as brakes on the spinning Earth, slowing it by friction in much the way that the brakes of a vehicle slow its wheels. As a result of this tidal friction, the rotation of the Earth is slowing down and the day is becoming longer at a rate of 2 milliseconds, or 0.002 s, per century. In other words, the days are getting longer at the rate of 1 second every 50,000 years, and tomorrow will be 60 billionths of a second longer than today.

As the Earth slows down, the angular momentum it loses is transferred to the Moon, which speeds up in its orbit around us. When we do the arithmetic, we find that the change of 0.002 s per century in the length of a day implies an outward motion of the Moon, amounting to about 0.04 m each year. Small as it is, this value is just measurable with pulses of laser light sent from the Earth to tiny corner mirrors placed on the Moon by the *Apollo* astronauts. When the speed of light is multiplied by the roundtrip travel time for the laser signals to go to the Moon and back, twice the Earth–Moon distance is obtained. Such measurements indicate that the Moon is moving away from the Earth at a rate of 0.038 m per year.

Because the Moon is moving away from the Earth, its tides are weakening, whereas the Sun's tide-producing force remains unchanged. In 2.96 billion years, the Sun's tides will be equal to the Moon's tides.

(continued)

Focus 2.2 *(continued)*

Tearing a Moon Apart

A planet's strong differential gravitational attraction, also known as *tidal interaction*, explains why large planets have rings and why they are usually closer to a planet than its large moons. The rings normally are confined to an inner zone, where the planet's gravitational forces would stretch a large natural satellite, or moon, until it fractured and split apart while also preventing small bodies from coalescing to form a larger moon. The outer radius of this zone in which rings are found is called the *Roche limit*, after the French mathematician Édouard A. Roche, who described it in 1849 (Fig. 2.5).

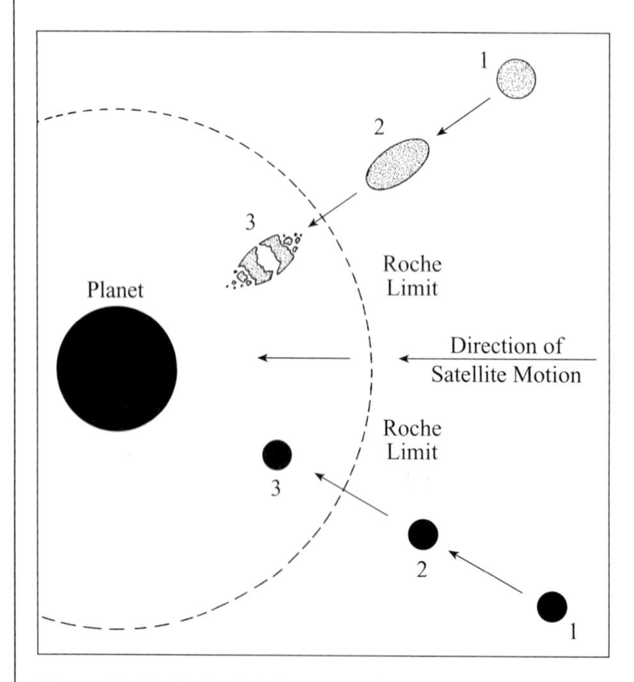

Figure 2.5. Roche limit. A large satellite *(top)* that moves well within a planet's Roche limit *(dashed curve)* will be torn apart by the tidal force of the planet's gravity, in an effect first investigated in 1849 by the French mathematician Edouard Roche (1820–1883). The side of the satellite closer to the planet feels a stronger gravitational pull than the side farther away, and this difference works against the self-gravitation that holds the body together. A small solid satellite *(bottom)* can resist tidal disruption because it has significant internal cohesion in addition to self-gravitation.

For a satellite with no internal strength and the mass density of which is the same as the planet, the Roche limit is 2.446 times the planetary radius, or about 147,000 km for Saturn and about 15,584 km for the Earth. If our Moon ever ventured within this distance from the Earth's center, it would be pulled apart by differential gravitational forces between the sides of the Moon facing toward and away from the Earth, and our planet would have rings. Not to worry: The mean Earth–Moon distance is much farther away, approximately 384,400 km.

2.4 Motion Holds Up the Planets

We might say that motion defines existence, for there is nothing in the universe that is completely at rest. All that exists, from atoms to planets and stars to galaxies, moves through space, all surely going somewhere. It is this motion that opposes the inward pull of gravity and shapes the universe, giving it form, structure, and texture. This motion explains why the stars have not all pulled together and coalesced under their mutual gravitational attraction and why the planets continue to revolve about the Sun rather than falling into it.

As proposed by Galileo, a moving body will continue in the same direction at constant speed unless disturbed; this principle was incorporated into *Newton's first law of motion*. It states that a moving object will continue in motion with the same speed and in the same direction unless an external force is applied to it. This means that a planet would continue going the way it started, moving along a straight line, if it were not for the solar gravitational force that deflects the planet into a curved orbit around the Sun.

Therefore, it is the Sun's gravitational attraction that keeps the planets in their orbits and holds the solar system together. But why does not the immense solar gravity pull all of the planets into the Sun? Motion holds up the planets, opposing the relentless pull of the Sun's gravity and keeping the planets from falling into the Sun.

Each planet is moving in a direction perpendicular to an imaginary line connecting it to the Sun, at exactly the speed required to overcome the Sun's gravitational pull, and maintaining an equilibrium between motion and gravitation that keeps the planet in perpetual motion.

The critical velocity required for this delicate balance between motion and gravitation is known as the *escape velocity*. When the speed of a small object is just enough to escape the gravity of a big object, we say that the former is moving at the escape velocity of the latter. A more massive body has a greater escape velocity due to its stronger gravitational pull; that velocity decreases with increasing distance from the object because its gravity weakens at greater distances.

For a planet orbiting the Sun, the escape velocity depends only on the Sun's mass and the distance of the planet. If a planet at a given distance moved any faster than the escape velocity, it would leave the solar system, moving off into interstellar space; if the planet moved any slower, it would be pulled into the Sun and consumed by it. The orbital speed of a planet therefore depends on the Sun's mass and is independent of the planet's mass, which is why the planetary realm is called the solar system, dominated by the central Sun.

2.5 The Massive Sun

Within the solar system, the dominant mass is that of the Sun, which far surpasses the mass of any other object there. As Newton demonstrated, we can determine the Sun's mass from the Earth's orbital parameters. He used Kepler's relationship between a planet's orbital period and distance to show that the force of gravity falls off as the inverse square of the distance from the center of the main source of gravity, the Sun. Once we know the

Earth's orbital period and mean distance from the Sun, we can weigh the Sun from a distance, determining its mass from Newton's formulation of Kepler's third law.

The Sun's mass is proportional to the cube of the mean Earth–Sun distance, the AU, and inversely proportional to the square of the Earth's orbital period of 1 year. The constant of proportionality is related to Newton's constant of gravitation, which is measured by gravitational effects on the Earth.

When we do the arithmetic, the mass of the Sun is determined, and it has a value of $M_{\odot} = 1.989 \times 10^{30}$ kg. The Sun's mass is denoted by the symbol $M_{\odot}$, or an uppercase M with a subscript of a circle and a dot in the center. This is the benchmark unit for specifying the mass of the stars and the galaxies. The Sun is 333,000 times more massive than the Earth and contains more than 99.9 percent of the mass of the entire solar system; therefore, the assumption that our planet's motion is controlled by the Sun is amply justified.

2.6 What Causes Gravity?

We cannot see the force of gravity and Newton did not know how it was exerted. Albert Einstein subsequently explained it by supposing that a massive body like a star bends nearby space into the curvature of an embrace, giving that space a shape and form. This bending, twisting, and distorting of space is the cause of the star's gravity. However, such curvature effects are noticeable only in extreme conditions near a very massive cosmic object like a star, and the differences between the Newton and Einstein theories of gravity are indistinguishable in everyday life.

One result of the Sun's curvature of nearby space is that planetary orbits are not exactly elliptical. This solved a perplexing problem with the motion of Mercury, the nearest planet to the Sun. Instead of returning to its starting point to form a closed ellipse in one orbital period, Mercury moves slightly ahead in a winding path that can be described as a rotating ellipse (Fig. 2.6). As a result, the point of Mercury's closest approach to the Sun, the perihelion, advances by a small amount – only 43 seconds of arc per century – beyond the location predicted using Newton's theory.

Although this anomalous twist in Mercury's motion was discovered in 1854, it was not explained for more than a half-century, when Einstein proposed in 1915 that the planet is directed along a path in curved space-time (Fig. 2.7), making the planet overshoot its expected location. Observations of Mercury trace out the invisible curvature, like watching the slow arc of a bird gliding on unseen winds.

In the very paper that explained Mercury's anomalous motion, Einstein showed that the curvature of space near the Sun also deflects the path of light from other stars about to pass behind the Sun. The otherwise straight trajectory of starlight is bent by the Sun's gravity.

Newton previously had speculated that massive bodies might bend nearby light rays if light has mass; however, when Einstein took space curvature into account, the expected deflection was doubled. The effect can be measured during a solar eclipse when stars pass behind the darkened Sun. The apparent positions of these stars are displaced from their locations in the night sky when they are nowhere near the Sun.

The successful measurement of this deflection of starlight during the total solar eclipse on May 29, 1919 made Einstein famous, practically overnight. The November 7, 1919 headline

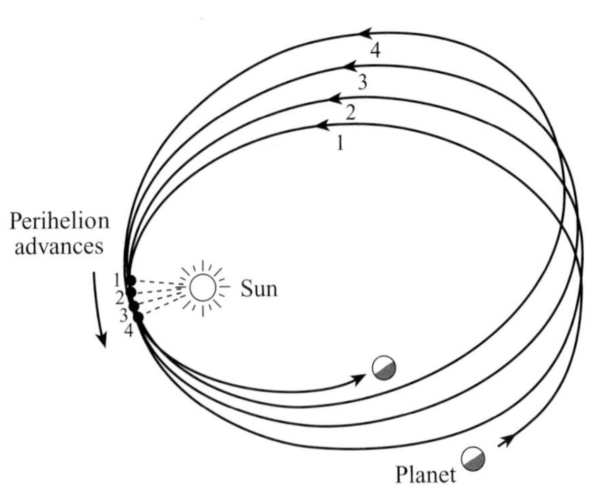

Figure 2.6. Precession of Mercury's perihelion. Instead of always tracing out the same ellipse, the orbit of Mercury pivots around the focus occupied by the Sun. The point of closest approach to the Sun, the perihelion, is slowly rotating ahead of the point predicted by Newton's theory of gravitation. This at first was explained by the gravitational tug of an unknown planet called Vulcan that was supposed to revolve about the Sun inside Mercury's orbit, but we now know that Vulcan does not exist. Albert Einstein (1879–1955) explained Mercury's anomalous motion in 1915 by inventing a new theory of gravity in which the Sun's curvature of nearby space makes the planet move in a slowly revolving ellipse.

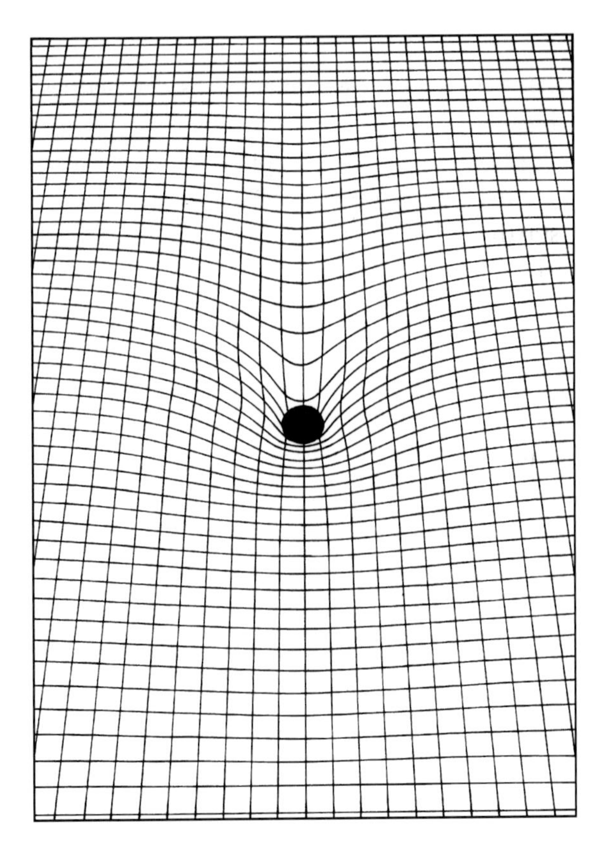

Figure 2.7. Space curvature. A massive object creates a curved indentation on the "flat" space described by Euclidean geometry, which applies in our everyday life on the Earth, where we do not directly encounter astronomical amounts of matter. Notice that the amount of space curvature is greatest in the regions near a cosmic object like a star, whereas farther away, the effect is lessened.

of the London *Times* read, "Revolution in Science – New Theory of the Universe – Newtonian Ideas Overthrown," and news of the discovery hit the front pages of every major newspaper in the world.

All of this excitement was over a tiny factor of two in a bending of starlight that was imprecisely measured. Nevertheless, the Sun's curvature of nearby space has now been measured with increasingly greater precision for nearly a century, confirming Einstein's prediction, to 2 parts in 100,000, or to the fifth decimal place. His *General Theory of Relativity*, which replaces the Sun's gravity with geometry, has been verified by so many solar experiments that it now is widely accepted.

All of the stars bend the transparent space around them, imprinting an invisible curvature within it. As a star's radiation carries energy through space and time, the starlight resembles an enormous flowing river, bending around any star in its path.

3

Atomic and Subatomic Particles

3.1 Inside the Atom

What is matter made of? To find out, we might try breaking any material object into increasingly smaller pieces until we reach a stage when the smallest piece cannot be broken apart. The last step in this imaginary, successive division of matter suggests the existence of unseen *atoms*, a Greek word meaning "indivisible," or something that cannot be divided further.

In the fifth century BC, for example, the ancient Greek philosopher Democritus and his mentor Leucippus proposed that all matter is composed of combinations of a small number of separate atoms coming together in different ways. They supposed that all substances are composed of four types of elemental atoms: air, earth, fire, and water. Mud, for example, could be made from earth and water, and fire could turn water into vapor, or air.

Then, about two millennia ago, the Roman poet Titus Lucretius Carus (Lucretius) wrote a wonderful epic poem, in Latin *De Rerum Natura* (*On the Nature of the Universe*), which described these indestructible atoms that are so exceedingly small that they are invisible and infinitely vast in number. To Lucretius, these atoms were the building blocks of all that exists; the fundamental idea persists to this day. Everything that we see, from a friend to a tree, to the Sun and the stars, consists of innumerable atoms, all moving randomly about, colliding, gathering together, and breaking apart again. The atoms are immortal, the former ingredients of all that existed in the past and the seeds of everything that will exist in the future.

All ordinary matter indeed is composed of elemental atoms, and there is a limited number – ninety-four naturally occurring atoms – detected directly on the Earth or in astronomical spectra. These atoms are known also as chemical elements because they cannot be decomposed by chemical means. They are very small and exceedingly numerous. A simple drop of water contains about 100,000 billion billion, or 10^{23} atoms – close to the number of stars in the universe.

Atoms combine to form molecules, and there are many more kinds of molecules than there are atoms. The vast numbers of molecules differ only in the kind and relative number of the atoms of which they are constructed. A molecule may be a combination of single chemical atoms, such as oxygen molecules, O_2, that we breathe, which consist of two oxygen atoms, each designated by O. A molecule also may contain different elements, as in water, designated H_2O, which is composed of two atoms of hydrogen, H, and one of oxygen, O. The Earth's transparent atmosphere consists mainly of molecules of oxygen, O_2 (21 percent), and nitrogen, N_2 (78 percent), with trace amounts of carbon dioxide, CO_2, and water vapor. Methane, CH_4, the natural gas used in a stove, consists of one atom of carbon, C, and four of hydrogen, H. Organic molecules contain more complex combinations of carbon, hydrogen, and other atoms.

The elemental atoms can be broken into smaller subatomic pieces. The first such particle was located through investigations of electricity. The English physicist Michael Faraday found that the electrical charge carried by different atoms is always an integer multiple of a basic amount, an atomicity of electrical charge, the electron. The concept of such an indivisible quantity of charge was proposed to explain the chemical properties of atoms. Interactions between electrons hold the atoms in a molecule together in a chemical bond. Similar but much weaker interactions among electrons hold many molecules together.

In 1894, the Irish physicist George Johnstone Stoney introduced the designation *electron* to describe the fundamental unit of electricity, which is transferred by the flow of electrons. A few years later, the English physicist Joseph John Thomson and his colleagues identified the electron as a particle and determined its charge-to-mass ratio. Thomson was studying cathode rays, which carry electrons between the electrodes of a tube of gas. He showed that the electrons are deflected when an external magnetic or an electrical field is applied, which meant that they are electrically charged. Using these curved trajectories, Thomson showed that electrons are very light, roughly 1/1,000th of the mass of the least massive atom, hydrogen. It would take 30 billion billion billion, or 3×10^{27}, electrons to make a total mass of just 1 ounce, which is only 28 grams.

Thomson received the 1906 Nobel Prize in Physics for his investigations in the conduction of electricity by gases. A few years later, the American physicist Robert A. Millikan determined the elementary charge of the electron by measuring the electrical force on charged droplets of oil suspended against gravity between two metal electrodes. He was awarded the 1923 Nobel Prize in Physics for this and related work.

The electron, which carries a negative electrical charge and has no known components or substructure, is believed to be a truly elementary particle. It is small, tinier than could be seen with the best microscope, and its charge is also an exceedingly minute quantity – but no one knows why an electron has its mass and charge. When the electrical current in a house is turned on to light a lamp, about 1 million trillion, or 10^{18}, electrons are flowing through the wires every second.

In the early twentieth century, the New Zealand–born British physicist Ernest Rutherford and his colleagues showed that radioactivity is produced by the disintegration of atoms, and they discovered that radioactive material emits energetic subatomic particles. When bombarding gold leaf with beams of these particles, they found to their astonishment that about 1 in 20,000 particles bounced right back from where it originated, whereas all of the others passed through the gold. This meant that atoms are largely empty space and that most of the mass of an atom is concentrated in a nucleus that is 100,000 times smaller than an atom.

Within a decade, Rutherford was able to show that the nucleus of different atoms contains various amounts of the nucleus of the simplest atom, hydrogen. He named this nuclear building block a *proton*, from the Greek for "first" because it was the first nuclear particle to be discovered.

A proton is positively charged, with a charge equal in amount to that of an electron but opposite in sign; particles with an opposite sign to their electrical charge attract one another. The negatively charged electrons surround the positively charged protons in an atom, in which the total positive charge of the protons is equal to the total negative charge of the electrons. An atom has no net electrical charge and it is electrically isolated from external space.

Due to their charged similarity, particles with the same electrical charge are driven apart by an electrical repulsion. Rutherford therefore postulated the existence of an uncharged nuclear particle, later called the *neutron*, to help hold protons together in the atomic nucleus and prevent the protons from dispersing as they repelled one another. After an 11-year search, the English physicist James Chadwick discovered the neutron in 1932.

Protons and neutrons collectively are named *nucleons* because they are the two constituents of the atomic nucleus. Because they consist of yet smaller components, known as *quarks*, the proton and the neutron are not truly elementary particles. Nevertheless, the nuclear-fusion reactions that make stars shine can be adequately understood by assuming that all atomic nuclei are composed of protons and neutrons.

These nucleons are bound together by an exceptionally strong force, the *nuclear force*, which allows them to cling tightly to one another and build the dense, compact atomic nucleus. Although powerful, this attractive force has a short range, operating over very close, limited distances. The strong force decreases to insignificance at distances greater than about 1 million billionths, or 10^{-15}, of a meter and closes the nucleus at an atom's center.

Therefore, an atom is largely empty space, like the room in which we are sitting. A tiny, heavy, positively charged nucleus lies at the heart of an atom, surrounded by a cloud of relatively minute, negatively charged electrons that define most of an atom's size and govern its chemical behavior.

A proton and a neutron have about the same mass, which is nearly 2,000 times that of an electron. To be exact, the mass of the proton and the mass of the neutron are, respectively, 1,836 and 1,839 times the mass of an electron.

The mass of an atomic nucleus is always less than the sum of the masses of its protons and neutrons because they have expended energy to bind themselves together. This binding energy is released in nuclear reactions. The physical properties of basic atomic and nuclear particles are listed in Table 3.1.

What holds familiar solid objects together? Why does not a chair fall apart when one sits in it? All durable material objects that surround us consist of atoms and, given the emptiness of an atom, we might wonder why we cannot easily crush them into smaller entities.

When we push any two pieces of material together, the forces of electrical repulsion between the atomic electrons in their adjacent surfaces resist the action. This prevents us from pushing them into each other, and that is why we cannot jam our finger into the chair in which we are sitting.

The simplest and lightest atom consists of a single electron circling around a nucleus composed of a single proton without any neutrons; this is an atom of hydrogen. Most of

Table 3.1. *Physical properties of the electron, proton, neutron, and atom*

Electron

m_e = mass of electron = 9.1094×10^{-31} kg

e = elementary charge = electron charge = 1.602×10^{-19} C

r_e = classical electron radius = 2.817940×10^{-15} m

σ_e = Thomson cross section = 0.665246×10^{-28} m^2

Atomic Nucleus (proton and neutron)

m_P = mass of proton = mass of hydrogen nucleus = 1.67262×10^{-27} kg

m_n = mass of neutron = 1.67493×10^{-27} kg

Z = total number of protons in nucleus = atomic number

A = total number of protons and neutrons in nucleus = atomic mass number

R = nuclear radius = $r_0 A^{1/3}$ for an atom with mass number A

$r_0 = 1.25 \times 10^{-15}$ m = 1.25 fm

Atom

a_0 = Bohr radius = 0.5292×10^{-10} m

u = atomic mass unit = $1.6605389 \times 10^{-27}$ kg

m_H = mass of hydrogen atom = 1.007825 u = 1.6739326 kg

m_{He} = mass of helium atom = 4.00260 u = 6.646473×10^{-27} kg

the universe, and the majority of the stars, is composed mainly of hydrogen. The nucleus of helium, the next most abundant atom in the cosmos, contains two neutrons and two protons; therefore, the helium atom has two electrons (Fig. 3.1).

So, an atom is composed of a dense and massive nucleus containing protons and neutrons and surrounded by electrons. An electrically neutral atom has as many electrons as there

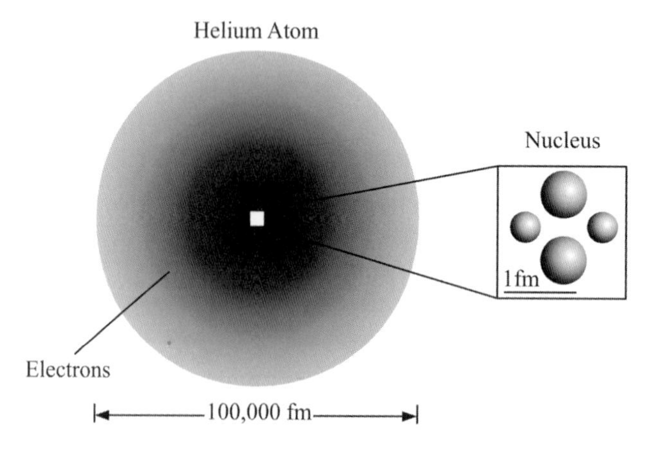

Figure 3.1. The helium atom. An atom of helium contains two electrons that swarm about the atom's nuclear center in a cloud of largely empty space. The shading shows that the electrons can be anywhere but are most likely to be found near the center of the atom. The magnified nucleus of the helium atom consists of two protons and two neutrons bound together by a strong nuclear force. The nucleus and each of its four particles are spherically symmetrical. The size of the helium nucleus is about 1 fermi, or 1 fm, which is equivalent to 10^{-15} m. The atom is about 100,000 times bigger than the nucleus, with an atom size of about 10^5 fm, or 10^{-10} m.

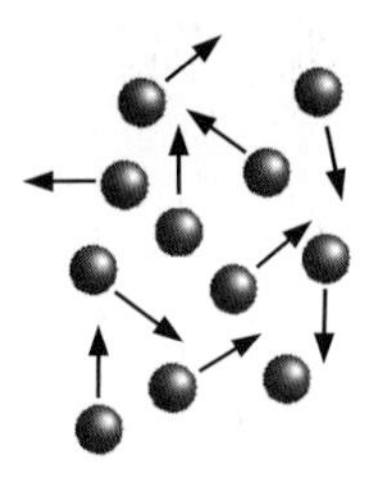

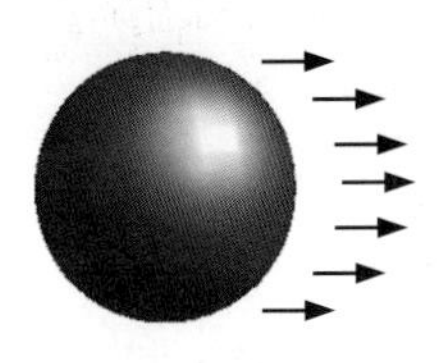

Figure 3.2. Random and ordered motion. Particles within a hot gas (*left*) move here and there in random directions that continually change as the result of collisions between particles. This supports the gas against inward gravitational forces. A planet or star (*right*) moves along a well-defined, ordered trajectory determined by external gravitational forces on it. When a large number of stars has gathered together and is confined within a star cluster, the stars also move in random directions, supporting their combined gravitational pull.

are protons and is therefore without net charge. An ionized atom has fewer electrons than protons and therefore has a positive charge. The atomic number, designated by the symbol Z, is equal to the number of protons in an atom's nucleus. Hydrogen has an atomic number of 1, helium 2, carbon 6, nitrogen 7, oxygen 8, lead 82, and uranium 92. The atomic mass number, designated A, is equal to the total number of protons and neutrons in the nucleus.

3.2 Heat, Motion, and Pressure

From Where Does Heat Come?

Heat is a form of energy caused by the motion of tiny unseen particles, such as the molecules in a gas, which are in a state of ceaseless motion and permanent restlessness. These particles move randomly in all directions and do not contribute to the overall motion of the gas in which they reside.

The energy of motion is called *kinetic energy*, after the Greek word *kinesis* meaning "motion" – the word *cinema* has the same root, referring to motion pictures. Individual gas particles are changing direction constantly in an irregular zigzag trajectory within a star, and the kinetic energy arising from this haphazard motion is heat energy. In contrast, all of the particles in a star move together in the same overall direction, and they are responsible for the bulk kinetic energy of star motion (Fig. 3.2).

The molecules of every gas always are moving, and the hotter they become, the faster they move and the greater their kinetic energy. The lowest possible temperature is absolute zero, or zero on the kelvin scale, denoted K. At absolute zero, molecules cease to move and are completely at rest; they have no kinetic energy. At this temperature, the constituent particles stick together and behave as a frozen solid resembling ice.

Raise the temperature above absolute zero and the molecules begin to move about and collide, turning ice into liquid water. Further increase the temperature and the molecules can move so fast that they overcome the cohesive forces that bind them together. A gas is formed, capable of nearly unlimited expansion in all directions, as when water evaporates; lower the temperature to produce rain or snow.

Table 3.2. *Range of temperatures*

Location	Temperature (kelvin)
Absolute zero	0 K
Cosmic microwave background radiation	3 K
Water's freezing (triple) point	273 K
Water's boiling point	373 K
Incandescent light bulb	2,500 K
Visible solar disk	5,780 K
Center of the Sun	1.56×10^7 K
Atomic bomb	3.5×10^8 K
Particle accelerator (anti-proton creation)	10^{13} K
"Big bang"	10^{32} K

A metal can be changed into liquid and even vaporized if it is heated and made hot enough; if molten metal is cooled enough, it will solidify. Thus, by adding more heat, we can transform a solid into a liquid and then a gas; we can reverse the process by removing heat and lowering the temperature. All familiar objects exist in one of these three fundamental states: solid, liquid, or gas. At higher temperatures within cosmic objects, there is a fourth state of matter known as *plasma*, in which the atoms are torn into their subatomic ingredients. The range of temperatures found in the universe is illustrated in Table 3.2.

The German physician Jules Robert Mayer realized that heat is a form of energy, generally called "force" in his time, and that this energy can change form. This had a crucial role in the discovery of the conservation of energy. The heat energy might be produced by or transferred into another type of energy, but the energy never disappears. The total energy is conserved and the balance remains unchanged.

The English physicist James Prescott Joule provided experimental verification of this principle, in particular cases, in a lecture titled "On Matter, Living Force, and Heat." Today, we identify Joule's "living force" with kinetic energy. The joule unit of energy, abbreviated J, is appropriately named after him.

If we bring a hot body into contact with a colder one, the fast-moving particles in the hot body will collide, at the common boundary, with the adjacent slower-moving particles, transferring to them a part of the kinetic energy. The fast-moving particles gradually slow down, the slow ones accelerate, and a state of thermal equilibrium is obtained. A single temperature characterizes this thermal equilibrium.

Thermal Energy and Thermal Velocity

The portion of an object's internal energy that is responsible for its temperature is called the *thermal energy*, and the average thermal energy is proportional to the temperature. A particle is said to move at the *thermal velocity* when its kinetic energy is equal to its thermal energy. As might be suspected, this thermal velocity increases with temperature and decreases with particle mass. Hotter particles move faster and more massive ones move more slowly. The thermal speed of nitrogen molecules at sea level in air is about 506 m s^{-1},

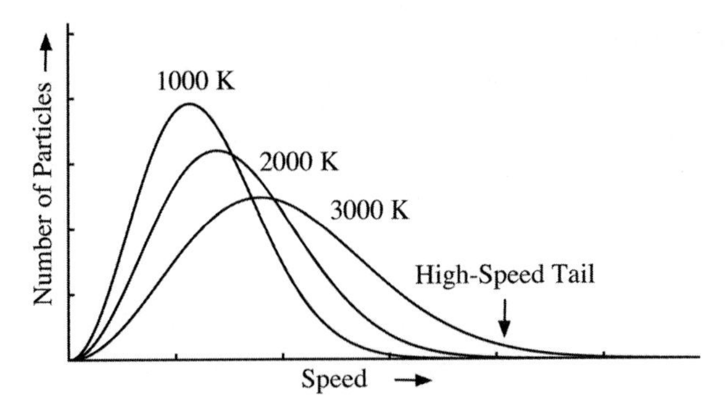

Figure 3.3. Maxwell distribution of particle speeds. The speeds of particles with the same mass and three different temperatures. The peak of this distribution shifts to higher speeds at higher temperatures. There is a small fraction of particles having high speed, residing in the high-speed tail of the distribution, and this fraction increases with temperature. The fraction of particles with low speed becomes smaller at higher temperatures but does not vanish. The peak also shifts to higher speeds at lower mass when the temperature is unchanged. The Scottish scientist James Clerk Maxwell (1831–1879) derived this distribution in 1860.

so they are rushing about at a speed faster than a vehicle moves – when the speedometer reads 60 miles per hour, the vehicle is moving at about 27 m s^{-1}. The abundant hydrogen atoms in the Sun's visible solar disk move at even faster speeds of about 12,000 m s^{-1} because the disk is hotter, at 5,780 K, and because hydrogen atoms are about 28 times less massive than nitrogen molecules.

Statistically Speaking, There Is a Distribution of Speeds

More than 150 years ago, the Scottish physicist James Clerk Maxwell introduced a statistical approach to the kinetic theory of gases, which recognizes that every gas particle has a different speed and that each collision between particles changes the speeds of those particles. He proposed that the numerous collisions between the large number of molecules in a gas produce a statistical distribution of speeds, in which all of the speeds might occur with a different and known probability.

In thermal equilibrium, the average value of the kinetic energy of particles in a gas is distributed equally among all of the particles, but this equality is only statistically true. Most particles move with the same average speed, but not all of them. Some move with faster than average speed and others move more slowly. That is, gas particles can gain or lose speed by collisions with one another, so they do not all move at the same average speed. In any given instant, the speed and kinetic energy of most of the particles are close to the average value, but there is always a small percentage that moves faster or slower than the average.

The *Maxwell speed distribution*, shown in Figure 3.3, gives the fraction of gas molecules, or other particles, moving at a particular speed at any given temperature and particle mass. The most probable speed is close to the average in thermal equilibrium; therefore, the most likely speed for any given type of particle increases with its temperature. However, there is a range of speeds, both higher and lower than the average value, and this range also increases with the temperature. In other words, the Maxwell distribution function becomes broader and its peak shifts to higher speeds when the temperature rises.

Table 3.3. *Range of pressures*

	Pressure[a] (pascal)
Interstellar medium	10^{-13} Pa
Beneath leg of a spider	1 Pa
Visible solar disk[b]	10 Pa
Atmospheric pressure on Mars	10^3 Pa
The Earth's atmosphere at sea level[c]	10^5 Pa
Inside a champagne bottle	5×10^5 Pa
Surface pressure of atmosphere on Venus	9×10^6 Pa
Inside a fully charged scuba tank	10^7 Pa
Center of the Earth	4×10^9 Pa
Center of Jupiter	7×10^{12} Pa
Center of the Sun	2×10^{16} Pa

[a] The SI unit of pressure is the pascal, denoted Pa, where 1 Pa = 1 N m^{-2}. The bar unit of pressure is used for planetary atmospheres, where 1 bar = 100,000 Pa.
[b] The gas pressure at the visible solar disk, known as the photosphere, is about the same as the vacuum pressure inside an incandescent light bulb.
[c] The standard atmosphere of the Earth has a sea-level pressure of 1.01325 bar = 101,325 Pa.

Although this Maxwell distribution function appears to be symmetrical, it is enhanced in a high-speed tail. Particles in the high-speed tail have greater kinetic energy than other particles in the distribution, and they have an important role in nuclear-fusion reactions that make the Sun and other stars shine.

The Maxwell speed distribution applies to all sorts of objects in thermal equilibrium, as long as there are many of them. In addition to atoms or molecules, it can be used to describe the speeds of numerous subatomic particles inside the Sun, which have been freed from their atomic bonds at very high temperatures. The distribution also describes the speeds of millions of stars that are collected together in star clusters, in which the stars behave like gas particles.

Gas Pressure

Why isn't the sky falling, as Chicken Little believed, because the Earth's gravity is relentlessly pulling it down? Because the atmosphere is warmed by the Sun, its atoms and molecules are in continuous motion and collide with one another, producing a pressure that prevents them from collapsing to the ground.

Moving particles exert a gas pressure, which increases with the temperature and the number of particles per unit volume – that is, number density. When a gas is compressed and the particles are pushed closer together, the gas particles collide more frequently, the gas heats up, and the temperature rises. When particles are crowded together, both the temperature and the gas pressure rise. In other words, closeness brings warmth.

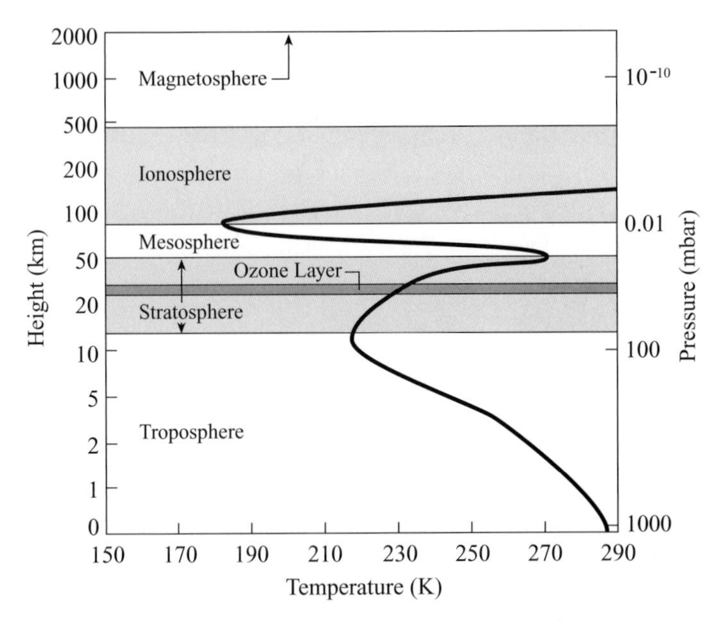

Figure 3.4. Earth's layered atmosphere. The pressure of the atmosphere (*right scale*) decreases with altitude (*left scale*). This is because fewer particles are able to overcome the Earth's gravitational pull and reach higher altitudes. The temperature (*bottom scale*) also decreases steadily with height in the ground-hugging troposphere, but the temperature increases in two higher regions that are heated by the Sun: the *stratosphere*, with its critical *ozone layer*, and the *ionosphere*. The stratosphere is heated mainly by ultraviolet radiation from the Sun, and the ionosphere is created and modulated by the Sun's x-ray and extreme ultraviolet radiation that breaks apart the atmospheric molecules, and strips electrons off their component atoms to produce ions. The process of ionization by the Sun's invisible rays releases heat to warm the ionosphere, so the temperature rises with altitude in it. In the ionosphere, at about 100 to 500 km above the ground, the temperatures skyrocket to higher values than anywhere else in the entire atmosphere. At higher altitudes, the atmosphere thins out into the exosphere, or the "exit to the outside sphere." The temperature is so hot out there, and the particles move so fast, that some atoms and molecules may slowly evaporate.

The SI unit of pressure is the *pascal*, abbreviated Pa, which is a force per unit area. Atmospheric pressure on the Earth and other planets often is measured in the bar unit of pressure, where 1 bar = 100,000 Pa. The Earth's standard atmospheric pressure at sea level is defined as 1.01325 bar. Representative pressures, which display the enormous cosmic range, are given in Table 3.3.

Any hot gas exerts gas pressure, and the gas pressure will vary with distance from whatever is heating the gas. The Earth's atmosphere, for example, is heated from below at the warm ground and from above by the Sun's radiation (Fig. 3.4 and Focus 3.1). Unlike our atmosphere, the Sun is heated from the inside, at the center of its hot dense core.

Radiation, wind, or a magnetic field also can produce a pressure, which is known as *radiation pressure*, *wind pressure*, or *magnetic pressure*. The radiation pressure of sunlight can be used to propel spacecraft using a solar sail, and both solar-radiation pressure and solar-wind pressure push different types of comet tails away from the Sun.

The inside of the Sun resembles the Earth's atmosphere in that it is dense at the bottom and rarefied at the top, except that the bottom is at the center of the Sun, which is much hotter

Focus 3.1 Density, Pressure, and Temperature of the Earth's Atmosphere

Our thin atmosphere is pulled close to the Earth by its gravity and suspended above the ground by molecular motion. Because air molecules are mainly far apart, our atmosphere is mostly empty space, and it always can be squeezed into a smaller volume. The atmosphere near the ground is compacted to its greatest density and pressure by the weight of the overlying air. Yet, even at the bottom of the atmosphere, the density is only about 1/1,000th of that of liquid water; an entire liter of this air weighs only 1 gram.

At greater heights, there is less air pushing down from above, so the compression is less and the pressure and density of air gradually fall off into the vacuum of space. At a height of 10 km, slightly higher than Mount Everest, the density of air has dropped to 10 percent of its value near the ground. No insects and few birds can fly in such rarefied air. At altitudes above about 50 km, the air is too thin to support a jet airplane.

We can infer the thinning of the air at greater heights by watching hawks circling above a meadow or open field. As the ground is heated by the Sun, the adjacent air warms up, expands, and becomes less dense, rising into the thinner atmosphere above. The rising air carries heat from the ground and distributes it to higher levels, giving free rides to the soaring birds. Hawks riding the currents of heated air sometimes rise so abruptly that it appears as if they were lifted and jerked up by strings.

The decrease in air pressure with height accounts for the ascent of balloons. When filled with a light gas, a balloon is buoyed upward by the greater pressure of the air beneath it. If the upward force of buoyancy exactly matches the downward weight of the balloon and its contents, the balloon will remain suspended at the same altitude, moving neither up nor down.

Not only does the atmospheric pressure decrease as we go upward, but the temperature of the air also changes; however, it is not a simple fall-off. It falls and rises in two full cycles as we move off into space (Fig. 3.4). An external force is energizing the atmosphere from outside, heating its outer layers; the invisible force is ultraviolet and x-ray radiation from the Sun.

and more dense. The weight of overlying particles compresses those beneath, increasing the density, temperature, and gas pressure toward the solar center. The moving particles are energized and the heat is sustained by nuclear-fusion processes occurring at the center of the Sun.

3.3 The Density and the Temperature within the Sun

What Supports the Sun?

Our Sun is a giant sphere of extremely hot gas, rarefied on the outside and compacted on the inside; unlike the Earth, it has no solid surface. The Sun's gases can expand into space when set free of solar gravity, and they can be noticeably compressed under the action of the Sun's gravitational force.

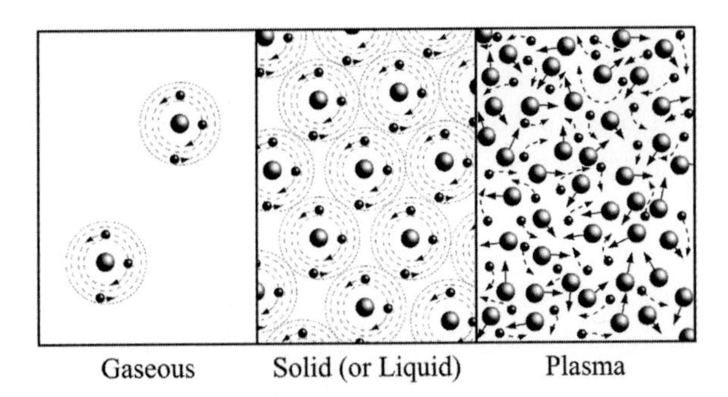

Figure 3.5. States of matter. The locations of atoms, large dashed circles; their component electrons, small filled circles; and central nucleus, large filled circles, for the gaseous (*left*), liquid or solid (*center*), and plasma (*right*) states of matter. In the gaseous state, the atoms are widely separated and free to move about. The atoms are practically touching one another in the solid and liquid states. At sufficiently high temperature and pressure, the atoms cease to exist and the plasma state is created. The atoms are torn into their constituents by frequent collisions at the high temperatures. The plasma consists of bare nuclei and unattached electrons moving about in random directions within the former empty space of atoms. In the plasma state, matter regains the compressibility of the gaseous state and plasma behaves like a gas.

Why does not the enormous mass and gravity of the Sun pull it all together into a tiny ball, and what holds up the gas within a gigantic sphere? The high compressibility of the solar gas brings about a rapid increase in density proceeding from its visible disk to its center. As a result of this crowding, the gas particles collide more frequently with higher speeds than elsewhere in the Sun. The compacted gas particles also push more vigorously outward, producing strong gas pressure that keeps the Sun from collapsing.

To understand how this works, imagine 100 mattresses stacked into a pile. The mattresses at the bottom must support those above, so they will be squeezed thin. Those at the top have little weight to carry, and they retain their original thickness. The particles at the center of the Sun similarly are squeezed into a smaller volume by the overlying material, so they become hotter and more densely concentrated. This results in increased pressure that supports the Sun from within.

Atoms Are Torn Apart within the Sun

Whole atoms are found only in the outer visible layers of the Sun, where the temperature is a relatively cool 5,780 K. Raise the temperature by only a factor of three, to about 17,000 K – which happens just beneath the solar disk we see with our eyes – and the Sun's atoms are stripped bare, losing their identity.

The hot atoms move rapidly here and there, colliding at high speeds, and the violent force of these collisions is enough to fragment the atoms into their subatomic constituents. Because the Sun is mostly hydrogen, its interior consists mainly of protons, the nuclei of hydrogen atoms, and free electrons that have been torn off the atoms by innumerable collisions and set free to move throughout the Sun.

What is left is plasma, a seething mass of electrically charged particles: the electrons and the protons (Fig. 3.5). The electrical charge of the protons balances and cancels that of the electrons, which have been removed from atoms to also release the protons; therefore, the plasma has no net electrical charge. The Sun is simply one huge mass of incandescent plasma, compressed on the inside and more tenuous farther out.

Plasma has been called the *fourth state of matter* to distinguish it from the gaseous, liquid, and solid states. The interiors of most stars are plasma, but protons and electrons still behave like a gas that is described by thermal equilibrium and the Maxwellian speed distribution.

With their electrons gone, the hydrogen nuclei (i.e., protons) can be packed more tightly than complete atoms. This is because all atoms are largely empty space, with their electrons located at relatively remote distance from their nuclei. The bare hydrogen nuclei, the protons, can be squeezed together within the empty space of former atoms.

A Hot, Dense Core

The Sun's heat is sustained from inside, at its center, and because heat flows from hot to cold places, the central solar furnace must be hotter than its visible disk. The amount of heat supplied at the Sun's center is just sufficient to hold up the overlying material, so the central gas pressure supports the inward pull of the entire star.

In this condition, which is known as *hydrostatic equilibrium*, the central pressure needed to resist the weight of the overlying gas is 223,000 million times the pressure of the Earth's atmosphere at sea level. Under this extraordinary pressure, the solar material is compressed to a density that can exceed that of normally solid or liquid bodies. Deep down inside, within the Sun's dense central core, the density has increased to more than 10 times greater than the density of solid lead but still behaves like a gas.

The central temperature can be estimated by assuming that each proton down there is hot enough and moving fast enough to counteract the gravitational compression that it experiences from the rest of the star. This balanced condition occurs at a central temperature of 15.6 million K. If the temperature of a kitchen stove were brought to this value, its thermal radiation would burn up half of a good-sized town.

The center of the Sun is slightly more than 100 times denser than the Sun taken as a whole, at a mean value near that of water. In contrast to both the central and mean densities, the outer layers of the Sun are quite rarefied because there is less overlying material to support as distance from the center increases. The compression is less, so the gas becomes thinner and cooler (Fig. 3.6). Halfway from the center of the Sun to the visible disk, the density is the same as that of water; about nine tenths of the distance from the center to the Sun's apparent edge, there is material as tenuous as the transparent air that we breathe on the Earth.

At the visible solar disk, the rarefied gas is about 1,000 times less dense than the atmosphere at sea level. Out there, in the more rarefied outer parts of the Sun, the pressure is less than that beneath the leg of a spider, and the temperature has fallen to 5,780 K. Examination of this outer, cooler solar atmosphere indicates the elemental constituents of the Sun.

3.4 What Is the Sun Made Of?

When we glance at the Sun, it looks like a great featureless yellow disk. Our eyes are looking through the transparent outer atmosphere of the Sun, to a level where the solar gas becomes opaque enough to see. It resembles looking into the distance on a foggy day; the fog builds up with distance until it blocks any further view. When we closely scrutinize the visible Sun, we can determine the composition of the opaque solar fog.

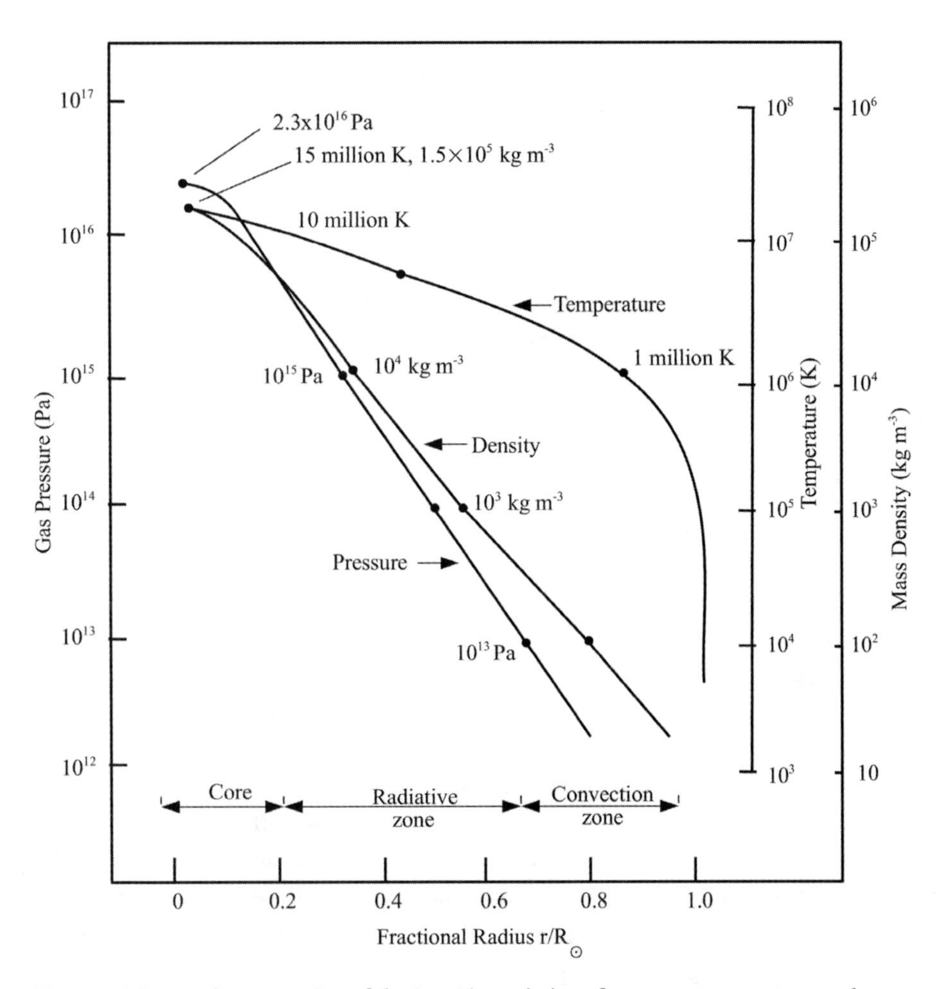

Figure 3.6. Internal compression of the Sun. The variation of pressure, temperature, and mass density with fractional radial distance from the Sun's center *(left)* to its visible disk *(right)*. At the Sun's center, the temperature is 15.6 million K and the mass density is 151,300 kg m^{-3}; the central pressure is 2.33 × 10^{16} Pa, or 233 billion times that of the Earth's atmosphere at sea level (one bar is equivalent to 100,000 Pa). Nuclear reactions occur only in the central core to about 25 percent of the Sun's radius. The energy produced in the core is transported by radiation to 71 percent of the star's radius, where the temperature has dropped to about 2 million K and the density has fallen to about 200 kg m^{-3}. The energy then is transported by convection out to the Sun's visible disk, known as the *photosphere*, where the temperature is 5,780 K, and the pressure and density have dropped off the scales of the graph.

The visible disk of the Sun is called the *photosphere*, which simply means the sphere in which sunlight originates – from the Greek *photos* for "light." When the photosphere's light is spread out into its different colors or wavelengths, it is cut by several dark gaps. They were first noticed by the English astronomer William Hyde Wollaston and then investigated in greater detail by the German astronomer Joseph von Fraunhofer, who detected and catalogued more than 300 gaps, assigning upper and lowercase Roman letters to the most prominent.

By directing the incoming sunlight through a slit and then dispersing it with a prism, Fraunhofer was able to overcome the blurring of colors from different parts of the Sun's disk, discovering numerous dark features in this spectral display. When coarse wavelength

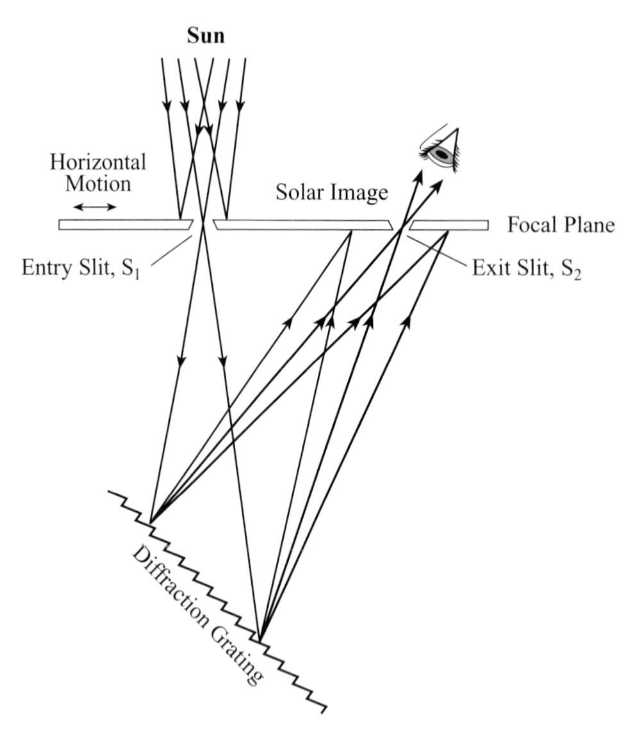

Figure 3.7. Spectroheliograph. A small section of the Sun's image at the focal plane of a telescope is selected with a narrow entry slit, S_1, and this light passes to a diffraction grating to produce a spectrum. A second slit, S_2, at the focal plane selects a specific wavelength from the spectrum. If the plate containing the two slits is moved horizontally, then the entrance slit passes adjacent strips of the solar image. The light leaving the moving exit slit then builds up an image of the Sun at a specific wavelength.

resolution is used, adjacent bright emission obscures the dark places, which are no longer found. An instrument similar to the one that Fraunhofer used to observe these features is known as a *spectroscope*, a joining of *spectro* for "spectrum" and *scope* for "telescope."

Fraunhofer's early life was grim. He was the eleventh child of an impoverished father, who died when the boy was just 12 years old. Two years later, the slum building in which Fraunhofer lived collapsed, trapping him and killing everyone else inside. When the Bavarian prince, Maxmilian Joseph, heard of the ordeal, he rescued Fraunhofer from his misery, giving him enough gold ducats to follow an interest in optics, lens making, and the Sun.

The dark gaps of missing colors found in a display of the Sun's radiation intensity as a function of wavelength, or in its spectrum, are now called *lines* because they each look like a line in the spectral display. They are designated further as *absorption lines* because they are produced when atoms in a cool, tenuous gas absorb the radiation of hot, dense underlying material. The term *Fraunhofer absorption line* also is used, in recognition of his investigations. Such lines also can appear in emission under different circumstances.

The Sun is so bright that its light can be spread out into small wavelength intervals with enough intensity to be detected. The instrument used to make and record such a spectrum is called a *spectroheliograph*, a composite word consisting of *spectro* for "spectrum," *helio* for the "Sun," and *graph* for "record" (Fig. 3.7). It uses the grooves of a diffraction grating to reflect

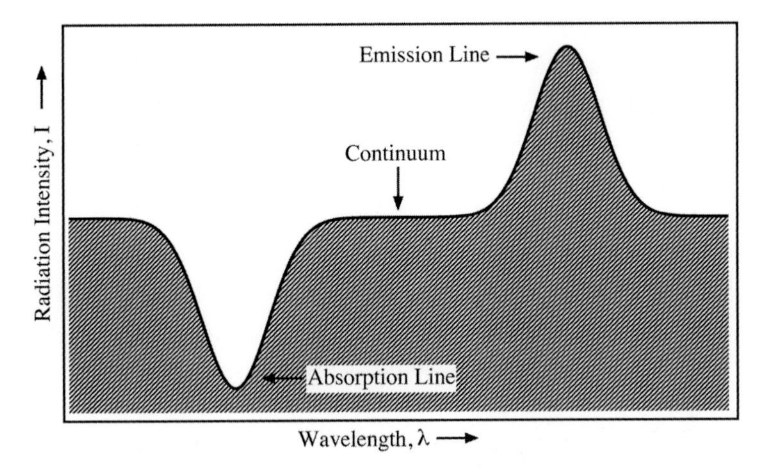

Figure 3.8. Absorption and emission lines. The spectrum of a star or other cosmic object displays the intensity of its radiation as a function of wavelength, denoted by the Greek symbol, λ. Any hot gas radiates at all wavelengths, producing a continuum spectrum with an intensity that varies with the wavelength and depends on the temperature. When this thermal radiation passes through an outer, cooler layer of a star, the atoms in this layer can produce absorption at a specific wavelength. This feature is called an *absorption line* because it looks like a line in the spectrum. When atoms are excited at high temperatures, they can radiate an *emission line*. The line wavelength indicates which atom is responsible for the absorption or emission. The intensity of a stellar line is related to both the number of atoms and the physical conditions in the star's atmosphere. The lines have a width and a shape that can be observed under close scrutiny in wavelength, and they provide information about local physical conditions. The motion of the absorbing or emitting atoms, for example, broadens the lines (see Fig. 3.13).

sunlight into different locations according to color or wavelength, similar to what a rainbow and compact disk do.

The Sun's absorption lines provided the first clues to the composition of the stars. In the mid-nineteenth century, the German physicist Gustav Kirchhoff and his chemist colleague Robert Bunsen showed that every chemical element, when burned and vaporized as a gas, emits brightly colored lines the unique wavelengths of which coincide with those of the dark absorption lines in the Sun's spectrum. By comparing the Sun's absorption lines with the emission lines of elements vaporized in the laboratory, Kirchhoff identified in the solar atmosphere several elements known on the Earth. As Bunsen wrote:

> At the moment I am occupied by an investigation with Kirchhoff, which does not allow us to sleep. Kirchhoff has made a totally unexpected discovery, inasmuch as he has found out the cause for the dark lines in the solar spectrum and can produce these lines artificially intensified both in the solar spectrum and in the continuous spectrum of a flame, their position being identical with that of Fraunhofer's lines. Hence the path is opened for the determination of the chemical composition of the Sun and the fixed stars.[4]

There are both absorption and emission lines (Fig. 3.8), and either type can be used to fingerprint an element and to identify the ingredients of the cosmos. For stars, it is mainly absorption lines that are used to determine the composition of their relatively cool, outer atmospheres, whereas the ingredients of the hotter, rarefied emission nebulae are established from observations of emission lines.

The lines designated by Fraunhofer with the letters H and K were associated with calcium, and vaporized iron was assigned the letter E. The letter b was attributed to magnesium, and the close pair of dark lines in the yellow, specified by the letter D, were attributed to sodium; they produce the distinctive yellow color of the sodium-vapor streetlights used at the time. This suggested that the Sun – and presumably all of the other stars – are composed of terrestrial elements and it unlocked the chemistry of the universe.

Each chemical element – and only that element – produces a unique set, or pattern, of wavelengths at which the dark lines fall. It is as if every element has its own characteristic barcode that can be used to identify it, as a fingerprint or DNA sample might identify a criminal. Every one of the numerous absorption lines found in the Sun's spectra were identified with a specific chemical element or compound.

When Kirchhoff and Bunsen vaporized an individual element in a flame, the hot vapor produced a distinctive pattern of sharply defined, bright lines. Moreover, when the light produced by a hot radiating object (e.g., a tungsten lamp) was passed through the cooler vaporized gas, dark lines were produced at exactly the same locations. Kirchhoff generalized this into a law stating that the powers of emission and absorption of a body at any particular wavelength are the same at any given temperature. He also concluded that the visible solar disk was hot and incandescent, producing a continuum spectrum (i.e., the type without lines), which became crossed by the dark Fraunhofer lines when passing through cooler overlying gas.

The Swedish spectroscopist Anders Jonas Ångström subsequently identified hydrogen in the solar spectrum; it is associated with Fraunhofer's letters C and F. Ångström published a comprehensive atlas of more than 1,000 absorption lines in the Sun's spectrum, identifying them with hydrogen, sodium, calcium, barium, strontium, magnesium, copper, iron, chromium, nickel, cobalt, zinc, and gold. He also established a scale of wavelengths for measuring the spectral lines, now known as the Ångström, abbreviated Å, where $1 \text{ Å} = 10^{-10}$ m $= 0.1$ nm.

The Fraunhofer lines designated A and B are unrelated to the composition of the Sun; they only appear in spectra gathered beneath the Earth's atmosphere. Molecular oxygen in the terrestrial atmosphere absorbs sunlight at the wavelengths of the A and B Fraunhofer lines, creating the dark lines that are superposed on the Sun's spectrum.

Some of the Sun's absorption lines are very strong, extracting great amounts of energy from sunlight. They are produced by hydrogen, sodium, magnesium, calcium, and iron (Table 3.4), but iron accounts for more lines than any other element. Because abundant heavy iron accounts for the Earth's high average mass density, and also because most of the other solar lines corresponded to elements known on the Earth, it was initially supposed that the Sun is made of the same material as the Earth – but this is only partly true. Many of the Sun's visible spectral lines were associated with hydrogen, a terrestrially rare element.

The Earth is a tiny dirt speck in the cosmos, an anomaly, for it is primarily made of heavy elements that are relatively uncommon in the Sun and the rest of the universe. Hydrogen is

Table 3.4. *Prominent absorption lines and elements detected in sunlight*[a]

Wavelength (nm)	Fraunhofer Letter	Element Symbol and Name
393.368	K	Ionized Calcium, Ca II
396.849	H	Ionized Calcium, Ca II
410.175	h	Hydrogen, H_δ, Balmer delta transition
422.674	g	Neutral Calcium, Ca I
431.0	G	CH molecule
434.048		Hydrogen, H_γ, Balmer gamma transition
438.356	d	Neutral Iron, Fe I
486.134	F	Hydrogen, H_β, Balmer beta transition
516.733	b_4	Neutral Magnesium, Mg I
517.270	b_2	Neutral Magnesium, Mg I
518.362	b_1	Neutral Magnesium, Mg I
526.955	E	Neutral Iron, Fe I
588.997[b]	D_2	Neutral Sodium, Na I
589.594	D_1	Neutral Sodium, Na I
656.281	C	Hydrogen, H_α, Balmer alpha transition
686.719	B	Molecular Oxygen, O_2, in the Earth's atmosphere
759.370	A	Molecular Oxygen, O_2, in the Earth's atmosphere

[a] The photosphere is the visible solar disk. The wavelengths are in nanometer units, where 1 nanometer $= 10^{-9}$ m $=$ 1 nm. Astronomers often use the Ångström unit of wavelength, where 1 Ångström = 1 Å = 0.1 nm. Joseph von Fraunhofer used the letters to designate the spectral lines before they were chemically identified, and the subscripts denote components that were not resolved by Fraunhofer. A Roman numeral I after an element symbol denotes an electrically neutral, or un-ionized, atom, with no electrons missing; whereas the Roman numeral II denotes a singly ionized atom with one electron missing. The lines A and B are produced by molecular oxygen in the terrestrial atmosphere.

[b] Fraunhofer's D line includes the two sodium lines, designated D_1 and D_2, and the helium line at 587.6 nanometers, designated D_3.

about 1 million times more abundant than iron in the Sun, but iron is a main constituent of the Earth, which cannot even retain hydrogen gas in its atmosphere for any significant length of time.

Helium, the second-most abundant element on the Sun, is so rare on the Earth that it was first discovered in the Sun. The French astronomer Pierre Jules César Janssen observed an unidentified yellow emission line in the solar spectrum during the solar eclipse of August 18, 1868, which he observed from India. The emission originated in the chromosphere, a thin, slightly hotter layer of gas that lies just above the visible solar disk, or photosphere, and became visible when the Moon blocked the bright glare of the photosphere.

The British astronomer Sir Joseph Norman Lockyer observed the same yellow line in the solar spectrum without a solar eclipse, and subsequently was knighted for this discovery. It was probably not until 1869 that Lockyer convinced himself that the yellow line could not

Table 3.5. *The five most abundant elements in the solar photosphere*

Element	Symbol	Atomic Number, Z	Abundance[a] (logarithmic)	Discovery on the Earth
Hydrogen	H	1	12.00	1766
Helium	He	2	[10.93 $\pm$ 0.01]	1895[b]
Carbon	C	6	8.43 $\pm$ 0.05	(ancient)
Nitrogen	N	7	7.83 $\pm$ 0.05	1772
Oxygen	O	8	8.69 $\pm$ 0.05	1774

[a] Logarithm of the abundance in the solar photosphere, normalized to hydrogen H = 12.00, or an abundance of 1.00×10^{12}. The indirect solar estimate for helium is marked with [ ]. The data are from Apslund, Grevesse, Sauval, and Scott, "The Chemical Composition of the Sun," *Annual Review of Astronomy and Astrophysics* 47, 481 (2009).

[b] Helium was discovered on the Sun in 1868, but it was not found on the Earth until 1895.

be identified with any known terrestrial element. He named the element "helium" after the Greek Sun God, *Helios*, who daily traveled across the sky in a chariot of fire drawn by four swift horses.

Helium was not found on the Earth until 27 years after its discovery, when the Scottish chemist Sir William Ramsay discovered its spectral signature in a gaseous emission given off by a heated uranium mineral, cleveite. Helium is one of the noble, or inert, gases that include helium, neon, argon, krypton, xenon, and radon. These so-called noble gases do not combine with most other chemical elements, similar to people of nobility who are unwilling to associate with ordinary, common folk.

Today, helium is used on the Earth in a variety of ways, including inflating balloons and, in its liquid state, to keep sensitive electronic equipment cold. Although plentiful in the Sun, helium is almost nonexistent on the Earth. It is so terrestrially rare that we are in danger of running out of helium during this century. Japanese scientists plan to mine helium from the Moon's surface, where the terrestrially rare element has been implanted by winds from the Sun.

Altogether, 92.1 percent of the atoms of the Sun are hydrogen atoms, 7.8 percent are helium atoms, and all of the other heavier elements comprise only 0.1 percent. In contrast, the main ingredients of the rocky Earth are the heavier elements such as silicon and iron, which explains the Earth's high mass density – about four times that of the Sun, which is only about as dense as water.

By mass, hydrogen accounts for 71.54 percent of the Sun; the helium amounts to 27.03 percent by mass because it is more massive than hydrogen. All of the heavier solar elements amount to only 1.4 percent by mass.

The abundance of the five most abundant elements in the Sun are listed in Table 3.5, normalized to a hydrogen abundance of 1 million million, or 10^{12}. There is a systematic decrease in the abundance of solar elements with increasing atomic number (Fig. 3.9), but

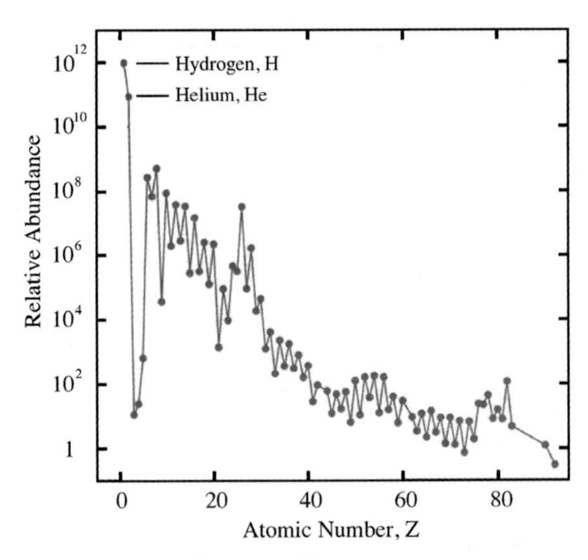

Figure 3.9. Elemental abundance in solar photosphere. The relative abundance of the elements in the Sun's visible disk, the photosphere, plotted as a function of atomic number. The atomic number, denoted by Z, is the number of protons in an atom's nucleus, or roughly half the atomic weight. Heavy elements, with high atomic numbers, are less abundant than light ones, with low atomic numbers; the most abundant element in the Sun is the lightest element, hydrogen. Helium is the second most abundant solar element. The abundance data are plotted in a logarithmic scale normalized to 1 million million, or 1.0×10^{12}, for hydrogen. (Adapted from Martin Asplund, Nicolas Grevesse, A. Jacques Sauval, and Pat Scott, "The Chemical Composition of the Sun," *Annual Review of Astronomy and Astrophysics* 47, 481–522 [2009].)

with a noticeable gap of unexpectedly low abundance for the light elements between helium and carbon.

3.5 Quantization of Atomic Systems

As discussed in Section 3.1, most of the mass of an atom is concentrated in its relatively small nucleus, which is surrounded by electrons. The nucleus has a positive charge due to the protons in it and is about 100,000 times smaller than the atom. The negatively charged electrons keep the atom distended, enlarging its shape; as a result, the atom is mostly empty space.

According to Rutherford's model of the atom, the electrons revolve around the central nucleus, somewhat like the planets that endlessly whirl around the Sun. Unlike a planet, however, an electron is electrically charged, and a revolving charge emits electromagnetic radiation. That is how radio signals are broadcast, by moving electrons through wires to generate the radiation.

An electron revolving in an atom-sized orbit will radiate light waves and, as a consequence of this emission, the electron will steadily lose its kinetic energy of motion. This means that electrons cannot be moving perpetually around the nucleus. Calculations indicate that any atomic electron should lose its orbital motion and spiral into the atom's nucleus in less than 1 second.

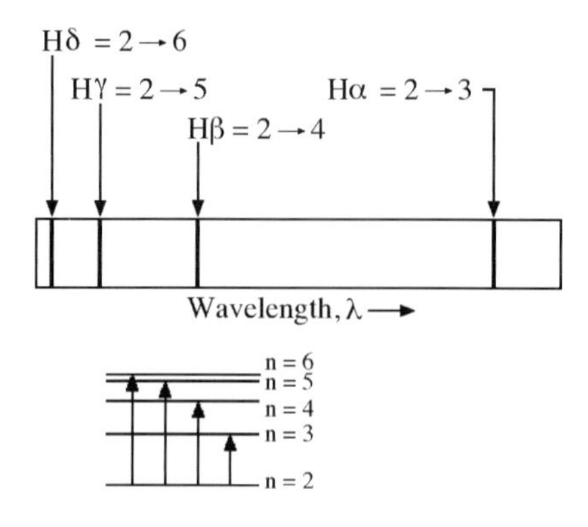

Figure 3.10. Balmer lines of hydrogen. The spectrum of the Sun's optically visible light exhibits four strong absorption lines that are attributed to hydrogen, whose line wavelengths are spaced closer together at shorter wavelengths (*top*). These lines are designated Hα at a red wavelength of 656.3 nm, Hβ at a wavelength of 486.1 nm, Hγ at the blue 434.1 nm, and Hδ at the violet 410.2 nm, where 1 nanometer = 1 nm = 10^{-9} m. These spectral features originate when an electron in a hydrogen atom moves from a low to a high electron orbit, the orbital energy of which is a function of the integer n (*bottom*). They were named Balmer lines after the Swiss mathematics teacher Johann Balmer (1825–1898), who first derived an equation in 1895 that describes their wavelengths in terms of integers.

Something was wrong with the basic assumptions of classical physics, which did not seem to apply to the small scales within an atom. Mysterious unknown forces seemed to be keeping the atomic electrons in perpetual motion around the nucleus, always remaining at a comparatively large distance from it. An unexpected feature of the Sun's spectral lines, which are absorbed or emitted by electrons, provided a resolution of this enigma. The spectral lines appear only at a set of well-defined wavelengths rather than at all wavelengths.

Moreover, the line wavelengths displayed an ordered arrangement, which helped describe the internal behavior of an atom. That is, the wavelengths of spectral lines from hydrogen indicated that electrons must follow certain "rules" if they want to belong to an atom. Not just any behavior is allowed and only certain orbits are permitted.

Adjacent hydrogen lines in the spectrum of the Sun or any other cosmic object systematically crowd together at shorter wavelengths (Fig. 3.10). The Swiss mathematics teacher Johann Balmer found a simple equation that describes their regular spacing.

The Danish physicist Niels Bohr explained Balmer's equation by a model of the hydrogen atom, now known as the *Bohr atom* (Fig. 3.11). In this model, a single electron in a hydrogen atom revolves about the nuclear proton in specific orbits with definite, quantized values of energy. An electron only emits or absorbs radiation when jumping between those allowed orbits, each jump associated with a specific energy and a single wavelength, like one pure note. If an electron jumps from a low-energy to a high-energy orbit, it absorbs radiation at this wavelength; radiation is emitted at exactly the same wavelength when the electron

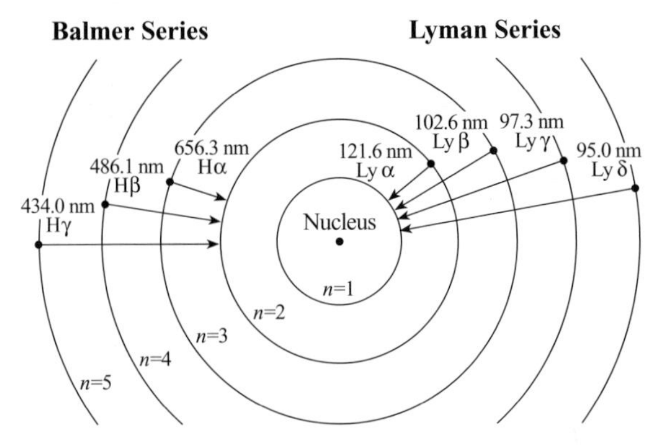

Figure 3.11. Bohr atom. In this model, proposed in 1913 by the Danish physicist Niels Bohr (1885–1962), a hydrogen atom's one electron revolves around the hydrogen nucleus, a single proton, in well-defined orbits described by the integer n = 1, 2, 3, 4, 5,.... An electron absorbs or emits radiation when it transitions between these allowed orbits. The electron can jump upward, to orbits with larger n, by absorption of a radiation photon of exactly the right energy, equal to the energy difference between the orbits; the electron can jump down to lower orbits, of smaller n, with the emission of radiation of that same energy and wavelength. Transitions that begin or end on the n = 2 orbit define the Balmer series observed at visible wavelengths. They are designated by Hα, Hβ, Hγ,.... The Lyman series, with transitions from the first orbit at n = 1, is detected at ultraviolet wavelengths. The orbits are not drawn to scale because the size of their radius increases with the square of the integer n.

jumps the opposite way. This unique wavelength is related to the difference between the two orbital energies. Bohr was awarded the 1912 Nobel Prize in Physics for these investigations of the structure of atoms and the radiation emanating from them.

Because only quantized orbits are allowed, spectral lines are produced only at specific wavelengths that characterize or identify an atom. An atom or molecule can absorb or emit a particular type of light only if it resonates to that light's energy. As it turns out, the resonating wavelengths or energies of each atom are unique.

A little more than a decade before Bohr introduced his model, Max Planck explained continuum thermal radiation, without lines, by supposing that the emission and absorption of light takes place only in the form of certain discrete portions, or *quanta*, of energy, now known as *photons*. He had quantized radiation. The energy of each separate light quantum, or photon, is proportional to the frequency of light, or inversely proportional to its wavelength.

Bohr went one step further and quantized the energy of motion of the electrons revolving in an atom. He proposed that the mechanical energy of any moving subatomic particle could take on only one of a certain set of discrete values, in an entirely new quantum mechanics. That is, he quantized the electron's angular momentum and energy. Because the quantum of an electron's orbital energy increases with the decreasing dimensions of the orbit, this suggested that quantum mechanics would become important only for very small, subatomic physical scales.

The permitted orbital energy increases with decreasing radius or size of the electron orbit. The closer an electron is to the nucleus, the greater is its allowed energy. Both the energy and radius of an electron's orbit vary as the inverse square of the quantum number.

3.6 Excited Atoms

Because atoms reside together in great numbers, we must use a statistical approach to determine their average properties, and their level of excitation depends on the temperature and the density, which influence how often the particles collide and become excited. The higher the temperature, the faster the particles move, on average, and the more frequent their collisions. When the particles are packed together in greater numbers per unit volume, with a greater density, the collision frequency also increases.

The number of atoms in the lowest possible energy state, called the *ground state*, is greater than the number in any other state of energy, essentially because it is easier to stay in the lowest energy state. Because it requires less energy, an atom prefers it; it is more difficult to exist and remain in an excited state of energy. Outside agencies are required to sustain the excitation, sometimes like a parent telling his teenage child to get out of bed.

The allowed energy levels of the electron orbits within an atom can be compared to the rungs of a ladder. Electrons can climb the ladder's energy rungs when an atom either collides with other atoms or absorbs radiation. Once an electron is up on a rung, it can jump downward, releasing the energy it attained to get there. It is easier to stay on the ground and never climb the ladder; therefore, most atoms are usually in the ground state. Because each type of atom or ion has a different type of ladder, with energy rungs located at different places, every atom or ion exhibits unique absorption or emission lines.

The number of atoms that exist in a given energy level depends on the energy of that level. Lower energy levels are naturally more populated than higher energy levels because it takes more energy to excite the higher states; the energy levels of an atom are populated inversely and exponentially as a function of the energy. When collisions are the dominant process that influences the energy-level population, then the ratio of the population of two energy levels of a given atom depends on the temperature. At higher temperatures, more atoms are pumped up to the more energetic states.

Under conditions of local thermodynamic equilibrium, the ratio of the number of atoms at two different energies depends on their energy difference divided by the temperature. At a higher temperature, there are more atoms with higher energy; there always are more atoms with the lowest possible energy, the ground state. The number ratio is known as the *Bolzmann distribution*, named after Ludwig Boltzmann who derived it in 1872.

Individual atoms in a collection of atoms are always moving about, colliding with other particles, becoming excited, and radiating away that excitation. As temperature increases, activity also increases. A measure of the energy emitted depends on the number of excited atoms, which varies as the temperature, the energy of the transition, and the transition probabilities tabulated in spectroscopic databases found on the Internet.

3.7 Ionization and Element Abundance in the Sun and Other Stars

Because a greater number of atoms will absorb more light, the relative darkness of the absorption lines in the Sun's spectrum should establish the relative abundance of the elements there. That is, darker, stronger absorption lines generally indicate greater absorption and therefore larger amounts of the absorbing element; however, the strength of an element's absorption lines depends only to some extent on the element's abundance. There were other mitigating circumstances, so unlocking the chemistry of the universe was not as straightforward as scientists initially supposed.

Atoms, for example, exist in altered physical states at the high temperatures that prevail within stars, which can result in a change in the wavelength and intensity of the spectral lines observed in stellar atmospheres. The Sun had to be hotter than typical laboratory temperatures or it would not shine so brightly, and different temperatures and pressures would prevail at various locations within the solar atmosphere. This suggested that an element would display different spectral lines depending on the physical conditions of the solar region in which it was located.

Moreover, some stars showed conspicuous lines other than the dominant lines in the Sun's spectrum, suggesting that different stars have different compositions; perhaps this was instead related to the stellar temperature.

To understand all of these differences, scientists needed to obtain a thorough knowledge of atomic structure, including the discoveries of the atomic electrons and the atomic nucleus, which indicated that the stellar temperature indeed would be important. At sufficiently high temperatures, electrons can be removed from atoms, thereby producing ions whose spectral lines can differ from their atomic counterparts.

The atoms in the Sun, for example, remain electrically neutral, or un-ionized, at only the relatively low temperatures of the visible photosphere, the temperature of which is 5,780 K. Immediately above the photosphere is the hotter *chromosphere*, the temperature of which can rise to 20,000 K. At this temperature, the thermal energy is roughly equal to the ionization energy of hydrogen.

When the theory of ionization in stellar atmospheres was developed, it became clear that the presence or absence of specific spectral lines did not necessarily indicate the chemical composition of a star's atmosphere. In 1920, Meghnad Saha, a young lecturer at the University of Calcutta, demonstrated that the spectral lines of different elements are excited under different conditions of temperature and pressure. This set the stage for demonstrating that many stars have similar compositions.

In his analysis, Saha demonstrated the analogy between the dissociation of molecules and the ionization of atoms. He replaced the mass of the atom with the mass of the electron in the expression for the degree of dissociation of a molecule, thereby obtaining his now-famous ionization equation. This formula, known as the *Saha equation*, relates the degree of ionization of an atom to temperature and pressure. It indicates that the relative intensities of a star's different spectral lines are caused, in part, by differences in the pressure and temperature of the stellar atmosphere.

Saha used his ionization equation in a physical theory for stellar spectra, specifying temperatures of stars of different spectral type. His results indicated that differences in stellar spectra are caused by differences in excitation rather than in chemical composition. The English astrophysicists Ralph A. Fowler and Edward Milne then showed that the number of atoms or ions responsible for the production of a spectral line could be estimated from the line intensity, once the temperature and pressure of the stellar atmosphere are known. This paved the way for the work of Cecilia Payne, who showed that stars with different spectra have essentially the same composition, and eventually for the realization that the

Focus 3.2 Hydrogen Is the Most Abundant Element in the Sun and Most Other Stars

In a brilliant doctoral dissertation, published in 1925, the American astronomer Cecilia H. Payne showed that the atmospheres of virtually every luminous, middle-aged star have the same ingredients. Payne, later Payne–Gaposchkin, eventually became the first female professor in the Faculty of Arts and Sciences at Harvard University, where she had studied. Her calculations also indicated that hydrogen is by far the most abundant element in the Sun and most other stars. However, she could not believe that the composition of stars differed so enormously from that of the Earth, where hydrogen is found rarely, so she mistrusted her understanding of the hydrogen atom. Prominent astronomers of the time also did not think that hydrogen was the main ingredient of the Sun and other stars, and this may have had a role in her considerations.

Subsequent detailed investigations of the Sun's absorption-line intensities indicated that the Sun is composed mainly of the lightest element, hydrogen, accounting for 92.1 percent of the number of atoms in the Sun.

The Danish astronomer Bengt Strömgren next calculated the hydrogen content in the interiors of stars, assuming that they are chemically homogeneous, and showed that their observed luminosities require that the entire star, not just its atmosphere, be composed predominantly of hydrogen.

We now know that very old stars have very few elements other than hydrogen and helium; these stars probably have existed since our Galaxy formed. Middle-aged stars like the Sun contain noticeable but still small amounts of heavier elements.

Hydrogen is the most abundant element in the stellar universe and there was nothing wrong with Payne's calculations. The Earth simply does not have sufficient gravity to retain hydrogen in its atmosphere for any length of time. Any hydrogen gas that our young planet once might have had must have evaporated while the Earth was forming and has long since escaped or become locked into water or surface rocks.

lightest element, hydrogen, is by far the most abundant element in most stars (Focus 3.2). This discovery also had a fundamental role in understanding how the Sun shines: by fusion reactions of the nucleus of the hydrogen atom, the proton.

In addition to specifying the compositions of stars, detailed observations of absorption or emission lines yield information about the temperature, density, motion, and magnetism of the Sun and other stars, as well as rarefied nebulae, interstellar matter, and various different cosmic objects.

3.8 Altering Spectral Lines

Radial Motion Produces a Wavelength Shift

Just as a source of sound can vary in pitch or wavelength, depending on its motion, the wavelength of electromagnetic radiation shifts when the emitting source moves with

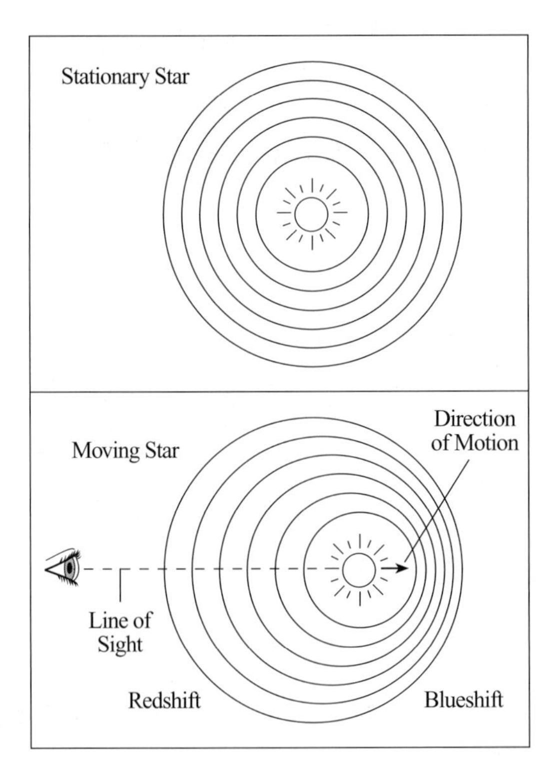

Figure 3.12. Doppler effect. A stationary source of radiation (*top*) emits regularly spaced light waves that get stretched out or scrunched up if the source moves (*bottom*). Here, a star moving away (*bottom right*) from the observer (*bottom left*) is shown. The stretching of light waves that occurs when the source moves away from an observer along the line of sight is called a *redshift* because red light waves are relatively long visible light waves. The compression of light waves that occurs when the source moves along the line of sight toward an observer is called a *blueshift* because blue light waves are relatively short. The wavelength change, from the stationary to moving condition, is called the *Doppler shift*, and its size provides a measurement of radial velocity, or the speed of the component of the source's motion along the line of sight. The Doppler effect is named after the Austrian physicist Christian Doppler (1803–1853), who first considered it in 1842.

respect to the observer. Such a shift is named after the Austrian scientist, mathematician, and schoolteacher Christian Doppler, who provided an explanation for it. We notice the effect on sound waves when listening to the changing pitch of a passing ambulance siren. The tone of the siren is higher as the ambulance approaches and lower when it moves away.

As a result of the Doppler effect, the wavelength of a spectral line that an astronomer observes can differ from the emitted line wavelength. The size of this wavelength shift depends on the relative speed of the radiating source along the observer's line of sight, known as the *radial velocity*. The greater the radial velocity, the larger the Doppler shift. In 1868, the English astronomer William Huggins was the first to use this method to determine the velocity of a star moving away from the Earth.

The Doppler shift is toward longer, redder wavelengths in the visible part of the electromagnetic spectrum and therefore also is known as a *redshift* (Fig. 3.12). When the motion is toward the observer, there is a blueshift to shorter, bluer wavelengths.

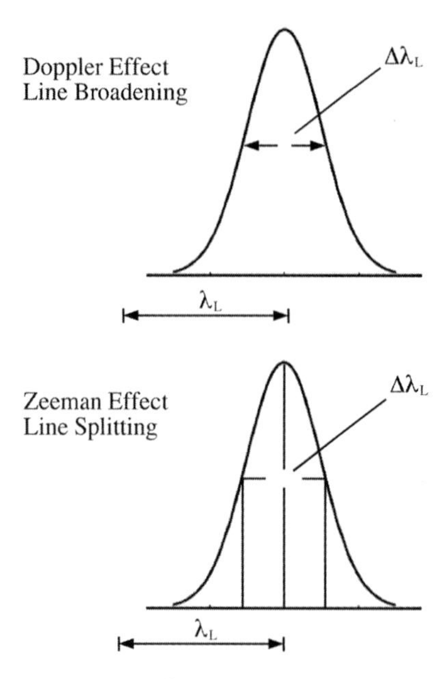

Figure 3.13. Effects that broaden a spectral line. The motion of absorbing or emitting atoms can broaden a line to the short-wavelength and long-wavelength sides of the resting, or nonmoving, wavelength, here denoted by λ_L (*top*). The Doppler effect describes the broadening. When the motion is due to the heat or temperature of the radiating atoms, the effect produces thermal broadening; when the average temperature of a collection of atoms increases, the thermal broadening becomes wider. An intense magnetic field can split a single line at wavelength λ_L into two components by the Zeeman effect (*bottom*). The wavelength difference $\Delta\lambda_L$ between the split lines is proportional to the strength of the magnetic field.

Gravitational Redshift

For massive, collapsed stars, there also is a detectable gravitational redshift caused by a radiation photon's loss in energy when overcoming the immense gravitational pull of a star. This has an insignificant effect for stars like the Sun but it increases for collapsed stars.

Thermal Motion Broadens Spectral Lines

Any observed spectral line is the superposition of the lines emitted by many individual atoms in different physical conditions. Rather than appearing at a single wavelength, the observed line therefore is broadened over a range of wavelengths (Fig. 3.13). The amount of wavelength broadening increases with the temperature of the source or, to be exact, it varies as the thermal velocity of the moving atoms and the square root of the temperature.

Rotation or Expansion of the Radiating Source Can Broaden Spectral Lines

If a source is rotating, the Doppler effect of the object's side rotating toward the observer produces a blueshift to shorter wavelengths; the other side, which is rotating away, shifts a line to longer wavelengths. The combined effect produces a line broadening that increases with the rotation velocity and that depends on the projected linear equatorial velocity or the observed rotational velocity of the line. A similar broadening applies to

Table 3.6. *Cosmic magnetic fields*

Object	Magnetic Field Strength (tesla)[a]
Earth (Equator to pole)	3×10^{-5} to 6×10^{-5}
Solar wind (at Earth orbit)	2.5×10^{-9}
Sunspot	0.3
Sun (global)	3×10^{-4}
Interstellar space	10^{-10}
White dwarf star	10^{2}
Pulsar	10^{8}

[a] 1 tesla = 1 T = 10^4 gauss = 10^4 G

an expanding source, which exhibits a line broadening that increases with the expansion velocity.

Magnetic Fields Split Spectral Lines

When an atom is placed in a magnetic field, it acts like a tiny compass, adjusting the energy levels of its electrons. If the atomic compass is aligned in the direction of the magnetic field, the electron's energy increases; if it is aligned in the opposite direction, the energy decreases. Because each energy change coincides with an alteration in the wavelength or frequency of the radiation emitted by that electron, a spectral line emitted at a single wavelength by a randomly oriented collection of atoms becomes a group of three lines of slightly different wavelengths in the presence of a magnetic field (see Fig. 3.13). The size of an atom's internal adjustments and the extent of its spectral division increase with the strength of the magnetic field.

This magnetic transformation was named the *Zeeman effect* after Pieter Zeeman, who first noticed it in the terrestrial laboratory. His Dutch colleague Hendrik A. Lorentz predicted the effect, and the pair received the 1902 Nobel Prize in Physics for their work.

The American solar astronomer George Ellery Hale made measurements of Zeeman splitting in sunspots, showing that they have magnetic field strengths of about 10,000 times the strength of the terrestrial magnetic field, which orients a compass. The split lines also are polarized circularly, and the direction of polarization indicates the direction of the magnetic field, pointing in or out of the Sun. The magnetic-field strengths of various cosmic objects are listed in Table 3.6.

4

Transmutation of the Elements

4.1 Things That Glow in the Dark

Near the end of the nineteenth century, university scientists in England, Germany, and France were actively investigating the light produced when electricity is passed through a glass tube or when certain substances shine after being exposed to sunlight. It must have been fascinating to study such things that can glow in the dark, at about the same time that the first long-lasting, practical electric light bulb was invented – by the American Thomas Alva Edison. In England and Europe, experiments with glowing tubes and shining materials led to discoveries of the electron, x-rays, and radioactivity.

When voltage was applied across the ends of a glass tube filled with rarefied gas, it emitted light, much like neon signs now used for advertising. An arc of light passed from the cathode (the negative end) of the gas tube to the other end (the positive anode) so rays passing through the gas were apparently producing the light. When vacuum pumps were used to reduce the gas pressure in the glass tubes, they ceased to glow inside, but the glass would shine where the invisible cathode rays apparently were striking it.

The French scientist Jean Perrin used a magnet to deflect the location of the glowing spot on the glass, suggesting that the unseen streams of "rays" were beams of negatively charged particles. A magnetic field deflects a negative charge in one direction and a positive charge in the other direction, and the size of the deflection depends on the mass of the charged particle, its velocity, and the strength of the magnetic field. The British physicist Joseph John Thomson subsequently used both electric and magnetic fields to measure the deflection of the beams of charged particles, now called *electrons*.

The deflection experiments worked this way: An electrical field produces a force that bends a beam of electrons in the direction of the field, toward positive and away from negative electrodes. By measuring the deviation from the original direction of motion, the charge to mass ratio can be determined. Because the deflection also depends on the velocity of the electrons, another measurement was needed, which was accomplished by placing a magnet near the electron beam (Focus 4.1). Combining both the electrical and magnetic experiments can determine the charge to mass ratio of the electron; from the fundamental unit of charge, the mass of the electron was determined. Thomson concluded that the electron is much less massive that any atom and roughly 1,000 times less massive than the lightest atom, hydrogen.

Focus 4.1 Charged Particles Avoid Magnetic Fields

When a moving charged particle encounters a magnetic field, it moves away from or around the magnetism. This is similar to changing direction to avoid an unpleasant situation. Humans have a choice, they can avoid or confront a situation, but a charged particle has no choice; it must swerve away from a magnetic field. It cannot move straight across a magnetic field but instead gyrates around it.

If the particle approaches the magnetic field straight on, in a perpendicular direction, a magnetic force pulls it into a circular motion about the magnetic field line. Because the particle can move freely in the direction of the magnetic field, it spirals around and along it with a helical trajectory.

Meanwhile, the German physics professor Wilhelm Röntgen inadvertently left some wrapped, unexposed photographic plates near a glowing gas tube that he was studying. Later, he found that the plates were fogged and that this was repeated when other new plates were left near the apparatus. Röntgen concluded that invisible rays were passing out of the tube and fogging the plates. At this time, photographs were taken with glass plates covered with a light-sensitive emulsion of silver salts. This preceded the use of photographic film, which has now been replaced by the CCDs used in digital cameras.

To remove any light, Röntgen enclosed the electrical discharge tube in black cardboard and noticed a glow coming from a nearby sheet of paper coated with a substance (i.e., barium platinocyande) known to shine in strong light – but there was no light to make it glow. The cause seemed to be the same invisible rays that fogged the photographic plates. Moreover, when Röntgen's wife placed her hand between the electrified gas tube and a photographic plate, the developed photograph showed the bones of her hand and the ring she was wearing.

The report of these previously unknown and penetrating rays caused a public sensation, for they could see inside humans and reveal the invisible. The rays were able to penetrate skin and muscle, detecting human bones, which revolutionized medicine.

No one knew what these mysterious, penetrating emissions were, but Röntgen preferred to call them *x-rays*, using the mathematical designation $\times$ for something unknown. Subsequent investigations eventually showed that x-rays are electromagnetic radiation of very short wavelength and high energy.

At about the same time, at the Sorbonne in Paris, Henri Becquerel was investigating fluorescent substances that could gather in the energy of light falling on them and remain luminous after the light source was removed. Uranium salts, for example, glowed in the dark, and Becquerel thought that they also might emit x-rays after being stimulated by sunlight. However, clouds shut out the Sun, as they often do in Paris, and he tossed the packet of uranium salts into a drawer in his worktable.

A box of unexposed photographic plates had been left in the drawer, wrapped in thick black paper and never opened; this led to another accidental discovery. When Becquerel took out the plates a few days later and developed the photographs taken with them, he found that they were ruined, as if they previously had been exposed to light. The uranium salts were emitting unseen rays that could affect the plates, even in pitch darkness. Becquerel called the invisible, highly penetrating phenomenon *uranic rays*; for a time, they were called *Becquerel rays*; and eventually they became known as *radioactive rays* – but for decades, no one knew exactly what they were.

The new type of rays was passing without difficulty through a covered box of photographic plates, but they might not penetrate metal. Becquerel repeated the experiment with an iron key placed between new plates and the uranium. When the plates were developed, they showed the silhouette of the key. This indicated that the uranium was emitting rays that were unable to pass through the iron, even though they could penetrate dark paper that blocks ordinary light. In this respect, the uranium rays resembled x-rays; but, unlike x-rays, the uranium was emitting rays spontaneously without previous excitation by sunlight or electricity.

Hearing of Becquerel's discovery, Pierre Curie, also a professor of physics at the Sorbonne, and the young graduate student he recently had married, Manya (Marie) Curie, began to investigate the new type of rays. Madame Curie wanted to know if uranium was the only element that emitted the mysterious rays, and she developed methods to measure the amounts being released. To her surprise, she found that impure uranium ores emitted more rays than could be explained in terms of the uranium they contained. The couple began a laborious two-year search for the unknown emitters; from 1 ton of uranium ore known as pitchblende, they extracted just a few grams of powerful new elements that had not been known previously. One was called *radium* and the other *polonium*, after Marie's native Poland. Only one year later, the French chemist André-Louis Debierne discovered the radioactive chemical element *actinium*, separating it from pitchblende residues left by Marie and Pierre Curie after they had extracted radium.

Radium is 1 million times more radioactive than uranium, which – in the terminology of the time – meant that radium is emitting the penetrating rays more intensely than uranium, not that either substance emits radio waves. Crystals containing radium can light up an otherwise dark room and burn the skin as well, as Pierre Curie discovered to his dismay.

4.2 Radioactivity

No one knew exactly what the radioactive rays were, where their energy came from, or why the radioactive materials kept pouring out energy, seemingly nonstop. Moreover, the amount of energy being released by radium was difficult to explain, for it far surpassed anything that had been achieved by chemical reactions. Respectable scientists, such as Lord Kelvin, even suggested that the radioactive energy was supplied from outside, perhaps being taken from the surrounding air by ether waves.

As proposed by the English physicist Ernest Rutherford, the source of radioactive energy must come from the interior of the radioactive atoms. These very heavy atoms apparently were unstable, disintegrating and falling apart by themselves. They were slowly leaking out energy from their atomic interiors in spontaneous transmutation. It is a natural process that happens all the time in the ground on which we stand.

Rutherford found that the radioactive rays emitted by uranium included at least two distinct types, termed *alpha rays* and *beta rays*. These rays are not waves of radiation; instead, they are beams of energetic, fast-moving particles. By using electrical and magnetic fields,

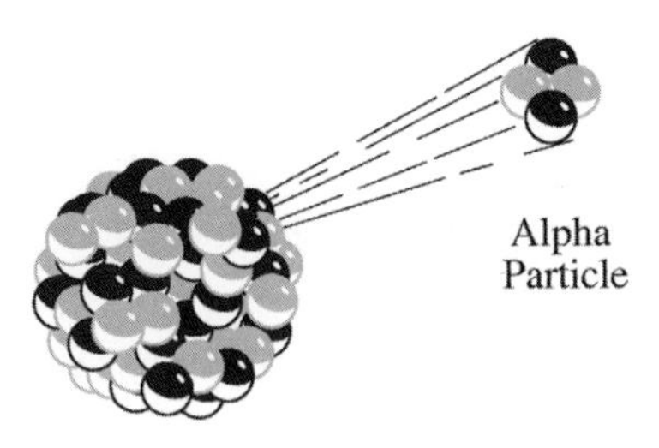

Figure 4.1. Radioactive alpha decay. An unstable, heavy nucleus of a radioactive element can disintegrate or decay into a stable, lighter nucleus, with the emission of an alpha particle that carries mass away from the heavy nucleus during its decay. The subatomic alpha particle consists of two protons and two neutrons. The nucleus of a helium atom is an alpha particle. Radioactive alpha decay of an individual heavy element such as uranium does not occur very often, on average. We would have to wait 704 million years for half of a rock of uranium to change into lead by emitting alpha particles.

the two types of particles could be separated and their physical properties examined. The directions in which the beams were deflected indicated the sign of their electrical charge, and the magnitude of the deflection provided a measure of both the charge and the mass.

Alpha particles carry a double dose of positively charged protons – the nuclei of hydrogen atoms – and they move at astonishingly high speeds of approximately 1/20th the speed of light. Rutherford and his colleagues eventually showed that an alpha particle is nothing more than the nucleus of the helium atom, containing two protons and two neutrons. By emitting alpha particles, a heavy, unstable atom was turning spontaneously into a slightly lighter atom, without any interactions with another particle or radiation from outside the atom (Fig. 4.1).

Radioactive decay occasionally is accompanied by the emission of charged beta rays, which make an electronic adjustment to an atom's nucleus without significantly changing its mass. Beta rays consist of negatively charged particles of low mass, which eventually were identified as high-speed electrons.

The ejection of alpha and beta particles often was accompanied by the emission of powerful electromagnetic radiation, akin to x-rays but with even shorter wavelength and greater energy. Because the energetic radiation, called *gamma rays*, is not charged, it is not deflected by electrical or magnetic fields.

Working with the young chemist Frederick Soddy, Rutherford found that radioactive atoms continued to disintegrate into other pieces after the emission of an alpha particle. Uranium, for example, initially turned into thorium, which also was radioactive and released other substances, including gaseous radon.

However, the progressive disintegration of heavy, unstable elements does not continue forever. As radioactive decay progresses, the inner parts of an atom rearrange into greater stability, eventually reaching an equilibrium that does not decay and fall apart. For uranium, this stable endpoint to successive transformation is lead.

A simplified notation, described in Focus 4.2, clarifies our understanding of these complex chains of radioactive decay.

Focus 4.2 Nuclear Nomenclatures

The number of protons in a nucleus is denoted by the atomic number Z. They account for the charge of the nucleus but not for all of its mass. A nucleus of any element except hydrogen has about twice the mass of the sum of its protons. The extra mass is due to neutral, or uncharged, particles called *neutrons*, each with about the same mass as a proton. The mass of the nucleus is specified by the mass number A.

Letters denote the nuclei and other subatomic particles. Both a letter and a superscript, the mass number A, designate a nucleus. An arrow $\rightarrow$ specifies the reaction. Nuclei on the left side of the arrow react to form products given on the right side of the arrow. The alpha decay of uranium, for example, is given by:

$$^{238}U \rightarrow \ ^{234}Th + \alpha$$

where U denotes a nucleus of uranium, Th indicates a nucleus of thorium, and the alpha particle α is the nucleus of the helium atom, also designated ^{4}He.

4.3 Tunneling out of the Nuclear Prison

Why do not the nuclei of all radioactive atoms completely and instantaneously disintegrate? Or, to ask a related question, how have the nuclei of so many uranium atoms managed to retain their alpha particles? After all, there is still plenty of uranium around billions of years after the Earth formed, continuing to make the rocks and soil around us radioactive. The reason is that it takes significant energy to break free of the strong forces that bind the protons and neutrons together in the nucleus of an atom. They are locked so firmly within the nucleus that exceptional force must be applied to dislodge them.

In fact, the escape of an alpha particle from an atomic nucleus seemed impossible from the viewpoint of classical physics. The forces holding the particle inside the nuclear prison are so strong that the energy required to overcome them is enormous. It is as if the atomic nucleus is surrounded by tall walls with energy much higher than that of the alpha particles.

The young Russian physicist George Gamow resolved this paradox in 1928, shortly after graduating from the University of Leningrad. When visiting the apparently dull university town of Göttingen, Germany, Gamow found little in the way of entertainment and therefore had time to use the uncertain, probabilistic nature of quantum theory to explain radioactive decay. Quantum theory indicated that the location of a tiny subatomic particle is not defined precisely. The particle instead acts like a spread-out entity with a set of probabilities of being in a range of places. As a result of this murky location uncertainty, a subatomic particle's sphere of influence is larger than was previously thought. It might be anywhere – although with decreasing probability at regions far from the most likely location.

This explains the escape of fast-moving, energetic alpha particles from the nuclei of radioactive atoms such as uranium. These particles usually lack the energy to overcome the nuclear barrier, but some have a small probability of escaping to the outside world. As Gamow described it, some of the alpha particles can "leak through" the nuclear walls, roughly like water that slowly leaks through a clogged drain. The rate of escape from the nuclear prison increases with the kinetic energy of the particle and decreases with its electrical charge.

In this surreal world of subatomic probability, we could relentlessly throw a ball against a wall, watching it bounce back countless times, until eventually the ball would tunnel under the wall or effectively pass through it. As Ahab said in *Moby Dick*, "How can the prisoner reach outside except by thrusting through the wall?"

Gamow's quantum-mechanical formula for the "transparency" of the nuclear walls was in good agreement with Rutherford's suggestion, in 1900, that the number of radioactive atoms in a rock changes with time, at a constant rate of decay. This disintegration is probabilistic, governed by the rules of quantum mechanics that describe a random, slow, and statistical decay. On the level of a single atom, it is impossible to predict when a given atom will decay, and the probability that a given unstable atom decays is the same for all atoms of that type, independent of age. For numerous identical atoms, the decay rate is predictable using quantum theory, and that rate depends on the radioactive element under consideration.

That is, the decay rates of various radioactive substances differ. The nuclei of uranium can retain their alpha particles for billions of years, whereas other radioactive nuclei eject them in a matter of seconds. These rates are quantified in terms of the nuclear half-life, which is the time needed for a given amount of a radioactive substance to decay to half of its initial value.

At almost the same time as Gamow, the English physicist Ronald W. Gurney and the American physicist Edward U. Condon developed a similar explanation for spontaneous radioactive decay with the emission of an alpha particle. In their communication, the two scientists wrote:

> It has hitherto been necessary to postulate some special arbitrary "instability" of the nucleus, but in the following note it is pointed out that disintegration is a natural consequence of the laws of quantum mechanics without any special hypothesis.... Much has been written about the explosive violence with which the alpha particle is hurled from its place in the nucleus. But from the process pictured above, one would rather say that the alpha particle slips away almost unnoticed.[5]

Even after being elected to the National Academy of Sciences and serving as both the director of the National Bureau of Standards and the president of the American Physical Society, Condon's security clearance was questioned by the House Un-American Activities Committee in the 1950s. A member of one loyalty review board noted that Condon had been at the forefront of a revolutionary movement in physics called quantum mechanics and therefore questioned whether he might not be at the forefront of another revolutionary movement, notably communism.

Radioactivity provides us with a way to measure the age of the solar system. The method is known as *radioactive dating*, and it works this way: The radioactive nuclei, known as *unstable parent isotopes*, decay at a constant rate into stable lighter isotopes, known as *daughters*. By measuring the amount of daughter material and knowing the rate of decay, the age of a rock can be estimated. We simply measure the relative amounts of radioactive parents and nonradioactive daughters. When this ratio is combined with the known rates of radioactive decay, the time since the rock solidified and locked in the radioactive atoms is found. The method is similar to determining how long a log has been burning by measuring the amount of ash and watching for a while to determine how rapidly the ash is being produced.

The daughter isotopes must be trapped in a rock and not allowed to escape or the estimated age will be too short. The daughters can escape easily when the rock is molten; only when it cools and solidifies do they start to accumulate. For this reason, the age determined for a rock is actually the time since the rock became solid. If the rock is remelted – for example, by the impact of a meteorite – its radioactive clock is reset and the age will measure the time since the last solidification.

Radioactive dating of primitive meteorites, ancient rocks returned from the Moon, and deep-ocean sediments indicates an age of about 4.6 billion years. These relics have remained unaffected by the geological erosion processes that removed the primordial record from most terrestrial rocks. If the solar system originated as one entity, then this also should be the approximate age of the Sun and the rest of the solar system.

Despite Gamow's tunneling discovery, our understanding of radioactivity was still incomplete; there was something wrong with the way the beta rays were behaving. This resulted in the discovery of an entirely new particle: the *electron neutrino*.

4.4 The Electron and the Neutrino

When first discovered, electrons emitted from radioactive elements were called *beta rays* to distinguish them from *alpha rays* (i.e., helium nuclei) and *gamma rays* (i.e., high-energy radiation) that also are emitted during radioactive-decay processes. From their measured charge and mass, it was discovered that the betas are not rays at all but instead ordinary electrons moving at nearly the speed of light. The emission of high-speed electrons by a radioactive element is known as *beta decay*.

Detailed measurements of the high-speed electrons, given off during radioactive decay, seemed to violate a fundamental principle of physics known as the *conservation of energy*. According to this rule, the total energy of a system must remain unchanged unless acted on by an outside force. We know of no process that disobeys this principle.

Nevertheless, the sum of the energy of the beta-decay nucleus and the energy of the emitted electrons sometimes turned out to be less energy than what was lost by the initial nucleus. Careful measurements failed to turn up the missing energy, which seemed to have vanished into thin air, suggesting that energy might not be conserved during beta decay. The eminent Danish physicist Neils Bohr proposed that the conservation of energy law was being violated on the atomic scale.

However, it turned out that a mysterious, invisible particle was spiriting away the missing energy. It was the elusive *neutrino*, the existence of which was postulated by Wolfgang Pauli, a brilliant Austrian physicist. Pauli proposed a "desperate way out" of the energy crisis, speculating that an electrically neutral particle, produced at the same time as the electron, carried off the remaining energy. The sum of the energies remains constant during the beta decay, so the energy is balanced and the principle of conservation of energy is saved. As Pauli expressed it:

> The conservation laws remain valid, the expulsion of beta particles [electrons] being accompanied by a very penetrating [energetic] radiation of neutral [uncharged] particles, which has not been observed so far.[6]

Pauli thought he had done "a terrible thing," for his desperate remedy postulated an invisible particle that could not be detected. Dubbed the *neutrino*, or "little neutral one," by the Italian physicist Enrico Fermi, the new particle could not be observed with the technology of the day because the neutrino is electrically neutral, has almost no mass, and moves at nearly the speed of light. Therefore, the hypothetical neutrinos were removing energy that would never be seen again. (Even in Pauli and Fermi's time, the observed high-energy shape of the emitted electron's energy spectrum indicated that the mass of the neutrino is either zero or very small relative to the mass of the electron.)

Unlike light or any other form of radiation, neutrinos can move nearly unimpeded through any amount of material, even the entire universe. In the parlance of modern physics, neutrinos are characterized by a weak interaction with anything in the material world.

Fermi formulated the mathematical theory of beta decay in a paper that was rejected by the journal *Nature* because "it contained speculations too remote from reality to be of interest to the reader." As beautifully described by Fermi, the decay process occurs when the neutron in a radioactive nucleus transforms into a proton with the simultaneous emission of an energetic electron and a high-speed neutrino. When left alone outside of a nucleus, a neutron in fact will self-destruct in about 10 minutes into a proton plus an electron to balance the charge and a neutrino to help remove the energy.

As far as anyone could tell, an atomic nucleus consists only of neutrons and protons, so the electron and neutrino seemed to come out of nowhere. They do not reside within the nucleus and are created at the time of nuclear transformation. No one knew exactly how neutrinos were formed.

How do you observe something that spontaneously appears out of nowhere and is so close to being nothing at all? Calculations suggested that the probability of a neutrino interacting with matter, so that it might be seen, is so incredibly small that no one could ever detect it. To see one neutrino, we would have to produce enormous numbers at about the same time and build a massive detector to increase the chances of catching it. Although almost all of the neutrinos still would pass through any amount of matter unhindered and undetected, a rare collision with other subatomic particles might leave a trace.

Manmade nuclear reactors produce large numbers of neutrinos, and if a massive detector is placed near a large nuclear reactor, with appropriate shielding from extraneous signals, the telltale sign of the hypothetical neutrino may be barely observed.

The existence of the neutrino was finally proven by Project Poltergeist, an experiment designed by Clyde L. Cowan and Frederick Reines of the Los Alamos National Laboratory in New Mexico. They placed a 10-ton (10,000-liter) tank of water next to a powerful nuclear reactor engaged in making plutonium for use in nuclear weapons. After shielding the neutrino trap underground and running it for about 100 days, Reines and Cowan detected a few synchronized flashes of gamma radiation that signaled the interaction of a few neutrinos with the nuclear protons in water.

The neutrinos themselves were not observed, and they never have been. Their presence was inferred by an exceedingly rare interaction. One of every billion billion, or 10^{18}, neutrinos that passed through the water tank hit a proton, producing the telltale burst of radiation. Cowan and Reines telegraphed Pauli with the news:

> We are happy to inform you that we have definitely detected neutrinos from fission fragments by observing inverse beta-decay of protons![7]

Pauli promptly sent them a case of champagne in recognition of their accomplishment. The ghostly neutrino, which most scientists thought would never be detected, finally had been observed. Nearly four decades later, Reines received the 1995 Nobel Prize in Physics for the detection of the neutrino; by that time, however, Cowan had died and therefore could not share in the award.

As discussed in Chapter 5, the Sun emits copious amounts of neutrinos. Every second, trillions upon trillions of neutrinos that were produced inside the Sun pass right through the Earth without even noticing that it is there. The indestructible neutrinos interact so rarely with the material world that almost nothing ever happens to them. Billions of ghostly neutrinos from the Sun are passing right through us every second, even in our bedrooms at night, and they did not come through the door. The solar neutrinos travel right through the Earth, a building, and us, without our body noticing them or them noticing our body.

Moreover, when a minute number of the Sun's neutrinos were snared in massive underground detectors, fewer than expected were observed; this eventually led to a new understanding of the neutrino. For the time being, however, let us move on to energetic cosmic rays that are always entering the atmosphere from outer space.

4.5 Particles from Outer Space

High above the ground, our upper atmosphere is immersed within a cosmic shooting gallery of subatomic particles coming from all directions in interstellar space and moving at nearly the speed of light. The perpetual high-energy rain was discovered about a century ago, when the Austrian physicist Victor Franz Hess, an ardent amateur balloonist, measured the amount of ionization at different heights within our atmosphere.

It was already known that radioactive rocks at the Earth's surface were emitting energetic "rays" – the alpha and beta particles – which ionize molecules in the atmosphere near the ground; they split molecules in the air into positive and negative ions, making the air electrically conducting.

It was expected that the ionizing rays would be absorbed completely after passing through sufficient quantities of the atmosphere. The measured ionization at first decreased with altitude, as expected from atmospheric absorption of energetic particles emitted by radioactive rocks. However, the ionization rate measured by Hess increased unexpectedly at even higher altitudes to levels exceeding that at the ground. This meant that some penetrating source of ionization came from beyond the Earth. By flying his balloons at night and during a solar eclipse, when the high-altitude signals persisted, Hess showed that they could not come from the Sun but rather from some other source.

The American physicist Robert A. Millikan subsequently used high-altitude balloon measurements to confirm that the "radiation" comes from beyond the terrestrial atmosphere, and he gave it the present name of *cosmic rays*. Millikan believed the cosmic rays were gamma rays associated with the synthesis of heavy elements deep in space, the "birth cries" of new matter. We now know that cosmic rays are energetic charged particles, not radiation, and more likely the "death cries" of massive exploding stars.

Global measurements showed that cosmic rays are electrically charged. During an ocean voyage, the Dutch physicist Jacob Clay, for example, found lower cosmic-ray intensity near the

Earth's Equator than at higher terrestrial latitudes; his results were confirmed and extended by Arthur H. Compton of the University of Chicago. Compton conclusively demonstrated an increase in cosmic-ray intensity with terrestrial latitude and also made measurements at mountain altitudes, where the increase with latitude was even stronger. His results indicated that cosmic rays must be electrically charged particles deflected by the Earth's magnetic field toward its magnetic poles, which are near the geographic poles.

Unlike cosmic radiation, the charged cosmic-ray particles also are deflected and change direction during encounters with the interstellar magnetic field that winds its way among the stars. Therefore, we cannot look back along their incoming path and determine where cosmic rays originate; the direction of arrival shows only where they last changed course. The favored hypothesis, proposed by Walter Baade and Fritz Zwicky, is that cosmic rays are accelerated to their tremendous energy during the supernova explosion of massive stars that have run out of thermonuclear fuel.

Instruments carried by high-altitude balloons established that the most abundant cosmic-ray particles arriving in the Earth's upper atmosphere are protons – the nuclei of former hydrogen atoms – and the second most abundant particles are helium nuclei – the alpha particles. Cosmic-ray electrons arriving near the top of the atmosphere were eventually discovered; they are far less abundant than the cosmic-ray protons at a given energy.

Although they are relatively few in number, cosmic rays contain phenomenal amounts of energy. That energy usually is measured in units of electron volts, abbreviated eV. The greatest flux of cosmic-ray protons arriving at the Earth occurs at 1 to 10 GeV, or 1 billion (10^9) to 10 billion (10^{10}) eVs of energy. Cosmic rays do not all have the same energy, and some reach an energy of 10^{20} eV, or more than 10 billion (10^{10}) times the abundant ones; however, the flux of the most common, lower-energy cosmic rays at 10^9 to 10^{10} eV is greatest (Fig. 4.2).

At an energy of 10^9 eV, a cosmic-ray proton must be traveling at 88 percent of the speed of light. For comparison, a helium nucleus, or alpha particle, emitted during radioactive decay reaches no more than 1 million eV in energy, or 1,000 times less than that of a cosmic-ray proton.

Cosmic rays enter the atmosphere with such great energies that they act like colossal atom destroyers, hitting molecules and their component atoms in the upper atmosphere and producing showers of subatomic debris, known as *secondary cosmic-ray particles*. This eventually led to the first observation of an energetic particle that does not belong to the atom. But first, a method needed to be developed to detect the then-unknown particle.

Subatomic particles coming down through the atmosphere are detected near or at the ground by tracks in a cloud chamber, which creates a cloudy mist that precipitates as long thin bands of fog along the trajectory of the particles. This is similar to the white vapor trails of jet aircraft, which record an airplane's movement in the sky. Fine water droplets condense from the jet exhaust fumes and create the elongated clouds.

The first cloud chamber, invented by Charles Thomas Rees Wilson, was very simple, consisting of a metallic cylinder with a glass cover and a piston that could be moved up and down from below, permitting air filled with water vapor to enter the space above it. When the piston was lowered quickly, the sudden expansion cooled the gas so that a mist formed in the chamber, like the foggy mist found high in the mountains or nearer the ground

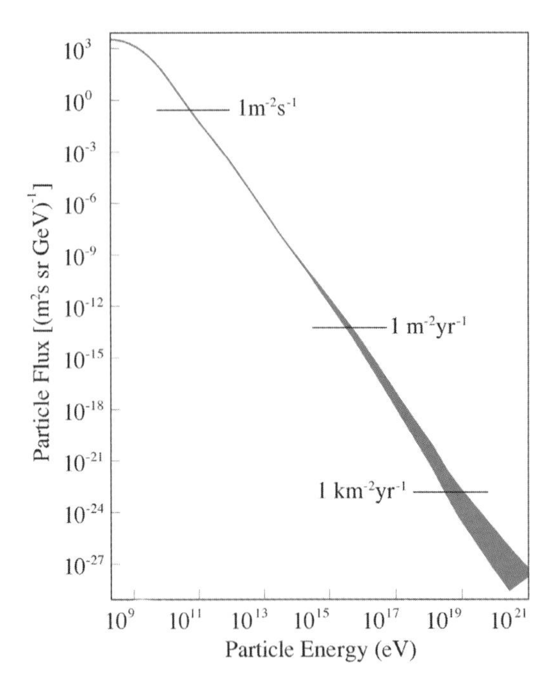

Figure 4.2. Flux of cosmic rays. The energy spectrum of cosmic-ray particles striking the outer atmosphere of the Earth. The particle flux is plotted as a function of the particle energy in units of electron volts, abbreviated eV, where 1 eV = 1.602×10^{-19} J and 1 GeV = 10^9 eV, or 1 billion eV. The most abundant cosmic-ray particles are protons with energies of about 1.5 billion eV. Every second, about 640 enter every square meter of the Earth's outer atmosphere. They probably are accelerated to high energy during the supernova explosions of massive stars. One cosmic-ray proton of 10 billion eV in energy enters each square meter of the Earth's outer atmosphere every second. The more energetic cosmic-ray particles of a million billion eV are less abundant, with one per square meter every year. Solar flares can emit protons with energies of 10 billion eV or less, and these solar energetic particles can strike the Earth when the solar active region is on the near side of the Sun. Cosmic rays with low flux and very high energy, greater than 1 million billion eV, may be of extragalactic origin.

on a winter day. The water vapor in the chamber condensed or precipitated out on any ions present, making the ionized tracks of cosmic rays visible and showing where they had moved.

When the cloud chamber is placed between the poles of a strong magnet, the magnetic field exerts a force on any charged particle entering the chamber, which produces a curved particle track. When Carl Anderson, who was Millikan's student at the California Institute of Technology, built such a device using a powerful electromagnet, he found that a few of the cosmic-ray showers produce two similar curved trajectories in opposite directions (Fig. 4.3). Further experiments revealed that an electron, which has a negative charge, was producing one of the curved tracks, whereas a particle with about the same mass as the electron and a positive charge of the same amount but opposite sign as the electron was producing the track curved in the opposite direction. Anderson had discovered the *positron*, short for "positive

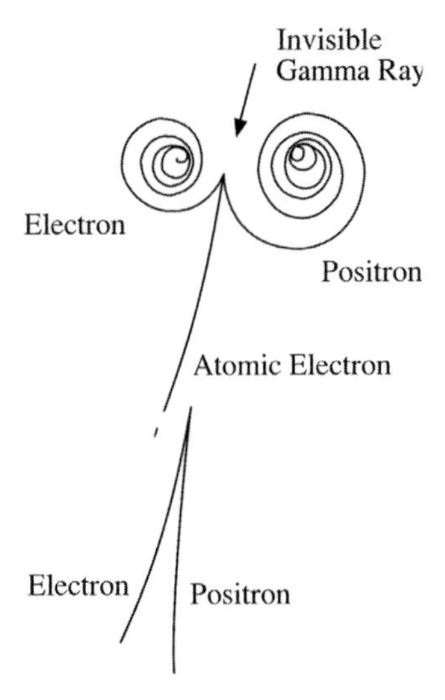

Figure 4.3. The electron and the positron. An invisible gamma-ray photon (*top*) produces an electron and a positron (short for positive electron), seen by curved tracks in a bubble chamber. Both the electron and the positron are bent into circular tracks by the instrument's magnetic field, moving in opposite directions because of their opposite electrical charge and spiraling into a smaller circular motion as they lose energy. In this upper pair, some of the photon's energy is taken up in displacing an atomic electron, which shoots off toward the bottom left. In the lower example, all of a gamma ray's energy goes into the production of the electron–positron pair. As a result, these particles are more energetic than the upper pair, and their tracks do not curve so tightly in the chamber's magnetic field. (Schematic of a Lawrence Berkeley Laboratory bubble-chamber image, reproduced by Frank Close, Michael Marten, and Christine Sutton in *The Particle Explosion*, New York: Oxford University Press, 1987.)

electron," and the anti-matter counterpart of the electron. The energetic cosmic rays had produced a new type of particle that had never been seen before!

As it turned out, Paul Adrien Maurice Dirac, then at Cambridge University, predicted the existence of anti-matter at about the same time that Anderson observed it. For Dirac, mathematical beauty was the most important aspect of any physical law describing nature. He noticed that equations that describe the electron have two solutions. Only one was needed to characterize the electron; the other solution specified a sort of mirror image of the electron – that is, an anti-particle, now called the positron. At the time of his discovery, Anderson nevertheless was unaware of Dirac's theoretical prediction of the positron.

Anderson received the 1936 Nobel Prize in Physics for his discovery of the positron, sharing the award with Hess for his discovery of cosmic rays. Wilson already had received recognition nine years earlier for his method of making visible the paths of electrically charged particles – the cosmic rays – by condensation of vapor.

Once created, anti-matter does not stay around for very long for any anti-matter will promptly self-destruct when it encounters ordinary matter. When an electron and positron meet, they annihilate one another and disappear in a puff of energetic radiation. As discussed subsequently, this electron–positron annihilation reaction helps to produce radiation in the core of the Sun and also is observed during explosive flares on the visible solar disk.

4.6 Nuclear Alchemy and Atomic Bombs

What happens if we turn radioactivity around and, instead of watching the nucleus of an unstable heavy atom decay, we bombard a perfectly normal, lighter nucleus with very energetic particles? Perhaps this normally stable element could be transformed artificially on the Earth through such a nuclear bombardment. After all, that is what the cosmic rays were doing, resulting in all kinds of surprises, turning some atoms into previously unknown, fundamental particles.

Medieval alchemists had been trying to change one element into another (e.g., lead into gold) for centuries, but they always failed in their attempts because the chemical and thermal reactions they employed were nowhere near energetic enough to crack open the nucleus of an atom. Their efforts were comparable to driving a vehicle into a concrete wall; the crash destroys only the vehicle and indicates nothing about the wall.

The first successful attempts to transform elements in the terrestrial laboratory occurred when Patrick Blackett, a recent graduate of Cambridge University, directed a beam of fast alpha particles, ejected by radioactive decay, into Wilson's cloud chamber. The chamber was filled with normal atmospheric air, which is composed mainly of nitrogen molecules. Blackett improved the cloud chamber so that the air expanded and cooled automatically, and he took automatic photographs of the tracks. Most of the alpha particles passed straight through the chamber. However, after more than 23,000 photographs of alpha particles bombarding nitrogen in the cloud chamber during a three-year period, Blackett finally succeeded in recording only eight head-on collisions of alpha particles with the nuclei of nitrogen atoms.

On each photograph, the track of an alpha particle suddenly stopped, being replaced with the fine, straight track of an ejected proton and the short, stubby recoil track of the struck nucleus (Fig. 4.4). However, there was no sign of the recoiling alpha particle. The collision had brought the alpha particle into the nitrogen nucleus, forming a nucleus of a form of oxygen. The reaction can be written as follows:

$$^{4}He + {}^{14}N \rightarrow {}^{17}O + {}^{1}H,$$

where the collision of an alpha particle, or helium nucleus ^{4}He, with a nitrogen nucleus, ^{14}N, gave rise to the nucleus of oxygen, ^{17}O, and a proton, the nucleus of hydrogen, ^{1}H. The old alchemist's dream finally had been realized in a laboratory on the Earth, in which nuclear transformation had been induced and recorded.

Enthusiastic scientists directed beams of alpha particles into many other elements, creating nuclear transformations similar to the one observed for nitrogen. When it came to heavier elements, with greater nuclear charge, however, a nuclear transformation could not be produced. The greater charged defense of nuclei with atomic number Z greater than 18 always withstood the bombardment by alpha particles.

That is when Gamow's 1928 paper on the decay of heavy radioactive nuclei had a decisive role. His calculations indicated that on rare occasions, alpha particles could tunnel through the positively charged wall of a nucleus but that fast protons would more easily overcome

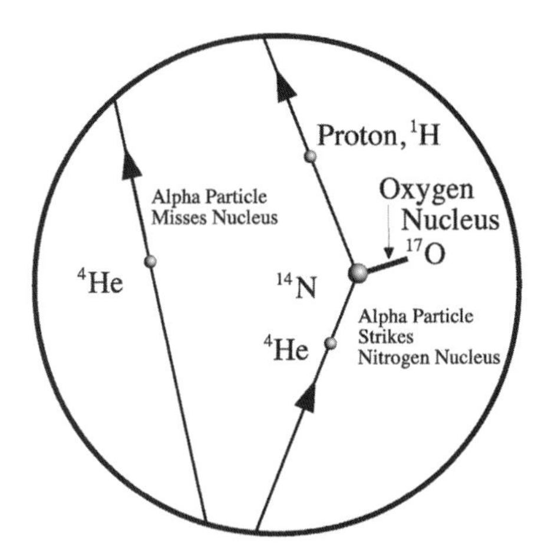

Figure 4.4. Nuclear transformation. When an alpha particle, or helium nucleus denoted ^{4}He, is sent through a cloud chamber, it usually passes right through it, with a trajectory that marks out a straight line. Occasionally, the alpha particle will strike the nucleus, ^{14}N, of a nitrogen atom in the air within the chamber, transforming it into the nucleus, ^{17}O, of an oxygen atom with the emission of a proton, the nucleus of a hydrogen atom and denoted ^{1}H. Such a nuclear transformation was first observed in cloud-chamber photographs taken in the early 1920s by Patrick Blackett (1897–1974).

the barrier. Because it has a smaller electrical charge, a proton suffers less nuclear repulsion when approaching a given nucleus and therefore has a greater probability of penetrating it. Moreover, because a proton is four times less massive than an alpha particle, it might be easier to accelerate it to high speed.

Rutherford's student John Cockcroft had studied electrical engineering and used his background to build a machine that would accelerate the hydrogen nuclei, the protons, in very intense electrical fields. When applying 500,000 volts, Cockcroft was able to produce a parallel beam of protons traveling at the speed of 1/30th the speed of light.

When the beam was directed at lithium nuclei, the protons split the nucleus apart, forming two alpha particles with the following reaction:

$$^1H + {}^7Li \rightarrow {}^4He + {}^4He.$$

The collisions between the proton, ^1H, and lithium, ^7Li, had completely transformed the two particles into pure helium nuclei, ^4He. The colliding atomic nuclei had disappeared, creating new nuclei.

This stimulated the construction of improved particle accelerators that created beams of protons moving at even faster speeds with greater kinetic energy. Ernest Lawrence at Berkeley University, for example, invented the cyclotron, which used magnets to bend the path of a charged particle into a circular orbit that passed across an alternating and accelerating voltage. As the radius of the spiraling orbit increased, so did the particle's speed; therefore, the time to complete each orbit remained constant, and the particles were repeatedly accelerated in equal time intervals before directing them into a collision with something else. This is similar to pushing a child in a swing at the same part of its swinging motion, pumping it up to greater and greater speed.

Lawrence built increasingly larger cyclotrons at Berkeley's Radiation Laboratory, whirling the protons to faster and faster speeds. His final cyclotron, with a diameter of nearly 5 meters, could accelerate deuterium nuclei to energies of 195 MeV.

In 1939, the Nobel Prize in Physics was awarded to Ernest Lawrence for the invention and development of the cyclotron and for results obtained with it, especially with regard to artificial radioactive elements. It wasn't until 1951 that John Cockcroft and Ernest Walton received the prize for their pioneer work on the transmutation of atomic nuclei by artificially accelerated atomic particles.

Eventually, accelerators were built that reached cosmic-ray energies. The Bevatron at the Lawrence Berkeley Radiation Laboratory began accelerating protons in 1954 up to an energy of 1 GeV. More than a half-century later, the 2-kilometer-diameter Tevatron, an accelerator at the Fermi National Accelerator Laboratory (the Fermi lab), used thousands of electromagnets to whirl protons up to 1,000 GeV, or 1 TeV, of energy. By 2009, the Large Hadron Collider at the European Organization for Nuclear Research (CERN, for Conseil Européan pour la Recherche Nucléaire) was using superconducting magnets in a circular tunnel 27 kilometers in circumference to produce two beams of protons moving in opposite directions and eventually directed into collision with one another (Fig. 4.5).

In the meantime, on the eve of World War II in 1938, the German radio chemist Otto Hahn, who had been working with Fritz Strassmann and Lise Meitner, showed that when uranium is bombarded with neutrons, it could be split into two nearly equal fragments. The process is known as *nuclear fission*, analogous to binary fission in the biological sciences. This was altogether different from the proton bombardment of much lighter nuclei, for the heavy uranium was broken in two and released enormous amounts of energy in the process. Moreover, it also freed additional neutrons, which in turn can trigger the fission of neighboring nuclei, resulting in a runaway chain reaction if not properly controlled. When controlled, a nuclear chain reaction can be used to generate electricity in nuclear reactors; when uncontrolled, it has applications in nuclear weapons.

Meitner, who was in exile in Copenhagen, confirmed the fission of uranium by neutron bombardment, and the Danish physicist Niels Bohr described it to Albert Einstein, who had emigrated to the United States. By the time Einstein heard about uranium fission, World War II (1939–1945) had begun, and scientists feared that Nazi Germany would use the discovery to build an atomic bomb to conquer the world. Einstein wrote to the President of the United States, Franklin Roosevelt, encouraging a program that would achieve a nuclear chain reaction and the United States to consider the development of "extremely powerful bombs" that the Germans also might be constructing. The concern was real, for it was later discovered that prominent German physicists, including Werner Heisenberg and Carl von Weizsäcker, helped the Germans investigate the feasibility of constructing nuclear weapons during World War II.

In 1942, the famed American physicist Jules Robert Oppenheimer invited a small group of theoretical physicists to the University of California at Berkeley to discuss how an atomic bomb might be assembled. Within a year, they had all moved to Los Alamos, the secret laboratory in New Mexico where thousands of scientists, technicians, and military personnel worked under Oppenheimer's enthusiastic direction to create the first atomic bomb.

Figure 4.5. Proton-proton collision. Two beams of protons have been whirled in opposite directions to nearly the speed of light, each with an energy of 7 trillion electron volts, or 7 TeV = 7×10^{12} eV = 1.12×10^{-6} J, and directed into collision with one another at CERN's Large Hadron Collider (LHC). This image displays the tracks of more than 100 charged particles as they fly away from the point of proton collision. By studying the collision-particle debris, including correlations among them, scientists hope to gain an improved knowledge of how subatomic particles interact at extremely high energies, including the hot, dense conditions only a small fraction of a second after the "big bang." At the point of proton impact, temperatures of more than 1 million million, or 10^{12}, K are generated, exceeding 100,000 times the temperature at the center of the Sun. Experiments with this instrument have provided evidence for a new, previously unseen particle, named the Higgs boson, in the mass-energy range of 10^{11} eV. CERN is a French acronym for the Conseil Européen pour la Recherche Nucléaire (the European Organization for Nuclear Research). The Compact Muon Solenoid (CMS) particle detector created this image. (Courtesy of CERN.) (See color plate.)

Some of the best scientific minds in the country were involved, including Hans Bethe, head of the Los Alamos theoretical division; Richard Feynman, who worked on numerical calculations of bomb performance; and the Italian immigrant Enrico Fermi, who helped produce the first self-sustaining nuclear chain reaction. Another famous physicist, Philip Morrison, accompanied the first bombs all the way to their final flight, caring for them before they were dropped on Japan.

The world's first nuclear device, a 21-kiloton bomb code-named Project Trinity, was detonated on July 16, 1945 at Alamogordo, New Mexico. Witnessing the explosion, Oppenheimer – who was deeply interested in Indian philosophy – apparently recalled a verse in Sanskrit from the *Bhagavad-Gita*, the English translation of which is as follows:

> If there should be in the sky
> A thousand suns risen all at once,
> Such splendor would be
> Of the splendor of that Great Being...
> I am Time, the mighty cause of world destruction,
> Here come forth to annihilate the worlds.[8]

Figure 4.6. Christmas Island. Photograph of the test of a nuclear bomb, code-named *Truckee*, delivered by air 16 kilometers south of Christmas Island (now Kiritimati), Pacific Ocean, on June 9, 1962 during Operation Dominic I. The explosion had a force equivalent to 210 kilotons of TNT, about 10 times that of the atomic bomb detonated over Nagaski, Japan, on August 9, 1945. *Truckee* was a prototype test of the W-58 warhead carried on the Polaris A-2 missile and deployed on submarine-launched ballistic missiles. Thermonuclear detonations in the humid Pacific created their own localized weather systems; visible in *Truckee's* cloud are multiple cloud-condensation structures known as "bells" and "skirts," as well as the more familiar "cauliflower" structure. The U.S. Air Force 1352nd Photographic Group, Lookout Mountain Station, took this image from Christmas Island. (Courtesy of Michael Light; also in his book, *100 Suns*, Alfred A. Knopf, 2003, Number 81.)

The last two lines also have been paraphrased as "I am become Death, the shatterer of worlds."

A few weeks later, two atomic bombs were dropped on the Japanese cities of Hiroshima and Nagasaki with devastating consequences. As far as the military was concerned, the bombs were a great success, ending the war and more than justifying their high cost.

Oppenheimer became a national hero, appearing on the covers of *Time* and *Life* magazines. Eventually, he confessed that "in some sort of crude sense, which no vulgarity, no humor, no overstatement can quite extinguish, the physicists have known sin, and this is a knowledge which they cannot lose."[9]

Nevertheless, many physicists subsequently condoned the creation of the more powerful hydrogen bomb as a deterrent to the Soviet Union's threat to the free world. The idea of such a superbomb apparently originated during the war, when Enrico Fermi used a stellar analogy to describe how it might be made. He knew that stars derive their energy from the fusion of the lightest element, hydrogen, into helium under extreme heat and compression in their core. He suggested that a manmade explosion might be used to implode and compact material, creating conditions close enough to those inside a star to start the fusion of deuterium, the heavy form of hydrogen nuclei, and releasing roughly 1,000 times more energy than an atomic bomb. The United States detonated the first such hydrogen bomb in 1953, and subsequent detonations of both atomic and hydrogen bombs by numerous countries have blasted their enormous destructive capability into our minds (Fig. 4.6).

Then, at the height of the anti-Communist crusade initiated by Senator Joseph R. McCarthy, the United States Government tried to remove Oppenheimer's security clearance. At the security hearings in 1954, he was accused of being a Soviet spy, perhaps because of his previous support of radical students at Berkeley, including both his one-time fiancé and his wife, who were members of the Communist Party. Hans Bethe defended his former boss, but Edward Teller, then associate director at the Lawrence Livermore Laboratory, strongly faulted Oppenheimer's judgment. Oppenheimer lost his security clearance and Teller lost the respect of many scientists.

However, to return to the trajectory of this book, some of the same scientists who developed the atomic bomb also showed how nuclear reactions deep inside the Sun and other stars makes them shine, while also synthesizing most of the elements heavier than helium that are now found in the universe.

5

What Makes the Sun Shine?

5.1 Awesome Power, Enormous Times

The Sun is so big, hot, and nearby that it is the brightest object in the daytime sky, warming our lands and lighting our days, even though the Earth intercepts only a modest fraction of the Sun's radiation. When we measure the total amount of sunlight that illuminates and warms our globe and then extrapolate back to the Sun, we find that it is emitting an enormous power of 385.4 million million million million, or 3.828×10^{26}, watts. In just 1 second, the energy output of the Sun equals the entire energy consumption of the United States for 1 million years.

The Sun's brilliance is far too great to be sustained perpetually, and no star can shine forever. All things wear out with time. We therefore wonder what heats the Sun and how long that heat will last.

No ordinary fire can maintain the Sun's steady supply of heat for long periods. If the Sun were composed entirely of coal, with enough oxygen to sustain combustion, it would be burned away and totally consumed in a few thousand years.

In the mid-nineteenth century, the German physicist Hermann von Helmholtz proposed that the Sun's luminous energy is due to its gravitational contraction. If the Sun were gradually shrinking, the compressed matter would become hotter and the solar gases would be heated to incandescence; in more scientific terms, the Sun's gravitational energy would be converted slowly into the kinetic energy of motion and heat up the Sun so that it would continue to radiate. This follows from the principle of the conservation of energy, which Helmholtz was one of the first to propose. It states that energy can be neither created nor destroyed; it can only change form.

As subsequently shown by the Irish physicist William Thomson, later Lord Kelvin, gravity might supply the Sun's energy for about 100 million years. In his article titled "On the Age of the Sun's Heat," published in 1862, he wrote the following:

> It seems, therefore, on the whole most probable that the Sun has not illuminated the Earth for 100 million years, and almost certain that he has not done so for 500 million years. As for the future, we may say, with equal certainty, that inhabitants of the Earth cannot continue to enjoy the light and heat essential to their life, for

> many million years longer, unless sources now unknown to us are prepared in the great storehouse of creation.[10]

To be precise, a contraction of 32.5 meters per year will power the Sun at its present rate. However, if the Sun continues to shine this way, it will shrink down to practically nothing and vanish from sight in 21.4 million years. The astonishing thing, which was not realized at the time Thomson wrote his influential article, is the Sun's durability. It has lasted much longer than he envisioned, so there are indeed other sources of energy "prepared in the great storehouse of creation."

In looking back at the Earth's history, we find that the Sun has been shining steadily and relentlessly for eons with a brilliance that could not be substantially less than it is now. The radioactive clocks in rock fossils indicate, for example, that the Sun was hot enough to sustain primitive creatures on the Earth 3.5 billion years ago, and an examination of the relative amounts of radioactive atoms and their decay products shows that the entire solar system is most likely 4.6 billion years old.

Even in the early twentieth century, no one had any clue as to why the Sun, or any other star, could shine so brightly for billions of years. That understanding had to await the discovery of subatomic particles and the realization that the Sun is composed mainly of hydrogen. Of equal importance was the fact that the center of the Sun is much hotter than an ordinary fire, enabling it to consume atomic nuclei.

5.2 How Hot Is the Center of the Sun?

The most abundant atom in the Sun is hydrogen, with a single proton at its nuclear center and one electron outside of the nucleus. It is so hot within most of the Sun, except its cool outer atmosphere, that all of the protons and electrons have been liberated from their atomic bonds and move about unattached to one another.

Protons are 1,836 times more massive than electrons; therefore, they dominate the gravitational effects inside the star. The temperature at the center of the Sun can be estimated by assuming that each proton is hot enough and moving fast enough to counteract the gravitational compression it experiences from the rest of the star. This indicates that a searing central temperature of 15.6 million K is required to support the Sun. The Sun's properties are summarized in Table 5.1.

Such extreme central conditions were recognized more than a century ago when Jonathan Homer Lane, an American astronomer and inventor working at the U.S. Patent Office, assumed that gas pressure supports the weight of the Sun. Although no one knew about nuclear protons at the time, Lane's basic reasoning still applies. The hot protons move about with high speeds, frequently colliding with one another and creating the gas pressure that holds up the Sun.

For the Sun, a central temperature of 15.6 million K establishes equilibrium between the outward pressure of moving protons and the inward gravitational pull at the Sun's center. In any layer within the Sun, the weight of the overlying gas must be equal to the outward-pushing pressure; otherwise, the Sun would expand or contract, which is not observed. At greater distances from the center, there is less overlying material to support and the

Table 5.1. *Physical properties of the Sun*

$M_\odot$ = mass of Sun = 1.989×10^{30} kg
$R_\odot$ = radius of Sun = 6.955×10^{8} m
$\rho_\odot$ = mean mass density of Sun = $3M_\odot/(4\pi R_\odot^3) \approx 1{,}409$ kg m^{-3}
$\rho_{C\odot}$ = central mass density of Sun = 151,300 kg m^{-3}
$T_{C\odot}$ = central temperature of Sun = $2Gm_pM_\odot/(3kR_\odot) \approx 1.56 \times 10^{7}$ K
$V_{esc\odot}$ = escape velocity from photosphere of Sun = $(2GM_\odot/R_\odot)^{1/2} \approx 6.177 \times 10^{5}$ m s^{-1}
$D_\odot$ = 1 AU = mean Earth–Sun distance = 1.4959787×10^{11} m $\approx 1.496 \times 10^{11}$ m
$\theta_\odot = R_\odot/D_\odot$ = angular radius of Sun = 959.63″ where 1″ = 1 second of arc
(At the Sun 1 second of arc = 1″ = 7.253×10^{5} m)
$P_{r\odot}$ = sidereal rotation period of visible solar disk at the equator = 25.67 days
$V_\odot$ = rotation velocity of visible solar disk at the equator = 1,971 m s^{-1}
$f_\odot$ = solar constant = 1361 J s^{-1}m^{-2}
$L_\odot$ = absolute luminosity of Sun = $4\pi f_\odot D_\odot^2 \approx 3.828 \times 10^{26}$ J s^{-1}
$T_\odot$ = effective temperature of visible solar disk = $[L_\odot/(4\pi\sigma R_\odot^2)]^{1/4} \approx 5780$ K
$m_{V\odot}$ = apparent visual magnitude of the Sun = −26.78 mag
$m_{bol,\odot}$ = apparent bolometric magnitude of the Sun = −26.83 mag
$M_{V\odot}$ = absolute visual magnitude of the Sun = +4.83 mag
$M_{bol\odot}$ = absolute bolometric magnitude of Sun = +4.74 mag
$B_\odot$ = magnetic field strength at visible solar disk = 100 G to 1000 G = 0.01 T to 0.1 T
X = mass fraction of hydrogen = 0.7154
Y = mass fraction of helium = 0.2703
Z = mass fraction of all other atoms = 0.0142
Age = 4.6×10^{9} yr

compression, pressure, and temperature are less, so the solar material becomes progressively thinner and cooler.

It is the moving particles inside any star, including the Sun, that hold up the star. This motion, pushing, and pressure of the particles prevent a star from collapsing under its enormous weight. What keeps the particles down there hot, and what sustains their rapid motion? It is nuclear-fusion reactions in the compact, dense core of the Sun that energize the particles there, sustaining their heat and making them move rapidly. Once the nuclear reactions begin, the subatomic energy that is liberated keeps the nuclei sufficiently hot to ensure continuation of the reactions.

5.3 Nuclear Fusion in the Sun's Core

Mass Lost Is Energy Gained

The only known method for keeping the Sun shining with its present luminosity for billions of years involves nuclear-fusion reactions under the intense pressures and exceptionally high temperatures at great depths within the Sun. They are termed *nuclear* because it is the interaction of atomic nuclei that powers the Sun. In nuclear-fusion reactions, two or more atomic nuclei fuse together to produce a heavier nucleus, releasing energy, subatomic particles, and radiation. For the Sun, it is protons, the nuclei of abundant hydrogen atoms that fuse together to make the nuclei of helium atoms, the next most abundant element in the Sun.

Energy can be derived only from energy, and the source of energy in nuclear fusion is mass loss. The basic idea was provided by Albert Einstein in his *Special Theory of Relativity*, which included the famous formula $E = mc^2$ for the equivalence of mass, m, and energy, E. Because the speed of light c, at about 300 million meters per second, is a very large number, only a small amount of mass is needed to produce a large amount of nuclear energy. Even a grain of sand holds an enormous quantity of energy locked up inside its atoms.

The nucleus of the atom was not discovered until 1920, when Ernest Rutherford, at the Cavendish Laboratory of Cambridge University in England, showed that the nuclei of all atoms are composed of hydrogen nuclei, which he named *protons*. In the previous year, the chemist Francis W. Aston, also working at the Cavendish, invented the mass spectrograph and used it to show that the mass of the helium nucleus is slightly less massive, by a mere 0.7 percent, than the sum of the masses of the four hydrogen nuclei, or protons, that enter into it.

At the same time that Rutherford and Aston were discovering the inner secrets of the atoms, the astronomer Arthur Stanley Eddington, director of the nearby Cambridge Observatory, was examining the internal workings of the Sun and other stars. Eddington was a Quaker who resisted military service in World War I (1914–1918) as a conscientious objector. Eddington also never married, which might explain his calm manner and extraordinary productivity.

Eddington also was an avid reader of mystery novels and once likened the process of understanding stars to analyzing the clues in a crime. He knew Rutherford and his colleagues were bombarding atoms of nitrogen, creating helium from them, and he reasoned "what is possible in the Cavendish Laboratory may not be too difficult in the Sun." In his view, the stars are the crucibles in which the heavier elements are made from the lighter ones.

Realizing that such stellar alchemy would release energy, Eddington proposed that hydrogen is transformed into helium inside stars, with the resultant mass difference released as energy to power them. What is obtained in making helium is less than what is put into it, like the usual outcome of any type of gambling. Mass disappears in the Sun and other stars, and the lost mass is converted into energy. Eddington rightly concluded that this could supply the Sun's current luminous output for an estimated 15 billion years.

In his farsighted 1920 essay on the internal constitution of the stars, Eddington wrote the following:

> Certain physical investigations in the past year make it probable to my mind that some portion of subatomic energy is actually being set free in the stars. F. W. Aston's experiments seem to leave no room for doubt that all the elements are constituted out of hydrogen atoms [nuclei] bound together with negative electrons. The nucleus of the helium atom, for example, consists of four hydrogen atoms [nuclei] bound with two electrons. But Aston has further shown conclusively that the mass of the helium atom is less than the sum of the masses of the four hydrogen atoms that enter into it.... Now, mass cannot be annihilated, and the

> deficit can only represent the mass of the electrical energy set free in the transmutation. . . . The total heat liberated will more than suffice for our demands, and we need look no further for the source of a star's energy.[11]

In the same article, he continued with the prescient statement that:

> If, indeed, the subatomic energy in the stars is being freely used to maintain their great furnaces, it seems to bring a little nearer to fulfillment our dream of controlling this latent power for the well being of the human race – or for its suicide.[12]

Physicists nevertheless were convinced that protons, the nuclei of hydrogen atoms, could not react with one another inside the Sun. Protons are positively charged, and like charges repel one another with an electrostatic force given by *Coulomb's law*, developed in 1785 by the French physicist Charles Augustin de Coulomb. This means that there is an electrified barrier that prevents protons from becoming too close.

The repelling force is proportional to the square of the electrical charge of the protons and inversely proportional to their separation. Therefore, the force of repulsion between two protons becomes increasingly larger as they are brought closer together. Even at the enormous central temperature of the Sun, two protons did not seem to have enough energy to overcome this electrical repulsion and move into each other. Stated another way, the protons were not moving fast enough, with enough kinetic energy, to overcome the electrical barrier and merge together.

Even in a high-speed, head-on collision at the enormous central temperature of the Sun, two protons did not have sufficient energy to overcome their mutual electrical repulsion and merge together. The velocity of the average proton at a temperature of 15.6 million K was far too slow, and the central temperature had to be at least 1,000 times hotter to raise the average kinetic energy of the protons above the electrical barrier. However, Eddington was certain that nuclear energy fueled the stars, remarking, "We do not argue with the critic who urges that the stars are not hot enough for this process; we tell him to go find *a hotter place*."

As it turned out, Eddington was correct, for the Russian physicist George Gamow already had provided an explanation for this paradox. While at the University of Göttingen, in what is now West Germany, Gamow showed how a subatomic particle could escape from the nucleus of a radioactive atom. He used the quantum theory of the very small, in which a particle can act like a spread-out wave with no precisely defined position, to determine the penetrability of the barrier surrounding a nucleus during radioactive decay. This tunnel effect works in the opposite way when nuclear particles merge rather than separate, and it explains why nuclei can fuse together inside a star.

With Gamow's encouragement, two young physics students, the English astronomer Robert d'Escourt Atkinson and the Austrian physicist Fritz Houtermans, applied and extended his quantum-tunneling theory to the process of nuclear fusion in stars. Atkinson had just received his Ph.D. degree at Göttingen and Houtermans in the previous year. By combining Gamow's penetration probability with the Maxwellian distribution of speeds, they were able to provide the first attempt at a theory of nuclear-energy generation within stars. Atkinson and Houtermans showed that fusing light nuclei into heavier ones

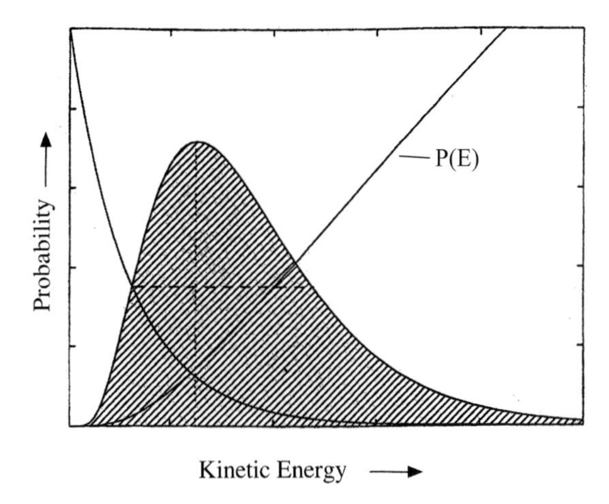

Figure 5.1. Nuclear tunneling in the Sun's core. The high-speed tail (*left*) of the Maxwell distribution of nuclear-particle speeds plotted as a function of kinetic energy, E, for protons near the center of the Sun. The P(E) function (*right*) describes the quantum-mechanical probability of two protons overcoming the electrical repulsion between them. The shaded area (*center*) is the product of the speed and penetration function, and it determines the nuclear-reaction rate in the core of the Sun.

could create energy in accordance with Einstein's formula connecting mass lost to energy gained.

The great stellar abundance of hydrogen had been established recently, and Atkinson also hit on the idea that the observed relative abundances of the elements might be explained by the synthesis of heavy nuclei from lighter ones within stars. It was immediately clear that the most effective nuclear interactions were those involving light nuclei with low charge and that only a few particles in the high-speed tail of the Maxwell speed distribution would be able to penetrate nuclei. For this reason, nuclear reactions proceed slowly in the Sun and other stars. Atkinson and Houtermans also demonstrated that the rate of nuclear reactions substantially increases with the increasing central temperature of stars.

The probability of penetration depends on the kinetic energy, or speed, and the electrical charges of the colliding particles. They have a greater impact when moving at faster speeds, but the electrical repulsion increases with the charge. For this reason, the lightest nuclei, the protons, are more likely to fuse together than the heavier ones, which have greater nuclear charge and mass. That is, the lightest elements carry the smallest charge, with less electrical repulsion between them, and they also move faster than heavier nuclei at a given temperature. Therefore, the initial nuclear reaction within stars would probably be the collision of two protons to form a deuteron and a positron.

Even with this enhanced penetration probability, the average kinetic energy of two colliding protons is not enough for fusion to occur at the center of the Sun. However, the particles in a hot gas do not all move at the average speed. There is a relatively small number, in the high-speed tail of the Maxwell speed distribution, that moves at much faster speeds, permitting fusion once the penetration probability also is considered.

The number of high-speed protons decreases exponentially with increasing speed and kinetic energy, whereas the tunneling probability increases exponentially with the energy (Fig. 5.1). In the overlap region, where the exponential decline meets the exponential rise, there are protons that can participate in the nuclear-fusion reactions that make the Sun shine. Thus, protons sometimes get close enough to move into each other and fuse together, even though their average energy is well below that required to overcome their electrical repulsion.

At a central temperature of 15.6 million K, the protons are darting about so fast within the dense solar core that each proton collides with other protons about 17 million times every second. Despite the exceptionally large number of collisions, a fusion reaction between two colliding protons does not happen very often. The protons nearly always bounce off one another without triggering a nuclear reaction during a collision. Even with the help of tunneling, the average proton must make about 10 trillion trillion, or 10^{25}, collisions before nuclear fusion can happen. It only occurs when the collision is almost exactly head on and between exceptionally fast protons; this explains why the Sun does not expend all of its nuclear energy at once, like an immense hydrogen bomb.

Hydrogen Burning

It was not until the late 1930s that nuclear physicists knew enough about subatomic particles to delineate the Sequence of nuclear reactions that makes the Sun shine; no one knew about neutrons, positrons, or neutrinos during the previous decade.

The German physicist Baron Carl Friedrich von Weizsäcker proposed that the solution to the solar-energy problem lay in the fusion of protons, which Atkinson previously regarded as the most likely explanation. Then Gamow, who had immigrated to the United States, suggested to one of his graduate students, Charles Critchfield, that he calculate the details of the reaction. The results were sent to the German-born American physicist Hans A. Bethe at Cornell University, who found them to be correct; the two published a joint paper titled "The Formation of Deuterons by Proton Combination." Weizsäcker also showed how other nuclear reactions could fuel stars by using carbon as a catalyst in the synthesis of helium from hydrogen.

Gamow, who was teaching physics at George Washington University in Washington, DC, organized a conference to bring astronomers and physicists together to discuss the problem of stellar-energy generation. The Danish astronomer Bengt Strömgren reported that because the Sun was predominantly hydrogen, it would have a central temperature of about 15 million K rather than 40 million K, as estimated by Eddington under the assumption that the Sun had approximately the same chemical composition as the Earth. The lower temperature meant that the calculations of Bethe and Critchfield correctly predicted the Sun's luminosity. Bethe, who attended the conference, was so stimulated by the meeting that within six months he had published a paper titled "Energy Production in Stars," which explains how the Sun fuses hydrogen into helium, releasing energy to heat the Sun's core and generate the radiation that makes it shine.

Bethe moved to Los Alamos, New Mexico, to use his knowledge of nuclear physics in support of the construction of the first atomic bomb. He eventually received the 1967 Nobel Prize in Physics for his contributions to the theory of nuclear reactions, especially his discoveries concerning the energy production in stars.

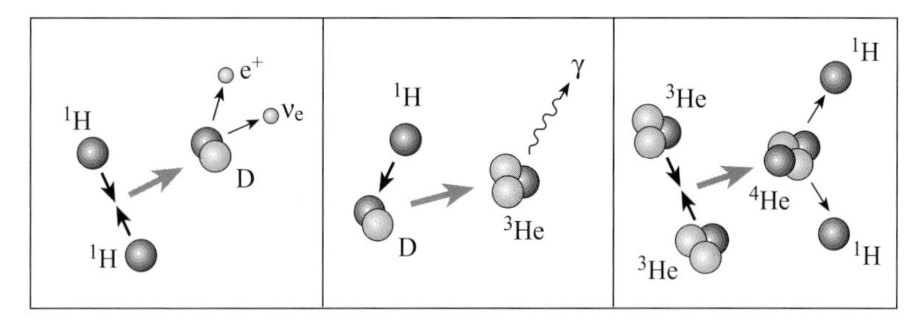

Figure 5.2. Proton-proton chain. Hydrogen nuclei, or protons, are fused together to form helium nuclei within the solar core, providing the Sun's energy. In 1939, the German-born American physicist Hans Bethe (1906–2005) described the detailed sequence of nuclear-fusion reactions, called the proton-proton chain. It begins when two protons, here designated by the letter ^{1}H, combine to form the nucleus of a deuterium atom, the deuteron that is denoted by ^{2}D, together with the emission of a positron, denoted by e^+, and an electron neutrino, designated by ν_e. Another proton collides with the deuteron to make a nuclear isotope of helium, denoted by ^{3}He, and then a nucleus of helium, designated by ^{4}He, is formed by the fusion of two ^{3}He nuclei, returning two protons to the gas. Overall, this chain successively fuses four protons together to make one helium nucleus. Even in the hot, dense core of the Sun, only rare, fast-moving particles can take advantage of the tunnel effect and fuse in this way.

In describing his life in astrophysics, Bethe recalls arriving in Stockholm to accept the award and asking a representative of the Nobel Foundation why there is no Nobel Prize in either astronomy or mathematics.[13] The representative answered that the wife of Alfred Nobel, the inventor of dynamite, had run off with a leading astronomer and mathematician of the time, and Nobel's endowment of the prize specified that neither field could be considered.

The sequence of nuclear reactions that make the Sun shine is called the *proton-proton chain* because it begins with the fusion of two protons. The complete chain of nuclear reactions also is known as the hydrogen-burning reaction for it is hydrogen nuclei, the protons, that are being consumed to make helium. However, it is not combustion in the ordinary chemical sense. In the proton-proton chain (Fig. 5.2), four protons are fused together to form a helium nucleus that contains two protons and two neutrons.

Nuclear reactions often are written in shorthand notation using letters to denote the nuclei and other subatomic particles. An arrow $\rightarrow$ specifies the reaction. Nuclei on the left side of the arrow react to form products given on the right side. The amount of energy released during the reaction also can be given on the far right side of the arrow. A letter and a preceding superscript designate a nucleus. For historical reasons, the nuclei of the hydrogen isotopes ^1H, ^2H, and ^3H also are named protons, deuterons, and tritons, and the nucleus of ^4He is called an *alpha particle*. A Greek letter γ denotes energetic gamma-ray radiation. A positron is denoted by e^+ and an electron is designated by e^-. The symbol ν_e denotes an electron neutrino.

The superscript that precedes a nucleus letter is the mass number, A, which is the sum of the neutrons and the protons and the total number of nucleons in the nucleus. Different isotopes of an element have the same number of protons and the same letter symbol but a different number of neutrons and a different superscript. For instance, a rare isotope of helium, designated ^3He, has two protons and one neutron in its nucleus, whereas the nucleus of the common form of helium, ^4He, has two protons and two neutrons.

In the first step of the proton-proton chain, two protons, each designated by either ^1H or p, meet head on and merge into one another, tunneling through the electrical barrier between them. The two protons combine to make a deuteron, ^2D, which is the nucleus of a heavy form of hydrogen known as *deuterium*.

Because a deuteron consists of one proton and one neutron, one of the protons must be neutralized. It is turned into a neutron, n, with the ejection of a positron – or positive electron denoted e^+ – to carry away the proton's charge and an electron neutrino, ν_e, to balance the energy in the reaction. This is the positive beta-decay reaction, denoted by the following:

$$p \rightarrow n + e^+ + \nu_e,$$

which applies to only one of the two protons making the deuteron. The initiating proton-proton reaction that involves both protons therefore is written as follows:

$$p + p \rightarrow {}^2D + e^+ + \nu_e.$$

Each proton inside the Sun is involved in a collision with other protons millions of times every second, but only exceptionally hot ones can tunnel through their electrical repulsion and fuse together. Only one collision in every 10 trillion trillion initiates the proton-proton chain.

The electron neutrinos produced in the first step of the proton-proton chain escape from the Sun without reacting with matter, carrying energy away; however, the positron – the anti-matter particle of the electron – is consumed immediately. Anti-matter and matter cannot coexist, and the reason why we live in a material world is simply because there is more matter than anti-matter. As soon as any anti-matter positron is produced, it is immediately wiped out of existence by colliding with an electron. The two subatomic particles annihilate one another in a flash of radiation at gamma-ray wavelengths.

The next step follows with little delay. In less than 1 second, the deuteron collides with another proton to form a nucleus of light helium, ^3He, and releases a gamma-ray photon, designated γ. In symbolic terms, the second step of the proton-proton chain is written as follows:

$$^2D + p \rightarrow {}^3He + \gamma.$$

This reaction occurs so easily that deuterium cannot be synthesized inside stars; it is consumed quickly to make heavier elements.

In the final part of the proton-proton chain, two such light helium nuclei meet and fuse together to form a nucleus of normal heavy helium, ^4He, and return two protons to the solar gas. This step takes about 1 million years on average and is written as follows:

$$^3He + {}^3He \rightarrow {}^4He + 2p.$$

This normal helium nucleus contains two protons and two neutrons; therefore, two of the protons that contributed to the formation of helium were converted into neutrons by the positive beta-decay reaction.

A total of six protons is required to produce the two ^3He nuclei that go into this last reaction, but two protons are returned to the solar interior to be reused later. Because two protons and a helium nucleus are produced, the net result of the proton-proton chain is as follows:

$$4p \rightarrow {}^4He + \text{ gamma-ray radiation} + \text{ neutrinos.}$$

The helium is slightly less massive, by a mere 0.007, or 0.7 percent, than the four protons that combine to make it, so there is energy released that eventually leaves the Sun as radiation. The part of this radiation that constitutes visible light is what makes the Sun shine. The subatomic energy that is liberated also keeps the core of the Sun hot, assuring continuation of the nuclear reactions.

The Sun is consuming itself at a prodigious rate. Every second, roughly 100 trillion trillion trillion, or 10^{38}, helium nuclei are created from about 700 million tons of hydrogen, where 1 ton is equivalent to 1,000 kg. In doing so, 5 million tons (0.7 percent) of this matter disappears, which is enough to keep the Sun shining with its present brilliant luminous output. However, the Sun's mass loss is trivial compared to its total mass, so the Sun has scarcely changed its mass even during the 4.6 billion years that it has been shining. It has lost only 1 percent of its original mass in all that time.

Why Doesn't the Sun Blow Up?

Hydrogen fusion has been used to create the powerful hydrogen bomb, which employs the fusion of the heavy isotopes of hydrogen to generate neutrons and increase the explosive yield of chain reactions. Inside the Sun, an energy equivalent to 2,000 million hydrogen bombs, of the 50-megaton variety, is being released every second.

So why doesn't the Sun blow up? Equilibrium between the inward pull of gravity and the outward pressure of the internal moving particles tames the hydrogen bomb inside the Sun, preventing detonation. The sheer weight of the gas surrounding the Sun's core keeps the lid on the nuclear cauldron and prevents it from blowing up.

Unlike a nuclear bomb on the Earth, the temperature-sensitive reactions inside the Sun act like a thermostat, releasing energy in a steady, controlled manner at exactly the rate needed to keep the Sun in equilibrium between the inward pull of gravity and the outward pressure of the moving subatomic particles. If the rate of the thermonuclear reactions in the central regions of the Sun rises as the result of a temperature increase, the nuclei move faster and create more pressure. The entire body of the Sun then would expand and thereby bring down its central temperature. If the rate of core nuclear reactions were to drop, gravity would pull the Sun inward, and the resulting increase in central temperature would bring the rate of energy production back into equilibrium. So, the pendulum continues to swing between gravity and fusion, with no winner. That is how the Sun harnesses its nuclear energy, which it has been doing for 4.6 billion years.

Nuclear-fusion reactions proceed at a slow and stately pace in the Sun, staying at just the right temperature to illuminate and heat the Earth for billions of years. If the central temperature of the solar crucible were much colder, the reactions would occur less often and make the Sun shine feebly, like a flashlight with a worn-out battery. If the temperature at the center were much higher, the frequent and rapid nuclear-fusion reactions would cause the entire Sun to blow up almost instantaneously, like a giant hydrogen bomb. It is the temperature sensitivity of the nuclear reactions that is the key to the Sun's stability, and it is the Sun's massive gravity that pulls and holds it all together, reining in all that nuclear energy.

5.4 Catching the Ghost

The Elusive Neutrino

Neutrinos, or "little neutral ones," are very close to being nothing at all. They are tiny, invisible packets of energy with no electrical charge and almost no mass, traveling at nearly the speed of light. These subatomic particles are so insubstantial and interact so weakly with matter that they streak through almost everything in their path, like a ghost that moves through walls. Unlike light or any other form of radiation, neutrinos can move nearly unimpeded through any amount of material, even the entire universe.

How do we know that such elusive, insubstantial particles even exist? They were proposed as an unseen agent to carry away energy that was missing in beta-particle radioactivity. They were first detected by placing a massive tank of water next to a nuclear reactor.

Neutrinos also are produced in much greater profusion by thermonuclear reactions in the Sun's core, removing substantial amounts of energy that is never seen again. Every second, trillions upon trillions of the solar neutrinos pass right through the Earth without even noticing that it is there. Billions upon billions are in our rooms, both night and day, and they did not enter through the door. At night, the solar neutrinos travel through the Earth before passing through the walls of our houses and even through our bodies, without us ever noticing them.

Neutrinos are the true ghost riders of the universe. As American writer John Updike put it:

> Neutrinos, they are very small.
> They have no charge and have no mass
> And do not interact at all.
> The Earth is just a silly ball
> To them, through which they simply pass,
> Like dust maids down a drafty hall.[14]

The Sun bathes our planet with a beam of neutrinos that is as steady as sunlight. Unlike radiation inside the Sun, which is diluted and transformed during its 170-thousand-year journey from the Sun's core, neutrinos pass quickly through the massive body of the Sun, bringing a unique message from deep within its hidden interior. They tell us what is happening in the center of the Sun right now – or, to be more exact, 499 seconds ago, which is the time required for a neutrino to move from the center of the Sun to the Earth at the speed of light.

By catching and counting solar neutrinos, we can open the door of the Sun's nuclear furnace and peer inside the energy-generating core. The flux of solar neutrinos expected at the Earth is calculated using supercomputers that have produced the Standard Solar Model, which best describes the Sun's luminous output, size, and mass at its present age. Once the Standard Solar Model has been used to specify the neutrino flux, the predictions are extended to specific experiments that detect neutrinos of different energies.

Solar Neutrino Detectors Buried Deep Underground

How do you observe something that seems to appear out of nowhere and interacts only rarely with other matter? Although the majority pass right through matter, there is a finite chance that a neutrino will interact with some of it. When this slight chance is multiplied by the prodigious quantities of neutrinos flowing from the Sun, we conclude that a few occasionally will strike an atom's nucleus squarely enough to produce a nuclear reaction that might signal the presence of an otherwise invisible neutrino. The neutrino detector must consist of large amounts of material, literally tons, to allow interaction with even a tiny fraction of the solar neutrinos and measure their actual numbers. Although almost all of the neutrinos still would pass through any amount of detector material unhindered and undetected, a rare collision with other subatomic particles might leave a trace.

Unlike a conventional optical telescope, which is placed as high as possible to minimize distortion by the Earth's obscuring atmosphere, a solar neutrino detector is buried beneath a mountain or deep within the Earth's rocks inside mines. This shields the instrument from deceptive signals caused by cosmic rays. There, beneath tons of rock that only a neutrino can penetrate, detectors unambiguously measure neutrinos from the Sun. If neutrino detectors were placed on the Earth's surface, they would detect high-energy particles and radiation produced by cosmic rays interacting with the Earth's atmosphere.

Thus, solar-neutrino astronomy involves massive subterranean detectors that look right through the Earth and observe the Sun at night or day. The first such neutrino telescope, constructed in 1967 by Raymond Davis, Jr., was a 615-ton tank located 1.5 km underground in the Homestake Gold Mine near Lead, South Dakota. The huge cylindrical tank was filled with 378,000 liters of cleaning fluid, technically called perchloroethylene (or "perc" in the dry-cleaning trade); each molecule of the stain remover consists of two carbon atoms and four chlorine atoms.

Most solar neutrinos passed through the tank unimpeded. Occasionally, however, a neutrino scored a direct hit with the nucleus of a chlorine atom, turning one of its neutrons into a proton, emitting an electron to conserve charge, and transforming the chlorine atom into an atom of radioactive argon. The new argon atom rebounds from the encounter with sufficient energy to break free of the perc molecule and enter the surrounding liquid. Because argon is chemically inert, it can be culled from the liquid by bubbling helium gas through the tank. The number of argon atoms recovered in this way measured the incident flux of solar neutrinos.

Every few months, Davis and his colleagues flushed the tank with helium, extracting about 15 argon atoms from a tank the size of an Olympic swimming pool. That was a remarkable achievement considering that the tank contained more than 1 million trillion trillion, or 10^{30}, chlorine atoms. The scientists persisted for nearly 30 years, capturing signatures of only 2,000 neutrinos in all that time. However, the consequences were enormous: The measurements implied not only that nuclear-fusion reactions indeed were providing the Sun's energy, making it shine, but there also was a small unexpected problem with the result that led to a new understanding of the physics of neutrinos.

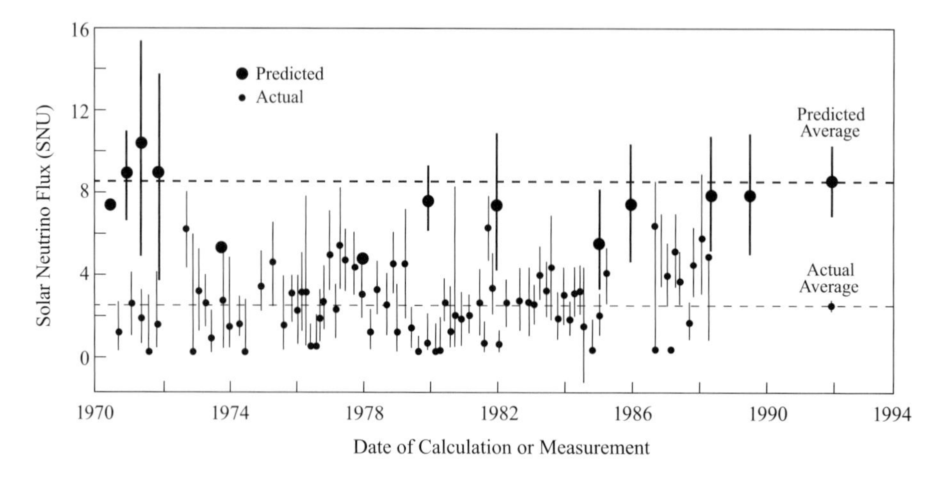

Figure 5.3. Solar neutrino fluxes. Calculated and measured solar neutrino fluxes have disagreed consistently for several decades. The fluxes are measured in solar neutrino units (SNU), which are defined as 1 neutrino interaction per trillion trillion trillion, or 10^{36}, atoms per second. Measurements from the chlorine neutrino detector (*small dots*) give an average solar neutrino flux of 2.6 ± 0.2 SNU (*lower broken line*), well below theoretical calculations (*large dots*) that predict a flux of 8.5 ± 1.8 SNU (*upper broken line*) for the Standard Solar Model. Other experiments also have observed a deficit of solar neutrinos, suggesting that either some process prevents neutrinos from being detected or there is an incomplete understanding of the nuclear processes that make the Sun shine.

The neutrino reaction rate with atoms in neutrino detectors is so slow that a special unit was invented to specify the experiment-specific flux. The Solar Neutrino Unit (abbreviated SNU and pronounced "snew") is equal to one neutrino interaction per second for every trillion trillion trillion, or 10^{36}, atoms. Even then, the predictions were only a few SNU per month for even the largest, most-massive detectors first constructed.

The Homestake detector always yielded results in conflict with the most accurate theoretical calculations. The final experiment value was 2.55 ± 0.25 SNU, where the $\pm$ value denotes an uncertainty of one standard deviation. (A *standard deviation* is a statistical measurement of the uncertainty of a measurement; a definite detection must be above three standard deviations and preferably above five.) In contrast, theoretical results using the Standard Solar Model predicted that the Homestake detector should have observed a flux of 8.5 ± 1.8 SNU. So, the tank full of cleaning fluid captured almost one third of the expected number of neutrinos (Fig. 5.3). The discrepancy between the observed and calculated values is known as the *Solar Neutrino Problem*.

In 1987, another giant underground neutrino detector, named Kamiokande, began to monitor solar neutrinos to confirm the neutrino deficit observed by Davis. This second experiment, located in a mine at Kamioka, Japan, consisted of a 4,500-ton, or 4.5-million-liter, tank of pure water. Nearly 1,000 light detectors were placed in the tank walls to measure signals emitted by electrons knocked free from water molecules by passing neutrinos.

The axis of the light cone gives the electron's direction, which is the direction from which the neutrino arrived. Because the observed electrons were preferentially scattered along the direction of an imaginary line joining the Earth to the Sun, the Kamiokande water experiment also confirmed that the neutrinos indeed are produced by nuclear reactions in the Sun's core.

The importance of the solar-neutrino measurements was confirmed in 2002, when Davis received the Nobel Prize in Physics; the Kamiokande scientist Masatoshi Koshiba shared the prize for the detection of cosmic neutrinos.

Solving the Solar Neutrino Problem

After almost 40 years of meticulous measurements and calculations, the neutrino count still came up short! Massive underground detectors always observed fewer neutrinos than theory states they should detect, and the Solar Neutrino Problem was a continuing embarrassment. Either the Sun did not shine the way we thought it did or our basic understanding of neutrinos was in error.

Is there something wrong with our understanding of the internal operations of the Sun? When the Solar Neutrino Problem first arose, some harried astrophysicists pondered deeply and lost sleep at night, coming up with all sorts of explanations, most of which involved a reduction of the central temperature of the Sun by about 3 percent. If the center of the Sun were about a half million degrees cooler than presently thought, the nuclear reactions would run at a slower rate, producing neutrinos in the observed amounts and resolving the dilemma.

A lower central temperature might result from a rapidly rotating core or a strong central magnetic field that would help hold up the Sun against the inward tug of its own gravity, thereby reducing the pressure and temperature at its center; or, the temperature might be reduced by mixing from the Sun's cooler outer layers.

However, these speculative solutions were unlikely, and most scientists agreed that an incomplete understanding of neutrinos was the cause of the Solar Neutrino Problem. It turned out that the neutrinos change into a different form during their journey from the center of the Sun, escaping detection by changing character. The reason that some neutrinos were not detected is because they were hiding in disguise, having metamorphosed into a different form, like a caterpillar turning into a butterfly.

Neutrinos apparently have an identity crisis! Each type of neutrino is not completely distinct, and the different types can be transformed into one another. In the language of quantum mechanics, neutrinos do not occupy a well-defined state; they instead consist of a combination or mixture of states. As neutrinos move through space, the states come in and out of phase with one another, so the neutrinos change form with time. The three possible types of neutrinos are called *electron neutrinos*, *muon neutrinos*, and *tau neutrinos*, each named for the type of particle with which it interacts.

The effect is called *neutrino oscillation* because the probability of metamorphosis between neutrino types has a sinusoidal, in and out, oscillating dependence on path length. The chameleon-like change in identity is not one way, for a neutrino of one type can change into another type and back again as it moves along. Like the Cheshire cat, the elusive neutrino can appear, disappear, and perhaps reappear.

The neutrino metamorphosis would happen extremely rarely in the vacuum of space but might be amplified in the dense interior of the Sun. Interactions between the electron neutrinos, made by nuclear reactions near the center of the Sun, and the densely packed solar electrons in the rest of the Sun can – when the density is right – alter the mass state of a neutrino traveling out through the Sun, thereby changing it into a muon neutrino. Once formed, the muon neutrino travels out into space and could remain invisible to the first solar neutrino detectors.

Such a transformation was suggested first by observations of nonsolar neutrinos using the Super-Kamiokande detector, which replaced the older, nearby Kamiokande instrument in 1996. Super-Kamiokande can observe both solar electron neutrinos and atmospheric muon neutrinos. The former are created by nuclear fusion at the center of the Sun, whereas the latter are created when fast-moving cosmic rays enter the Earth's atmosphere from outer space. Atmospheric neutrinos have higher energy and produce a tighter cone of light than solar electron neutrinos, which make a fuzzy, blurred, and ragged light pattern.

After monitoring light patterns for more than 500 days, the Super-Kamiokande scientists reported in 1998 that there were roughly twice as many muon neutrinos coming from the Earth's atmosphere directly over the Super-Kamiokande detector than those coming from the other side of the Earth. The muon neutrinos are produced in the atmosphere above every place on our planet, but some apparently disappeared while traveling through the Earth to arrive at the detector from below. Subsequent experiments using neutrinos generated by particle accelerators on the Earth confirmed the effect. They suggest that although all neutrinos produced by nuclear reactions in the Sun are electron neutrinos, they do not stay that way.

Nevertheless, the terrestrial neutrinos did not come from the Sun and are not directly related to nuclear-fusion reactions there. Therefore, the solution to the Solar Neutrino Problem was not known definitely until 2001, when a new underground detector in Canada, the Sudbury Neutrino Observatory, demonstrated that solar neutrinos are changing type when traveling to the Earth. The Sudbury Neutrino Observatory (abbreviated SNO and pronounced "snow") is located 2 km underground in a working nickel mine near Sudbury, Ontario. It is a water detector but, unlike Kamiokande and Super-Kamiokande, the SNO detector contains heavy water.

Heavy water is chemically similar to ordinary water and it does not appear or taste any different. In fact, heavy water exists naturally as a constituent of ordinary tap or lake water in a ratio of about 1 part in 7,000; expensive chemical and physical processes can separate it.

The hydrogen in heavy water has a nucleus, called a deuteron, which consists of a proton and a neutron. For ordinary water, the hydrogen is about half as light, with a nucleus that contains only a proton and no neutron. It is the heavier deuteron that makes the SNO sensitive to not just one type of neutrino but instead to all three known varieties.

One thousand tons, or 1 million liters, of heavy water was placed in a central spherical cistern with transparent acrylic walls. The heavy water was borrowed from Atomic Energy of Canada Limited, which stockpiled it for use in its nuclear power reactors – the heavy water moderates neutrons created by uranium fission in the reactors.

A geodesic array of about 10,000 photo-multiplier tubes surrounds the vessel to detect the flash of light given off by heavy water when it is hit by a neutrino. Both the light sensors and the central tank are enveloped by a 7,800-ton jacket of ordinary water (Fig. 5.4) to shield the heavy water from weak, unwanted signals from the underground rocks. As with other neutrino detectors, the overlying rock blocks energetic particles generated by cosmic rays.

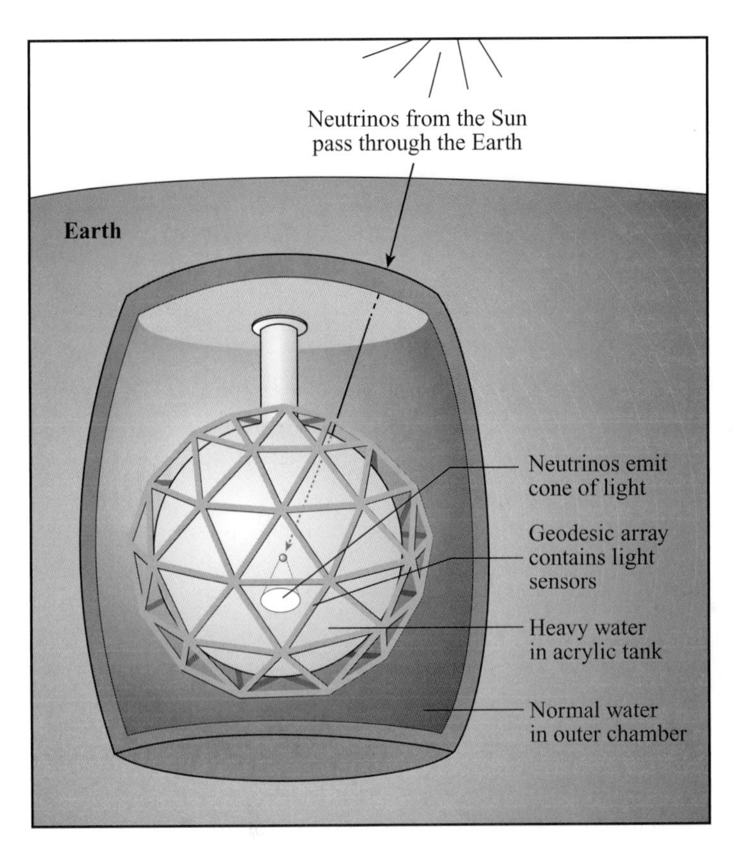

Figure 5.4. How Sudbury works. Neutrinos from the Sun travel through more than 2 kilometers of rock, entering the acrylic tank of the Sudbury Neutrino Observatory, which contains 1,000 tons (1 million liters) of heavy water. When one of these neutrinos interacts with a water molecule, it produces a flash of light that is detected by a geodesic array of photo-multiplier tubes. Some 7,800 tons (7.8 million liters) of ordinary water surrounding the acrylic tank block radiation from the rock, and the overlying rock blocks energetic particles generated by cosmic rays in our atmosphere. The heavy water is sensitive to all three types of neutrinos.

The SNO can be operated in two modes: one sensitive only to electron neutrinos and the other equally sensitive to all three types of neutrinos. Observations with both modes have confirmed that the Solar Neutrino Problem is caused by changes in the neutrinos as they travel from the Sun's core and, when this is taken into account, the total number of electron neutrinos produced in the Sun is as predicted. After 33 years, the Solar Neutrino Problem was solved!

5.5 How the Energy Gets Out

All of the Sun's nuclear energy is released deep down inside its high-temperature core, and no energy is created in the cooler regions outside of it. The energy-generating core extends to about one quarter of the distance from the center of the Sun to the visible solar disk, accounting for only 1.6 percent of the Sun's volume. However, about half of the Sun's mass is packed into its dense core.

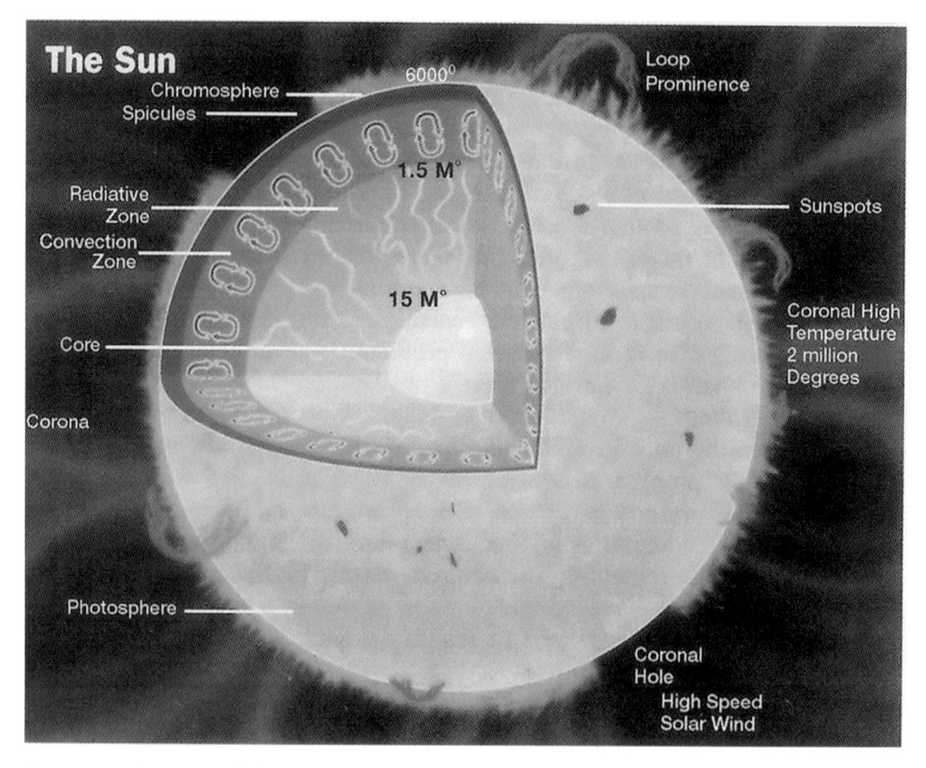

Figure 5.5. Anatomy of the Sun. The Sun is an incandescent ball of ionized gas powered by the fusion of hydrogen in its core. As shown in this interior cross section, energy produced by nuclear fusion is transported outward, first by countless absorptions and emissions within the radiative zone and then by convection. The visible disk of the Sun, called the *photosphere*, contains dark sunspots, which are Earth-sized regions of intense magnetic fields. A transparent atmosphere envelops the photosphere, including the low-lying chromosphere with its jet-like spicules and the 1-million-degree corona that contains holes with open magnetic fields, the source of the high-speed solar wind. Loops of closed magnetic fields constrain and suspend the hot million-degree gas within coronal loops and cooler material in prominences. (See color plate).

Because we cannot see inside the Sun, astronomers combine basic theoretical equations (e.g., for equilibrium and energy generation or transport) with observed boundary conditions (e.g., the Sun's mass and luminous output) to create models of the Sun's internal structure. These models consist of two nested spherical shells that surround the hot, dense core (Fig. 5.5).

The innermost shell, called the *radiative zone*, extends from the core to 71.3 percent of the Sun's radius. As the name implies, energy moves through this region by radiation. It is a relatively tranquil and placid place, analogous to the deeper parts of the Earth's oceans. The outermost layer is known as the *convective zone*, where energy is transported in a churning, wheeling motion called *convection*. It is a lively, turbulent, and vigorous place, encompassing the more serene radiative zone.

Although light is the fastest thing around, radiation does not move quickly through the solar interior. The center of the Sun is so dense that a single gamma ray produced by nuclear fusion in the solar core cannot move even a fraction of a millimeter before encountering a subatomic particle, where the radiation is scattered or absorbed and reemitted with less energy. This radiation quickly interacts with another particle in the radiative zone and is eventually reradiated at yet lower energy. The process continues countless times as the radiation moves outward on a haphazard, zigzag path, steadily losing energy at each encounter.

As a result of this continued ricocheting and innumerable collisions in the radiative zone, it takes about 170,000 years, on average, for radiation to work its way out from the Sun's core to the bottom of the convective zone, where the temperature has become cool enough for heavy nuclei to capture electrons and form atoms that absorb radiation. These atoms block the outward flow of radiation like dirt on a window, and the radiation heats the bottom of the convective zone.

This material becomes hotter than it otherwise would be, and it must find a way to release the pent-up energy. In response to heating from below, gases in the bottom layer of the convective zone expand, thereby becoming less dense than the gas in the overlying layers. Due to its low density, the heated material rises to the visible solar disk in about 10 days and then cools by radiation. The cooled gas then sinks because it is denser than the hotter surrounding gas, only to be reheated and rise again. Such convective motions occur in a kettle of boiling water or a simmering pot of oatmeal, with hot rising bubbles and cooler sinking material.

The convective zone is capped by the photosphere, where hydrogen atoms absorb some light while most of the sunlight travels out into interplanetary space; however, extra opacity was required to account for the detailed observations. They were explained when Rupert Wildt showed that collisions between un-ionized, or neutral, hydrogen atoms and free electrons lead to the formation of negative hydrogen ions. Despite their low concentration, they provide the absorption and extra opacity, beyond hydrogen atoms, needed to account for the observations of sunlight escaping from the photosphere.

As first noticed by William Herschel at the beginning of the nineteenth century, the photosphere contains a fine granular pattern. These closely packed granulation cells now can be examined using high-resolution images taken from ground-based telescopes under conditions of excellent observations (Fig. 5.6) or from spacecraft located outside of the Earth's obscuring atmosphere. The images reveal a host of granules with bright centers surrounded by dark lanes, exhibiting a nonstationary, overturning motion caused by the underlying convection.

The bright center of each granule, or convection cell, is the highest point of a rising column of hot gas. The dark edges of each granule are the cooled gas, which sinks because it is denser than the hotter gas. Individual granules last only about 15 minutes before they are replaced by another one, never reappearing in precisely the same location.

The mean angular distance between the bright centers of adjacent granules is about 2.0 seconds of arc, corresponding to about 1,500 km at the Sun. That distance seems significant, but an individual granule is about the smallest object we can see on the Sun when peering through our turbulent atmosphere.

At least 1 million granules are on the visible solar disk at any moment. They are evolving and changing constantly, producing a honeycomb pattern of rising and falling gas that is in constant turmoil, bubbling away and completely changing on time-scales of minutes.

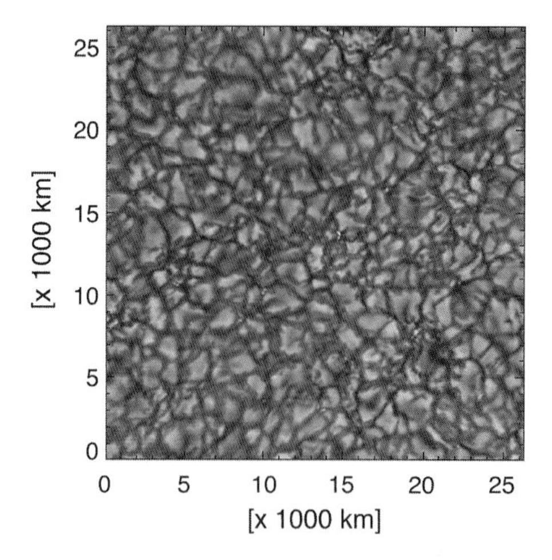

Figure 5.6. The solar granulation. Underlying convection shapes the photosphere, producing tiny, varying regions called *granules*. They are places where hot and, therefore, bright material reaches the visible solar disk. The largest granules are approximately 1,400 km across. They are not circular but rather angular in shape. This honeycomb pattern of rising (*bright*) and falling (*dark*) gas is in constant turmoil, completely changing on time-scales of minutes and never exactly repeating itself. This image was taken with exceptional angular resolution of 0.2 second of arc, or 150 km, at the Sun using the National Solar Observatory's Vacuum Tower Telescope at the Sacramento Peak Observatory. (Courtesy of Thomas R. Rimmele/AURA/NOAO/NSF.)

When studying a larger cellular convection, known as the *supergranulation*, Robert B. Leighton and his coworkers unexpectedly discovered vertical up and down motions in the subtracted difference between long- and short-wavelength solar images. They exhibited an oscillation with a period of about 5 minutes. These oscillations subsequently were used to investigate the unseen depths of the Sun.

5.6 Looking Inside the Sun

Taking the Pulse of the Sun

The interior of the Sun is as opaque as a stone wall, and there is no way we can see inside it! However, we can illuminate, or sound, the hidden depths of the Sun by recording oscillations of its visible disk, which are produced by a secret melody of sounds trapped inside the Sun.

On striking the visible solar disk and rebounding back down, the sound waves cause the gases there to move up and down, producing a widespread throbbing motion. These sounds are produced by vigorous turbulent motion in the convective zone, somewhat like the deafening roar of a jet aircraft or the hissing noise made by boiling water. Sound waves generated in the convective zone echo through the solar interior, causing the entire globe – or parts of it – to move in and out, slowly and rhythmically like the regular rise and fall of tides or a beating heart.

The sounds resonate within the convective zone like the plucked string of a guitar or the beat of a drum. When we move water regularly in a bathtub, the waves similarly grow in

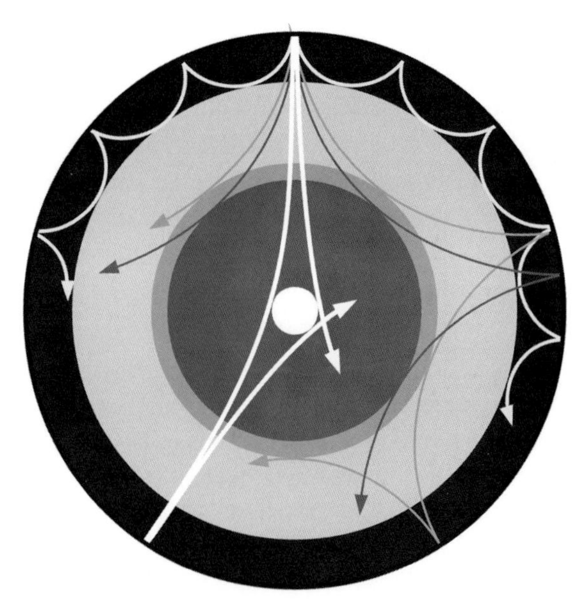

Figure 5.7. Sound paths. The trajectories of sound waves are shown in a cross section of the solar interior. The rays are bent inside the Sun, like light within the lens of an eye, and circle the solar interior in spherical shells called *resonant cavities*. Each shell is bounded at the top by a large, rapid density drop near the photosphere and bounded at the bottom at an inner turning point where the bending rays undergo total internal refraction due to the increase in sound speed with depth inside the Sun.

size, but when we swish the water randomly, it develops a choppy confusion of small waves. The resonating sound waves strike the photosphere and rebound back down, producing a localized motion that changes from moving outward to moving inward and back outward again in 5-minute intervals. Such 5-minute oscillations are imperceptible to the unaided eye because the photosphere moves a mere 1/100,000th (0.00001) of the solar radius.

A sound wave's path inside the Sun forms a regular sequence of loops (Fig. 5.7). When a sound wave angles up to the visible solar gases, in the photosphere, it strikes them with a glancing blow, turns around, and travels back into the Sun, like light reflected from a mirror. Above the photosphere, most of the sound waves are evanescent and cannot propagate.

The inner turning point of a sound wave's downward trajectory depends on the increase of sound speed, or wave velocity, with depth. Because the speed of sound is greater in a hotter gas, it increases in the deeper, hotter layers of the Sun. The deeper part of a wave front traveling obliquely into the Sun moves faster than the shallower part and pulls ahead of it. Gradually, the advancing wave once again is headed back up. Sound is similarly refracted down into the cool air above a mountain lake; the hotter, higher air bends the sound downward, permitting it to travel great distances across the lake's surface.

Some of the photosphere oscillations are created by sound waves that have moved just beneath the part of the Sun that we can see; others arise from sounds that have traveled deep into the Sun's interior. The information obtained from oscillations produced by sound waves that traveled to various levels within the Sun can be combined to create a picture of the Sun's large-scale internal structure, somewhat similar to an x-ray CAT scan that probes the inside of a human's brain.

Geophysicists similarly construct models of the Earth's interior by recording earthquakes, or seismic waves, that travel to different depths; this type of investigation is called *seismology*. Astronomers use the name *helioseismology* to describe somewhat similar studies of the Sun's interior; it is a hybrid name combining the Greek words *helios* for the "Sun" and *seismos* for "earthquake" or "tremor" – although there are nothing like earthquakes on the Sun.

Observations from space where night never falls and the Sun can be observed nonstop provide the best data for helioseismology. The SOlar and Heliospheric Observatory (SOHO) has provided them. Instruments aboard this spacecraft have observed solar oscillations 24 hours a day, every day, for more than 10 years. Analagous to terrestrial seismology, it is now possible to use these observations of the 5-minute solar oscillations to isolate sound waves penetrating to various depths. This has resulted in precise and detailed information about the properties of the solar interior, rivaling our knowledge of the inside of the Earth. A small but definite change in sound speed, for example, marks the lower boundary of the convective zone, which is located at a radius of 71.3 percent, or extends to a depth of 28.7 percent, of the radius of the visible Sun.

How Does the Sun Rotate Inside?

To a first approximation, the internal structure of the Sun has spherical symmetry, and the measured properties of the 5-minute oscillations depend on only radial variations in the Sun's internal properties. On close inspection, however, there are secondary effects that break this radial symmetry; the most pronounced symmetry-breaking agent is rotation.

Sound waves propagating in the direction of rotation tend to be carried along by the moving gas and move faster than they would without any rotation. A bird or a jet airplane similarly moves faster when traveling with the wind than against it. The resonating sound wave crests moving with the rotation therefore appear – to a fixed observer – to move faster and their measured periods are shorter. Waves propagating against the rotation are slowed down, with longer periods.

Thus, rotation imparts a clear signature to the oscillation periods, lengthening them in one direction and shortening them in the other. These opposite effects make the oscillation periods divide; such rotational splitting depends on both depth and latitude within the Sun.

Observations of sunspots indicate that the visible solar disk rotates with a period of approximately 25 days at the equator. The solar oscillations have a period of about 5 minutes, so the rotational splitting is roughly 5 minutes divided by 25 days, or about 1 part in 7,000. The oscillations must be measured 10 or a 100 times more accurately than this to determine subtle variations in the Sun's rotation, or as accurately as 1 part in 1 million.

It has been known since 1610 – when Galileo first used a telescope to observe the Sun – that sunspots near the solar equator rotate around the Sun faster than sunspots at higher solar latitudes; this was confirmed by Doppler-effect measurements of gas in the photosphere (Table 5.2). As on the Earth, latitude is the angular distance north or south of the equator. In contrast, the rotation rate of the solid Earth is the same at every latitude; therefore, all points of the globe take the same amount of time to complete a rotation, and a day lasts 24 hours everywhere on the planet.

Table 5.2. *Differential rotation of the Sun*[a]

Solar Latitude (degrees)	Rotation Period (days)	Rotation Speed (km h^{-1})	Rotation Speed (m s^{-1})	Angular Velocity (nHz)
0 (equator)	25.67	7,097	1,971	451
15	25.88	6,807	1,891	447
30	26.64	5,922	1,645	434
45	28.26	4,544	1,262	410
60	30.76	2,961	823	376
75	33.40	1,416	393	347

[a] Data from the Michelson Doppler Imager (MDI) instrument aboard the *SOHO* spacecraft.

The solar oscillation data indicate that this differential rotation, in which the equator spins faster than the poles, is preserved throughout the convective zone (Fig. 5.8). Within this zone, there is little variation of rotation with depth, and the inside of the Sun does not rotate any faster than the outside. At greater depths, the interior rotation no longer mimics that of sunspots, and differential rotation disappears. The internal accord breaks apart just below the base of the convective zone, where the rotation speed becomes uniform from pole to pole. Lower down, within the radiative zone, the rotation rate remains independent of latitude, acting as if the Sun were a solid body. Although gaseous, the radiative interior of the Sun rotates at a nearly uniform rate intermediate between the equatorial and high-latitude rates in the overlying convective zone.

Thus, the rotation velocity changes sharply at the top of the radiative zone, located nearly one third of the way to the core. There, outer parts of the radiative interior, which rotates at one speed, meet the overlying convective zone, which spins faster in its equatorial middle. The transition between these two different regimes takes place in a narrow region, where a strong radial rotational shear helps generate the large-scale solar magnetic field.

When the average differential rotation of the Sun is removed from the helioseismology data, rivers of gas are found sweeping around the Sun's equatorial regions in zonal bands, moving just slightly faster or slower than the average rotation. It is similar to watching water flow beneath an ice-covered stream in winter, when we do not notice that the Earth is rotating. The velocity of the faster zonal flows is about 5 to 10 m s^{-1} higher than gases to either side. This is substantially slower than the mean velocity of rotation, which is about 2,000 m s^{-1}; therefore, the fast zones glide along in the spinning gas, like a wide lazy river. These internal flows run deep, penetrating to a substantial fraction of the Sun's convective zone; they might involve the entire convective envelope of the Sun.

Ponderous, slow rivers of gas also circulate from the solar equator toward both poles. This persistent poleward flow, to the north and the south, is named the *meridional oscillation*. It also extends deep within the convective zone. The entire outer layer of the Sun, to a depth of at least 25,000 km, is slowly but steadily flowing from the equator to the poles at a speed of about 20 m s^{-1}. At this rate, an object would be transported from the equatorial middle

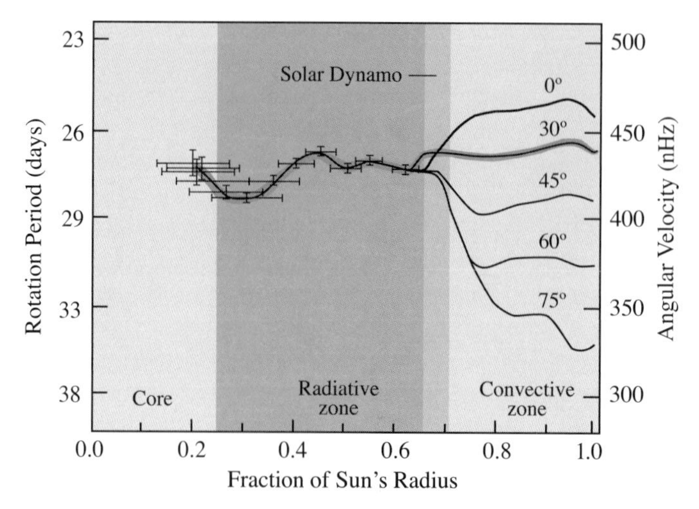

Figure 5.8. Internal rotation of the Sun. The rotation rate inside the Sun, determined by helioseismology using instruments aboard the *SOHO*. The outer parts of the Sun exhibit differential rotation, with material at high solar latitudes rotating more slowly than equatorial latitudes. This differential rotation persists to the bottom of the convective zone at 28.7 percent of the way down to the center of the Sun. The rotation period in days is given at the left axis, and the corresponding angular velocity scale is on the right axis in units of nanoHertz (nHz), where 1 nHz = 10^{-9} Hz, or 1 billionth, of a cycle per second. A rotation rate of 320 nHz corresponds to a period of about 36 days (*solar poles*) and a rate of 460 nHz to a period of about 25 days (*solar equator*). The rotation in the outer parts of the Sun is given at latitudes of 0 (*solar equator*), 30, 45, 60, and 75 degrees. Just below the convective zone, the rotational speed changes markedly, and shearing motions along this interface may be the dynamo source of the Sun's magnetism. There is uniform rotation in the radiative zone, from the base of the convective zone at 0.713 to about 0.25 solar radii. The sound waves do not reach the central part of the energy-generating core. (Courtesy of Alexander G. Kosovichev/convective zone/Sebastien Couvidat, Rafael García, and Sylvaine Turck-Chièze/radiative zone. *SOHO* is a project of international cooperation between ESA and NASA.)

to the polar top of the photosphere in a little more than one year. Of course, the Sun rotates at more than 100 times this rate, completing one revolution at the equator in about 25 days.

5.7 The Faint-Young-Sun Paradox

Nothing in the cosmos is fixed and unchanging or escapes the ravages of time. Everything moves and evolves, including the seemingly constant and unchanging Sun. It began about 4.6 billion years ago, when a spinning interstellar cloud fell in on itself. The center became increasingly dense until it became so packed, so tight, and so hot that protons came together and fused into helium, making the Sun glow. Ever since then, the Sun has grown slowly in luminous intensity as it aged, with a steady, inexorable brightening that is a consequence of the increasing amount of helium accumulating in the Sun's core.

As the hydrogen in the Sun's center slowly depletes and is steadily replaced by heavier helium, the core must continue producing enough pressure to prevent the Sun from collapsing. The only way to maintain the pressure and keep supporting the weight of a heavier material is to increase the central temperature. As a result of this slow rise in temperature, the rate of nuclear fusion gradually increases and so does the solar luminosity.

The luminosity, effective temperature, and radius of the Sun all have increased. The Sun is now 30 percent more luminous than it was 4.6 billion years ago. The brightening is sufficient to make the visible solar disk 300 K hotter and its radius 6 percent greater than when the Sun began shining. The luminosity has increased by only a miniscule 0.0000023 (2.3×10^{-6}) percent in the past 350 years, and there is no way that this small change ever will be directly measured. Yet, it has profound implications over cosmic periods.

Because the solar luminosity increases as time goes on, the Sun was significantly dimmer in the remote past, and the Earth should have been noticeably colder then. However, this does not agree with geological evidence. Assuming an unchanging terrestrial atmosphere, with the same composition and reflecting properties as today, the lower solar luminosity in the past would have caused the Earth's global surface temperature to be below the freezing point of water during the planet's first 2.6 billion years. The oceans would have been frozen solid, there would have been no liquid water, and the entire planet would have been locked into a global ice age.

Yet, sedimentary rocks – which must have been deposited in liquid water – date back to a time when the Earth was less than 800 million years old. There is fossil evidence in those rocks of living things at about that time. Thus, for billions of years, the Earth's surface temperature was not very different from today; conditions have remained hospitable for life on the Earth throughout most of the planet's history.

The discrepancy between the Earth's warm climatic record and an initially dimmer Sun is known as the *faint-young-Sun paradox*. It can be resolved if the Earth's primitive atmosphere contained about 1,000 times more carbon dioxide than it does now. Greater amounts of carbon dioxide would enable the early atmosphere to trap more solar heat near the Earth's surface, warming it by the greenhouse effect, which would prevent the oceans from freezing. Another possibility is that the Sun was more magnetically active in its youth, expelling strong winds, energetic particles, and radiation that might have kept the Earth warm.

What about the future? In only 1 billion years, the Sun will have brightened by another 10 percent. Calculations suggest that the Earth's oceans could evaporate then at a rapid rate, resulting in a hot, dry, uninhabitable Earth. If that does not do us in, any terrestrial life is doomed to fry in about 3 billion years from now. The Sun then will be hot enough to boil the Earth's oceans away, leaving the planet a burned-out cinder – a dead and sterile place.

5.8 When the Sun Dies

The Sun cannot shine forever because eventually it will deplete the hydrogen fuel in its core. Although it has converted only a trivial part of its original mass into energy, the Sun has processed a substantial 37 percent of its core hydrogen into helium in the past 4.6 billion years. There will be no hydrogen left within the solar core about 7 billion years from now. When that hydrogen is exhausted, the central part of the Sun will undergo a slow collapse, and the gradually increasing core temperature will cause the outer layers of the Sun to expand into a red giant star, with a dramatic increase in size and a powerful rise in luminosity (Fig. 5.9).

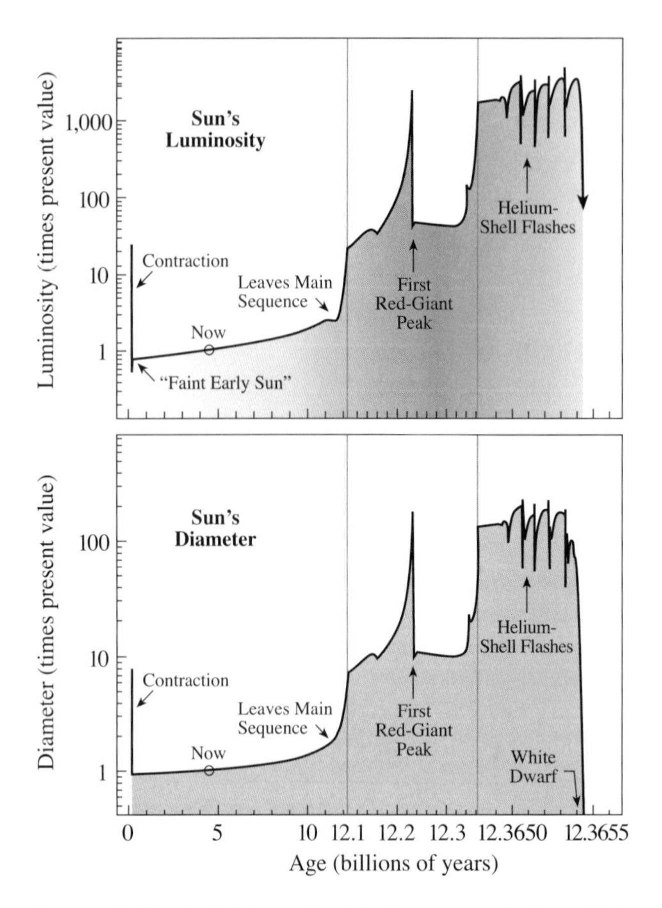

Figure 5.9. The Sun's fate. In about 8 billion years, the Sun will become much brighter (*top*) and larger (*bottom*). The time-scale is expanded near the end of the Sun's life to show relatively rapid changes. (Courtesy of I-Juliana Sackmann and Arnold I. Boothroyd.)

Mercury will become little more than a memory, being pulled in and swallowed by the swollen Sun. Eventually, the star will become 170 times larger than it is now. The Earth and Venus will probably remain outside the enlarged Sun, but their prospects are dim. The giant Sun will be 2,300 times more luminous than it is now, resulting in a substantial rise in temperatures throughout the solar system.

Meanwhile, the core of the Sun will continue to contract until the central temperature is hot enough to ignite helium burning – at about 100 million K. However, this conversion of helium into carbon will not last long compared to the Sun's 12 billion years of hydrogen burning. In about 35 million years, the core helium will have been used up and there will be no heat left to hold up the Sun. In a last desperate gasp of activity, the Sun will shed the outer layers of gas to produce an expanding "planetary" nebula around the star, and the core will collapse into a white dwarf star.

By this time, the intense winds of death will have stripped the Sun to about half of its original mass, and gravitational collapse will squeeze the remaining part into an insignificant cinder about the size of the present-day Earth. Nuclear reactions then will be a thing of the past, and there will be nothing left to warm the Sun or planets. The former Sun will gradually cool down and fade away, plunging all of the planets into a deep freeze.

Such events are in the very distant future, of course; but even now, the Sun threatens the Earth with its perpetually expanding atmosphere that envelops our planet and with explosive outbursts that can send energetic particles, intense radiation, and huge magnetic bubbles toward us.

6

The Extended Solar Atmosphere

6.1 Hot, Volatile, Magnetized Gas

The Million-Degree Solar Corona

The light we see coming from the Sun originates in the photosphere, a thin, tenuous layer of gas, only a few hundred kilometers thick, with a temperature of 5,780 K. We are looking at the round disk of the photosphere when we watch the Sun rise in the morning and continue on its daily journey across the sky. So it is natural to suppose that the photosphere is the surface of the Sun; however, being entirely gaseous, the Sun has no solid surface that divides the inside from the outside.

Moreover, the sharp outer rim of the Sun is illusory. A hot, transparent outer atmosphere envelops the photosphere and extends all the way to the Earth and beyond. Observing the Sun is like looking into the distance on a foggy day. At a certain distance, the total amount of fog we are looking through amasses to create an opaque barrier. The fog then becomes so thick and dense that radiation can penetrate no farther, and we can only see that far. When looking into the solar atmosphere, we similarly can see through only so much gas. For visible light, this opaque layer is the photosphere – the level of the Sun from which we receive our light and heat.

The unseen atmosphere just above the visible solar disk is far less substantial than a whisper and more rarefied than the best vacuum on the Earth. It is so tenuous that we see right through it, just as we see through the Earth's clear air. The diaphanous atmosphere of the Sun includes – from its deepest part outward – the underlying *photosphere*, from the Greek word *photos* for "light"; the thin *chromosphere*, from the Greek word *chromos* for "color"; and the extended *corona*, from the Latin word for "crown." We can observe the chromosphere and corona during a total solar eclipse, when the Moon blocks out the intense light of the underlying photosphere (Fig. 6.1).

Because of their very low densities and high temperatures, the chromosphere and corona produce bright spectral features called *emission lines*. Atoms and ions in a hot tenuous gas produce such emission features, heated to incandescence and shining at precisely the same wavelengths as the dark absorption lines produced by the same substance in the cooler

Table 6.1. *Strong coronal forbidden emission lines*[a]

Wavelength (nm)	Ion	Name	Wavelength (nm)	Ion
338.8	Fe XIII		670.2	Ni XV
423.2	Ni XII		789.2	Fe XI
530.3	Fe XIV	Green Line	802.4	Ni XV
569.4	Ca XV	Yellow Line	1,074.7	Fe XIII
637.4	Fe X	Red Line	1,079.8	Fe XIII

[a] The symbols Ca, Fe, and Ni denote, respectively, calcium, iron, and nickel. Subtract 1 from the Roman numeral to obtain the number of missing electrons. Thus, the ion Fe XIII is an iron atom missing 12 electrons. The wavelength is in units of nanometers, or 1 billionth of a meter.

photosphere. The corona's emission lines provided the initial evidence that it is hundreds of times hotter than the underlying photosphere.

The corona's searing heat was suggested by the identification of emission lines first observed during total eclipses of the Sun (Table 6.1). The intense green line, for example, was first observed during the solar eclipse of August 7, 1869. About 70 years later, Walter

Figure 6.1. Eclipse corona. The million-degree solar atmosphere, known as the corona, was seen around the shadowed disk of the Moon during the solar eclipse on July 11, 1991. The electrically charged gas was concentrated by magnetic fields into numerous fine rays as well as larger helmet streamers. (Courtesy of HAO/NCAR.)

Grotrian of Potsdam and the Swedish spectroscopist Bengt Edlén attributed it to emission from iron ions. These ions are iron atoms, denoted by Fe XIV, missing 13 of their 26 electrons.

The reason it took so long to identify the coronal emission lines is that no one realized the corona was so hot and also because such spectral features can arise only in the very tenuous corona. They are "forbidden transitions" that do not occur in terrestrial circumstances, where collisions between atoms keep them from happening even in the best vacuum.

The iron must be at a temperature of a few million K for atomic collisions to tear off so many electrons from the atoms. Edlén provided additional evidence for this hot temperature from the observed widths of the emission lines. Elements move at a faster speed in a hotter gas, broadening the observed spectral features as well as producing them. The million-degree temperature of the corona was subsequently confirmed by observations of the Sun's radio and x-ray radiation.

Because of its high temperature, the corona emits most of its energy, and its most intense radiation as x-rays. They can be used to image the hot corona all across the Sun's face with high spatial and temporal resolution. This is because the Sun's visible photosphere, being so much cooler, produces negligible x-ray radiation and appears dark under the million-degree corona. Because the Sun's x-ray radiation is absorbed totally in the Earth's atmosphere, it must be observed with telescopes lofted into space by rockets or in satellites.

To the casual observer, the Sun is simply a uniform ball of gas; however, close inspection of the Sun's its x-ray radiation shows that the star is in constant turmoil. The seemingly calm Sun is a churning, quivering, and explosive body, driven by intense, variable magnetic fields. This magnetism is responsible for dark sunspots that temporarily mark the visible face of the Sun.

Varying Sunspots and Ever-Changing Magnetic Fields

The solar corona is permeated by magnetic fields that are generated inside the Sun and rise up through the photosphere into the overlying atmosphere. The strongest magnetism protrudes to blemish the visible Sun with dark, Earth-sized sunspots (Fig. 6.2), which were seen by the unaided human eye up to 3,000 years ago (staring at the Sun can burn our eyes).

In the early twentieth century, the American astronomer George Ellery Hale first used the Zeeman effect to show that sunspots are regions of intense magnetism, thousands of times stronger than the Earth's magnetic field. The intense sunspot magnetism acts as both a valve and a refrigerator, choking off the outward flow of heat and energy from the solar interior and keeping the sunspots cooler and darker than their surroundings.

The strong magnetism exerts a pressure that tends to push apart the magnetic fields; however, by using helioseismology to look under the photosphere, astronomers discovered that flowing material pushes against the magnetic fields of sunspots, holding them in place (Focus 6.1).

Because the Sun's magnetism is forever changing and is never still, the sunspots are temporary, with a lifetime ranging from hours to months. Moreover, the total number of sunspots varies periodically, from a maximum to a minimum and back to a maximum, in about 11 years (Fig. 6.3). Samuel Heinrich Schwabe, an amateur astronomer in Dessau, Germany, discovered this periodic variation in the mid-nineteenth century. At the maximum

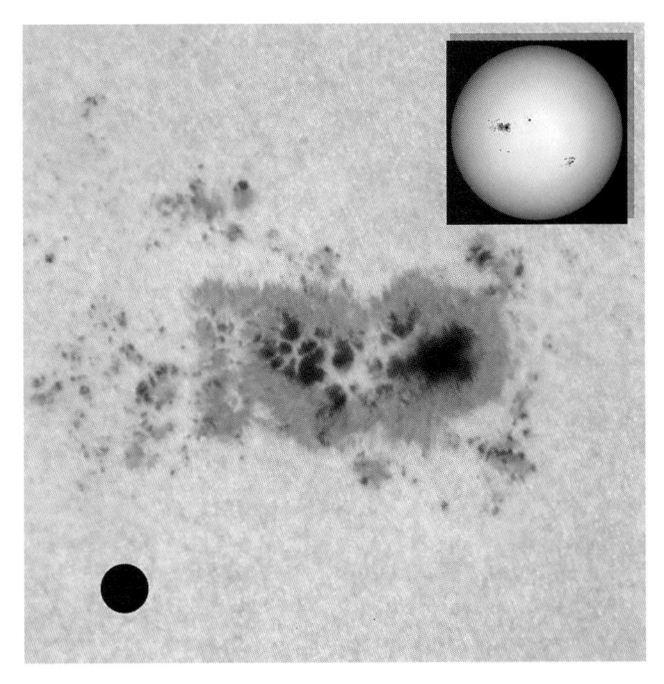

Figure 6.2. Sunspot group. Intense magnetic fields emerge from the interior of the Sun through the Sun's visible disk, the photosphere, producing groups of sunspots. The sunspots appear dark because they are slightly cooler than the surrounding photosphere gas. This composite image was taken in white light; that is, all of the colors combined. The enlarged image shows the biggest sunspot group, which is about 12 times larger than the Earth, the size of which is denoted by the black spot *(lower left)*. (Courtesy of *SOHO*/ESA/NASA.) (See color plate.)

in the sunspot cycle, there may be 100 or more spots on the visible hemisphere of the Sun at one time; at sunspot minimum, very few are seen and, for periods as long as a month, or more, none can be found. The locations where sunspots emerge and disappear also vary over the 11-year sunspot cycle, from mid-latitudes on the Sun to the solar equator (see Fig. 6.3).

The American astronomer Horace W. Babcock devised a conceptually simple model for the varying sunspots. His dynamo theory begins at sunspot minimum with a global, dipolar magnetic field that runs inside the Sun from south to north, or from pole to pole. Uneven, or differential, rotation – in which the equatorial regions rotate faster than the polar regions – shears the electrically conducting gases of the interior. As a result, the entrained magnetic

Focus 6.1 Looking into and Beneath Sunspots

Scientists were perplexed for decades about what holds sunspots together. Although sunspots can stay organized for several weeks at a time, their strong magnetic fields should repel each other and make the sunspots expand and disperse. Scientists have used instruments aboard the SOHO spacecraft to trace out the motions of hot, flowing gas in, around, and below sunspots. They found that sunspots are held together by powerful converging flows, which force the intense magnetic fields together.

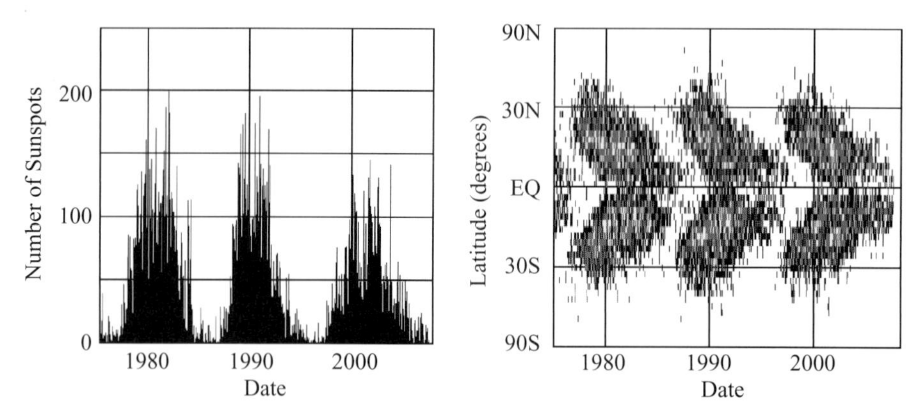

Figure 6.3. Solar magnetic activity cycle. The 11-year solar cycle of magnetic activity plotted from 1975 to 2007. Both the numbers of sunspots (*left*) positions of sunspots (*right*) wax and wane in cycles that peak every 11 years. Similar 11-year cycles have been observed for more than a century. At the beginning of each cycle, the first sunspots appear at about 30 degrees solar latitude and then migrate to 0 degrees solar latitude, at the solar equator (EQ), when the cycle ends. This plot of the changing positions of sunspots resembles the wings of a butterfly and therefore has been called the *butterfly diagram*. The cycles overlap with spots from a new cycle appearing at high latitudes when the spots from the old cycle persist in the equatorial regions. The solar latitude is the angular distance from the plane of the Sun's equator, which is very close to the plane of the Earth's orbit about the Sun, called the *ecliptic*. (Courtesy of David Hathaway/NASA/MSFC.)

fields are stretched out and squeezed together. The magnetism is coiled, bunched, and amplified as it is wrapped around the inside of the Sun. The surrounding gas buoys up the concentrated magnetism – just as a piece of wood is subject to buoyant forces when it is immersed in water – and eventually the magnetic fields become strong enough to rise to the surface and break through it in belts of bipolar sunspot pairs (Fig. 6.4).

The initial dipolar magnetic field is twisted into a submerged, ring-shaped field running parallel to the equator, or east to west. The dynamo generates two buried magnetic fields, one in the northern hemisphere and one in the southern hemisphere, but oppositely directed. They bubble up at mid-latitudes to spawn two belts of sunspots, symmetrically placed on each side of the equator. Thus, according to Babcock's scenario, we may view the solar cycle as an engine in which differential rotation and convection drive a periodic change in the magnetic fields hidden beneath the photosphere.

As the 11-year cycle progresses toward maximum activity, the internal magnetic field is wound increasingly tighter by the shearing action of differential rotation. The two sunspot belts slowly migrate toward the solar equator, where sunspots in the two hemispheres tend to merge.

Diffusion and poleward flows sweep the remnant magnetism into streams, each dominated by a single magnetic polarity, that slowly wind their way from the low- and mid-latitude belts to the Sun's poles. By sunspot minimum, the continued poleward transport of their debris may form a global dipole with reversed polarity. The north and south poles switch magnetic direction or polarity at the next sunspot minimum. When the Sun's magnetic

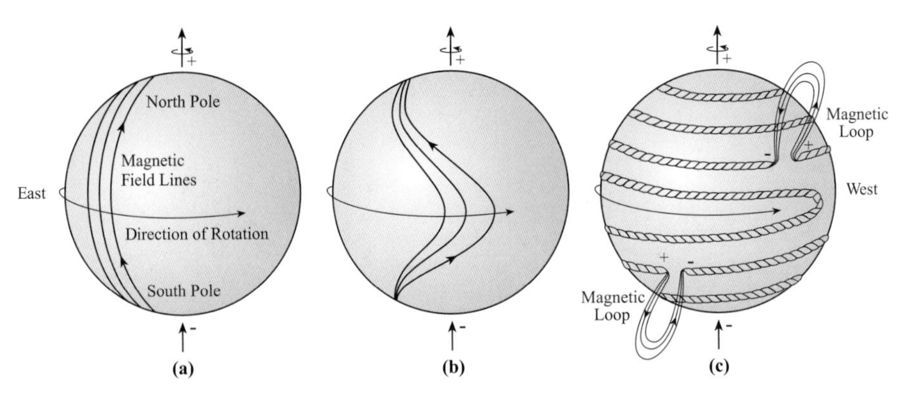

Figure 6.4. Winding up the field. A model for generating the changing location, orientation, and polarity of the sunspot magnetic fields. Initially, the magnetic field is supposed to be the dipolar or poloidal field seen at the poles of the Sun (*left*). The internal magnetic fields then run just below the photosphere from the Sun's south to north pole. As time proceeds, the highly conductive, rotating material inside the Sun carries the magnetic field along and winds it up. Because the equatorial regions rotate at a faster rate than the polar regions, the internal magnetic fields are stretched out and wrapped around the Sun's center, becoming concentrated and twisted together like ropes (*middle* and *right*). With increasing strength, the submerged magnetism becomes buoyant, rises and penetrates the visible solar disk – the photosphere – creating magnetic loops and bipolar sunspots that are formed in two belts, one each in the northern and southern hemispheres (*right*). (Adapted from Horace W. Babcock, "The Topology of the Sun's Magnetic Field, and the 22-Year Cycle," *Astrophysical Journal* 133, 572–587 [1961].)

flip is considered, we see that it takes two activity cycles, or about 22 years, for the overall magnetic polarity to return to where it began. The internal magnetism then has readjusted to its original submerged dipolar form, and the magnetic cycle begins again.

Coronal Loops

Magnetic fields are described by lines of force, like those joining the opposite poles of a bar magnet. The direction of the lines of force and the orientation of the magnetic fields can be inferred from the polarization of the spectral lines that have been split by the Zeeman effect. Magnetic field lines pointing out of the Sun have positive magnetic polarity, whereas inward-directed fields have negative polarity.

Sunspots usually appear in adjacent pairs or other close groupings of opposite magnetic polarity. Invisible magnetic arches loop between these oppositely directed magnetic regions in the photosphere, often emerging from a sunspot with one polarity and reentering a neighboring sunspot of opposite polarity.

The invisible magnetized bridges that join the ubiquitous pairs of opposite magnetic polarity are known as *coronal loops*. Although they remain unseen in optically visible sunlight, these coronal loops shine brightly in x-ray and extreme ultraviolet images of the Sun taken with telescopes in space (Fig. 6.5). Material is concentrated to higher densities and temperatures within these loops, so they emit this invisible radiation more intensely than their surroundings. This intense x-ray and extreme ultraviolet emission thus outlines the

Figure 6.5. Magnetic loops made visible. An electrified, million-degree gas, known as *plasma*, is channeled by magnetic fields into bright thin loops. The magnetized loops stretch up to 500,000 kilometers from the visible solar disk, spanning up to 40 times the diameter of planet Earth. The magnetic loops are seen in the extreme ultraviolet radiation of eight and nine times ionized iron, denoted Fe IX and Fe X, formed at a temperature of about 1.0 million K. The hot plasma is heated at the bases of loops near the place where their legs emerge from and return to the photosphere. Bright loops with a broad range of lengths all have a fine thread-like substructure with widths as small as the telescope resolution of 1 second of arc, or 725 kilometers at the Sun. This image was taken with the *Transition Region and Coronal Explorer* (*TRACE*) spacecraft. (Courtesy of the *TRACE* consortium, LMSAL, and NASA. *TRACE* is a mission of the Stanford–Lockheed Institute for Space Research, a joint program of the Lockheed-Martin Solar and Astrophysics Laboratory, [LMSAL] and Stanford's Solar Observatories Group.)

magnetic shape and structure of the Sun's outer atmosphere, indicating that the corona is stitched together by bright, thin, magnetized loops.

The magnetized atmosphere in, around, and above bipolar sunspot groups is called a *solar active region*. Active regions are places of concentrated, enhanced magnetic fields, sufficiently large and strong to stand out from the magnetically weaker areas. These disturbed regions are prone to awesome explosions, marking a location of extreme unrest on the Sun.

The number of active regions, with their bipolar sunspots and coronal loops, varies in step with the sunspot cycle, peaking at sunspot maximum when they dominate the structure of the inner corona. At sunspot minimum, the active regions are largely absent and the strength of the extreme-ultraviolet and x-ray emission of the corona is greatly reduced. Because most forms of solar activity are magnetic in origin, the sunspot cycle also is called the *solar cycle of magnetic activity*.

Unlike the Earth, magnetism on the Sun does not consist of only one simple dipole. The Sun is spotted all over and contains numerous pairs of opposite magnetic polarity. Powerful magnetism, spawned deep inside the Sun, threads its way through the solar atmosphere, creating a dramatic, ubiquitous, and ever-changing panorama of coronal loops (Fig. 6.6). Coronal loops provide the woven fabric of the entire corona.

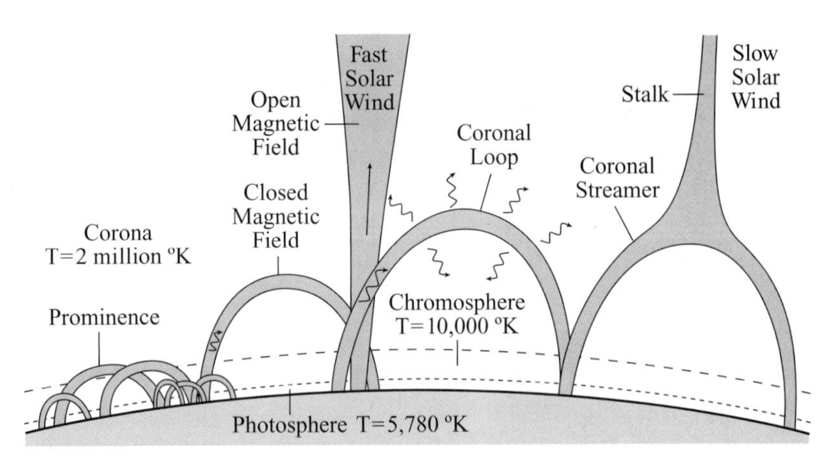

Figure 6.6. Coronal loops. The corona is stitched together with ubiquitous coronal loops that are created when upwelling magnetic fields generated inside the Sun push through the visible solar disk – the photosphere – into the overlying, invisible chromosphere and corona. These closed magnetic structures are anchored in the photosphere at footpoints of opposite magnetic polarity. Coronal loops can be filled with hot gas that shines brightly at extreme ultraviolet and x-ray wavelengths. Driven by motions in the underlying photosphere and below, the coronal loops twist, rise, shear, and interact, releasing magnetic energy that can heat the solar corona and power intense solar flares or coronal mass ejections. Large coronal loops are found in the bulb-like base of coronal streamers, whose long, thin stalks extend out into space. Magnetic fields anchored in the photosphere at one end also can be carried by the solar wind into interplanetary space, resulting in open magnetic fields and a channel for the fast solar wind.

Throughout the solar atmosphere, a dynamic tension is set up between the gas pressure of the charged particles and the pressure of the magnetic field (Fig. 6.7). In the photosphere and convective zone, the gas pressure dominates the magnetic pressure, allowing the magnetic field to be carried around by the moving gas. Because the churning gases are ionized and, hence, electrically conductive, they sweep the magnetic field along. The situation is reversed in the low corona within active regions. Here, strong magnetism wins and the hot ionized particles are confined within coronal loops that permeate the solar atmosphere (Fig. 6.8).

The coronal magnetic fields emerge from underneath the photosphere, where they are rooted, and they are continually displaced and replaced by convective motions just below the photosphere. As a result, the corona has no permanent features and it is never still, quiet, or inactive. It is always in a continued state of metamorphosis.

What Heats the Corona?

The visible solar disk, the photosphere, is closer to the Sun's center than the million-degree corona, but the photosphere is several hundred times cooler, with a temperature of 5780 K. This temperature difference is unexpected because energy should not flow from the cooler photosphere to the hotter corona any more than water should flow uphill. When we sit far away from a fire, for example, it warms us less.

The temperature of the corona is not supposed to be so much higher than that of the atmosphere immediately below it. It violates common sense, as well as the *second law of*

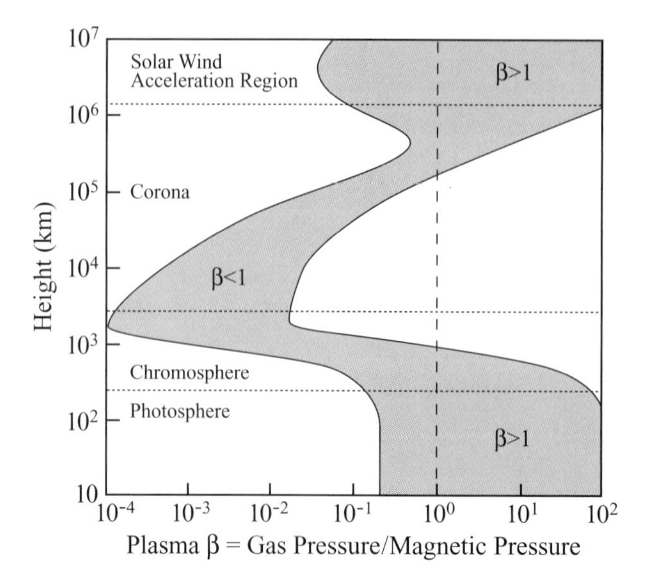

Figure 6.7. Gas and magnetic pressure. The ratio of gas to magnetic pressure, denoted by the symbol β, is plotted as a function of height above the photosphere. The magnetic pressure is greater than the gas pressure in the low corona, where β is less than 1, and magnetic fields dictate the structure of the corona. Farther out, the gas pressure can exceed the magnetic pressure, which permits the solar wind to carry the Sun's magnetic field into interplanetary space. In the photosphere, below the corona and chromosphere, the gas pressure also exceeds the magnetic pressure, and the moving gas carries around the magnetic fields. (Adapted from G. Allen Gary, "Plasma Beta above a Solar Active Region: Rethinking the Paradigm," *Solar Physics* 203, 71–86 [2001].)

thermodynamics, which states that heat cannot be continuously transferred from a cooler to a warmer body without doing work.

We know that visible sunlight cannot resolve the heating paradox. Radiation from the photosphere does not go into the corona; it goes through the corona. There is so little material in the corona that it is transparent to almost all of the photosphere's radiation. Therefore, sunlight passes right through the corona without depositing substantial quantities of energy in it, traveling out to warm the Earth and to keep the photosphere cool.

So, radiation cannot resolve the heating paradox. We must look for alternate sources of energy, and magnetic fields play an important role. The hottest and densest material in the low corona is located where the magnetic field is strongest; it is intense magnetism that molds the corona into loops, producing its highly structured, nonhomogeneous shape.

Motions down inside the convective zone twist and stretch the overlying magnetic fields, slowly building up their energy, and magnetic loops of all sizes always are being pushed up into the solar corona from below. When oppositely directed coronal loops are pressed together, they can merge and join at the place where they touch, releasing their energy to heat the corona.

Coronal Holes

In contrast to the dense, bright areas, the corona also contains less dense regions called *coronal holes*. These so-called holes have so little material in them that they appear as

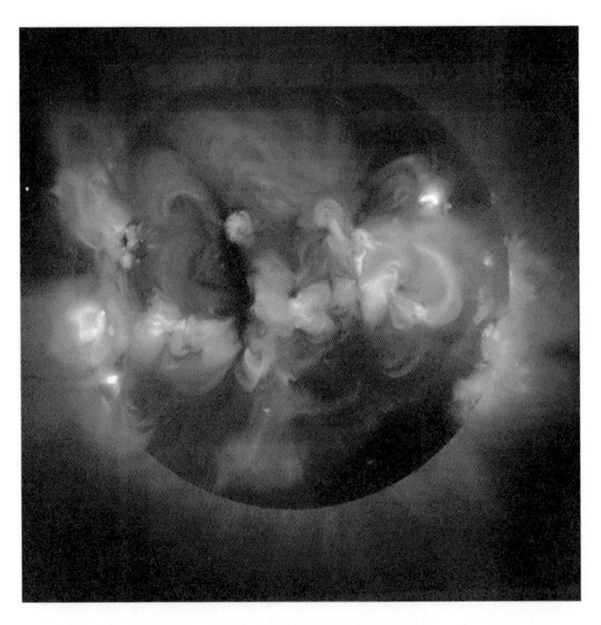

Figure 6.8. The Sun in x-rays. Ionized gases at a temperature of a few million K produce the bright glow seen in this x-ray image of the Sun. It shows magnetic coronal loops that thread the corona and hold the hot gases in place. The brightest features are called active regions and correspond to the sites of the most intense magnetic field strength. This image of the Sun's corona was recorded by the Soft X-Ray Telescope (SXT) aboard the Japanese *Yohkoh* satellite on February 1, 1992, near a maximum of the 11-year cycle of solar magnetic activity. Subsequent SXT images, taken about five years later near activity minimum, show a remarkable dimming of the corona when the active regions associated with sunspots have almost disappeared, and the Sun's magnetic field has changed from a complex structure to a simpler configuration. (Courtesy of NASA/ISAS/LMSAL/NAO Japan, University of Tokyo.) (See color plate.)

large dark areas on x-ray or extreme-ultraviolet images, seemingly devoid of radiation (see Fig. 6.8).

The coronal holes are neither constant nor permanent; they appear, evolve, and die away in periods ranging from a few weeks to several months, continuously changing in content, shape, and form.

At times of low solar activity, near the minimum of the Sun's 11-year magnetic activity cycle, coronal holes cover the north and south polar caps of the Sun. During more active periods, closer to the cycle maximum, the large coronal holes at the poles shrink and even disappear, and smaller coronal holes appear at all solar latitudes – even at or near the solar equator.

Coronal holes appear like a dark, empty void, as if there were a hole in the corona, but the rarefied coronal holes are not completely empty. The normally constraining magnetic forces relax and open up in the coronal holes to allow an unencumbered outward flow of electrically charged particles and magnetic fields into interplanetary space, keeping the coronal hole's density low and expelling a relentless high-speed wind.

Focus 6.2 Discovery of the Solar Wind

The notion that something is always being expelled from the Sun first arose from observations of comet tails. Comets appear unexpectedly almost anywhere in the sky, moving in every possible direction, but with tails that always point away from the Sun. A comet therefore travels head first when approaching the Sun and tail first when departing from it. Ancient Chinese astronomers concluded that the Sun must have a *chi*, or "life force," that blows away the comet tails. In the early 1600s, the German astronomer Johannes Kepler proposed that solar radiation pushes the comet tails away from the Sun.

Modern astronomers noticed that a comet could have two tails. One is a yellow tail of dust, which can litter the comet's curved path. The dust is pushed away from the Sun by the pressure of sunlight. The other tail is colored electric blue, shining in the light of ionized particles. The ions in comet tails always stream along straight paths away from the Sun with velocities many times higher than could be caused by the weak pressure of sunlight. In the early 1950s, the German astrophysicist Ludwig Biermann therefore proposed that streams of electrically charged particles, called *corpuscular radiation*, poured out of the Sun at all times and in all directions to accelerate and shape the comet ion tails.

Eugene Parker, a young astrophysicist at the University of Chicago, used the theory of hydrodynamics to show that the million-degree solar corona must expand rapidly outward because it is so extremely hot, and that as the outer corona disperses, gases welling up from below will replenish it. The expansion begins slowly near the Sun, where the solar gravity is the strongest; then accelerates outward into space until the winds break away from the Sun; and eventually cruises along at the roughly constant and supersonic velocities needed to account for the acceleration of comet tails. This creates a strong, persistent solar wind, forever blowing throughout the solar system.

Any doubts about the existence of the solar wind were removed by *in situ* (Latin for "in original place" or literally, "in the same place") measurements made by instruments on board the Soviet *Lunik 2* spacecraft on the way to the Moon in 1959 and by those aboard NASA's *Mariner II* spacecraft during its trip to Venus in 1962.

6.2 The Sun's Varying Winds

We Live inside the Expanding Sun

The Sun's radiation is not all that passes through the space between the planets. It is filled with electrons, protons, and magnetic fields emanating from the Sun in a ceaseless flow. These unseen particles and fields form a perpetual solar wind that extends all the way to the Earth and far beyond. It was inferred from comet tails, suggested by theoretical considerations, and fully confirmed by direct measurements from spacecraft in the early 1960s (Focus 6.2).

The eternal solar winds are always blowing out into space, never stopping for a rest or finding a place to call home; they just keep rolling along, like the tumbling tumbleweed.

Figure 6.9. Magnetic fields near and far. In the low solar corona, strong magnetic fields are tied to the Sun at both ends, trapping hot, dense electrified gas within magnetized loops. Far from the Sun, the magnetic fields are too weak to constrain the outward pressure of the hot gas, and the loops are opened up, allowing electrically charged particles to escape. They form the solar wind that carries the magnetic fields away from the Sun. (Courtesy of *Newton Magazine*, the Kyoikusha Company.) (See color plate.)

As with the winds in the vast Sahara Desert, we cannot see the solar wind in interplanetary space. However, we have found ways to watch the effects of the Sun's winds and to detect them in space.

The space between the planets is not completely empty; it contains an eternal solar wind – a rarefied mixture of electrons, protons, other ions, and magnetic fields that stream out radially in all directions from the Sun.

In the low corona, strong magnetic fields constrain the hot ionized gas within coronal loops. However, farther out in the corona, the magnetic fields decrease in strength and cannot restrain the outward flow of the million-degree gas; it also flows out unencumbered from the open magnetic fields in coronal holes (Fig. 6.9). The planets move through this wind as if they were ships at sea.

The solar gale brushes past the planets and engulfs them. The solar wind wraps itself around the Earth, for example, so that we live inside the Sun. The relentless wind blows on, carrying the Sun's corona out to interstellar space.

This radial, supersonic outflow creates a huge bubble of plasma, with the Sun at its center and the planets inside; this is called the *heliosphere*, from *Helios*, the "God of the Sun" in Greek mythology.

Table 6.2. *Mean values of solar-wind parameters at the Earth's orbit*

Parameter	Mean Value
Particle Density, N	$N \approx 10$ million particles per cubic meter
Velocity, V	Fast wind $V \approx 750$ km s^{-1}
	Slow wind $V \approx 375$ km s^{-1}
	Average $V \approx 600$ km s^{-1}
Mass Density, ρ	$\rho = 10^{-20}$ kilograms per cubic meter (protons)
Temperature, T	$T \approx 120{,}000$ K (protons)
	to 140,000 K (electrons)
Radial Magnetic Field, B_r	$B_r = 2.5 \times 10^{-9}$ T $= 2.5$ nT

The solar wind has been sampled for nearly a half-century, and it has never stopped. Contemporary solar spacecraft can observe the winds at their origin on the Sun and sample its ingredients near the Earth, within the Earth's orbital plane, and also far above the Sun's polar regions.

Properties of the Solar Wind

The million-degree corona is so hot that it cannot stand still. Indeed, the solar wind consists of an overflow corona, which is too hot to be entirely constrained by the Sun's inward gravitational pull and perpetually moves out into surrounding space. The hot gas creates an outward pressure that tends to oppose the inward pull of the Sun's gravity. At great distances, where the solar gravity weakens, the hot protons and electrons overcome the Sun's gravity and continue to accelerate, like water overflowing a dam. So, the solar corona is actually the visible, inner base of the solar wind, and the solar wind is simply the hot corona expanding into interplanetary space.

The Sun's continuous wind travels with two main velocities, like an automobile with one high gear and one low gear. There is a fast, uniform wind that blows at a speed of about 750 km s^{-1} and a variable, gusty slow wind that moves about half as fast. Both winds are supersonic, moving at least 10 times faster than the sound speed in the solar wind. It rushes on with little reduction in speed because there is almost nothing out there to slow it down. Both the fast and slow winds from the Sun are much more tenuous, hotter, and faster than any wind on the Earth.

Because the electrified wind material is an excellent conductor of heat, the temperature falls off only gradually with distance from the Sun, reaching between 120,000 K and 140,000 K at the Earth's distance. Because it is so hot out there, one wonders why astronauts do not burn up when they leave their spacecraft to walk in the solar wind. The reason is that the tenuous wind has been diluted to a rarefied plasma by the time it reaches the Earth's distance from the Sun, where there are approximately 5 million electrons and 5 million protons per cubic meter of solar wind. The density of the solar wind is so low that if we could go out into space and put our hands on it, we would not be able to feel it, and there are not enough particles to heat up an astronaut.

Physical properties of the solar wind at the Earth's distance from the Sun are listed in Table 6.2.

At large distances from the Sun, the charged particles in the solar wind drag the Sun's magnetic fields with them. Whereas one end of the interplanetary magnetic field remains firmly rooted in the photosphere and below, the other end is extended and stretched out by the radial expansion of the solar wind. The Sun's rotation bends this radial pattern into an interplanetary spiral shape within the plane of the Sun's equator, coiling up the magnetism like a tightly wound spring. Because it is wrapped into a spiral, the interplanetary magnetic-field strength falls off linearly with distance from the Sun, in contrast to the solar-wind number density, which decreases more rapidly, as the inverse cube of the distance, as it fills a larger volume.

Because the Sun is blowing itself away continuously, we might imagine that it would eventually vanish from view after expelling all of its substance into space. Every second, the solar wind carries about 1 billion kilograms, or 1 million tons, of the Sun into surrounding space. That seems significant, but in 4.6 billion years, the solar wind has carried away only 0.00005, or 5 millionths, of the Sun's mass. Moreover, this is five times less than the amount of mass turned into energy during this time by nuclear reactions near the center of the Sun.

Where Do the Two Solar Winds Come From?

Instruments aboard spacecraft have detected two solar winds with different physical properties. There is a fast wind that moves at a speed of about 750 km s^{-1} and a slow wind that blows at about half that speed. The high-speed wind is steady and uniform, whereas the slow-speed wind is variable and gusty.

The two solar winds do not blow uniformly from all points of the Sun; instead, they depend on solar latitude. The spatial distribution of the two types of winds also depends on the Sun's magnetic field configuration, which varies dramatically with the 11-year solar-activity cycle.

As suggested by Sir William Ian Axford, the steady, uniform, high-speed wind emanates from magnetically open configurations in coronal holes. They provide a conduit for the fast wind, like the express lane of a divided highway. In contrast, the slow wind – which is filamentary and transient – involves the intermittent release of material from previously closed magnetic regions.

The distribution of the open and closed magnetic regions on the Sun, and therefore the places of origin for the two solar winds, depends on the 11-year cycle of solar magnetic activity.

Near activity minimum, a steady torrent of high-speed wind rushes out of the open magnetic fields in large coronal holes located at the Sun's polar regions. A slow, gusty, and variable wind then moves away from closed magnetic regions near the Sun's equator.

The simple, bimodal distribution of fast and slow wind structures disappears near the maximum in the 11-year solar-activity cycle. The large polar coronal holes then shrink and even disappear and smaller coronal holes appear at all solar latitudes. A chaotic and complex mixture of varying solar-wind flows therefore is found at all solar latitudes near activity maximum. The slow winds still seem to be associated with closed magnetic structures, such as active regions, whereas the fast winds rush out of the interiors of coronal holes all over the Sun. Solar-active regions, with their explosive behavior, provide a noticeable third source for solar winds near the activity-cycle maximum.

6.3 Explosions on the Sun

Solar Flares

Suddenly, without warning, a part of the Sun will explode with awesome power and violence on a scale unknown on the Earth. Some of these solar flares are the biggest explosions in the solar system, releasing energy comparable in strength to 20 million nuclear bombs exploding simultaneously.

A substantial fraction of the flare energy goes into accelerating electrons and protons to nearly the speed of light. Some of these high-energy particles are hurled into the Sun, briefly raising the temperature of Earth-sized regions to more than 10 million K. Other accelerated particles are tossed out into interplanetary space and emit intense radio and x-ray radiation.

The short-lived solar flares unleash their energy in the vicinity of sunspots, covering just a few tenths of a percent of the solar disk. These incredible explosive outbursts become more frequent and violent when the number of sunspots is greatest; several solar flares can be observed on a busy day near the maximum of the sunspot cycle. However, they are not caused directly by sunspots; solar flares instead are powered by magnetic changes in the corona above sunspots.

Although it emits awesome amounts of energy, a solar flare usually releases less than 1/1,000th of the total energy radiated by the Sun every second, so they are only minor perturbations in the combined colors, or white light, of the Sun. The first record of a solar flare detected on the visible solar disk did not occur until the mid-nineteenth century, when the English astronomers Richard C. Carrington and Richard Hodgson independently noticed one.

A new perspective, which demonstrated the frequent occurrence of solar explosions, was made possible when flares were observed at radio and x-ray wavelengths. During World War II (1939–1945) it was discovered that sudden, intense radio outbursts from the Sun, associated with solar flares, could interfere with radio communications and radar systems. Soon after the war ended, J. Paul Wild's group of Australian radio astronomers used swept-frequency receivers to track the outward motions of the impulsive radio bursts. Some of the bursts moved up to half the speed of light, signaling the expulsion of high-speed electrons; others moved at about 1/100th of this speed and were attributed to shock waves.

The bulk of radiation from high-temperature solar flares is not emitted as radio waves but instead at extreme ultraviolet and x-ray wavelengths, where they can briefly outshine the entire Sun. This radiation is absorbed in the Earth's atmosphere; therefore, astronomers have observed it from outer space, beginning with primitive instruments aboard balloons or sounding rockets and continuing with increasingly sophisticated telescopes in many satellites, including the Yohkoh, Ulysses, Wind, SOHO, ACE, TRACE, Hinode, RHESSI, STEREO, and SDO spacecraft.

Why does a solar flare occur? What triggers the instability and suddenly ignites an explosion from magnetic fields that remain unperturbed for long intervals of time? They are triggered when magnetized coronal loops are pressed together, driven by motions beneath them, meeting to touch one another and merge.

Magnetic fields have a direction associated with them and, if oppositely directed magnetic fields are pushed together, they can interact. When these merging magnetic fields are closed coronal loops, they will break open to release magnetic energy in the form of flare heating and particle acceleration. The magnetic fields are not broken permanently; they simply reconnect to their closed state. For this reason, this merging and coupling is known as *magnetic reconnection*.

So, powerful solar flares stem from the interaction of coronal loops. These loops always are moving about, like swaying seaweed or wind-blown grass, and often are brought into contact by these movements. Magnetic fields coiled up in the solar interior, where the Sun's magnetism is produced, also can bob into the corona to interact with preexisting coronal loops. In either case, the coalescence leads to the rapid release of magnetic energy through magnetic reconnection.

The explosive instability of a solar flare has been compared to an earthquake, with the moving roots or footpoints of a magnetic loop resembling two tectonic plates. As the plates move in opposite directions along a fault line, they grind against one another and build up stress and energy. When the stress is pushed to the limit, the two plates cannot slide farther and the accumulated energy is released as an earthquake. That part of the fault line then lurches back to its original equilibrium position, waiting for the next earthquake. In this analogy, the moving magnetic fields become stressed to the breaking point and similarly regain their composure after an explosive convulsion on the Sun, fusing back together and becoming primed for the next outburst.

Because flares apparently originate in the low corona and the ubiquitous coronal loops dominate its structure, it is perhaps not surprising that solar-flare models involve a single coronal loop (Fig. 6.10). Magnetic reconnection triggers the release of magnetic energy just above the loop top, where electrons and protons are accelerated. In less than one second, electrons are accelerated to nearly the speed of light, producing intense radio signals. Protons likewise are accelerated to high speeds, and both the electrons and protons are hurled down into the Sun and out into space.

Nuclear reactions, with the creation of anti-matter, have been observed near the visible disk of the Sun during solar flares. When protons and heavier ions are accelerated during solar flares and beamed downward into the Sun, they slam into the dense, lower atmosphere, shattering the nuclei of atoms there. Some release neutrons and protons, and positrons – the anti-matter counterparts of electrons – are created. The positrons annihilate the electrons, producing radiation at 0.511 MeV, which is the energy contained in the entire mass of a single nonmoving electron.

Because the chromosphere has been heated very rapidly by the accelerated particles that were hurled down into it, that part of the chromosphere explodes, or evaporates, into the corona to release the excess energy. This process may include the more gradual release of energy when the coronal loop relaxes into a more stable configuration during the decay phase of a solar flare.

Coronal Mass Ejections

Coronal mass ejections are gigantic magnetic bubbles that can rush away from the Sun at supersonic speeds, expanding to become larger than the Sun in a few hours (Fig. 6.11). They carry about 10^{13} kg, or 10 billion tons, of material out into space, produce intense shock waves, and accelerate vast quantities of energetic particles in interplanetary space. A

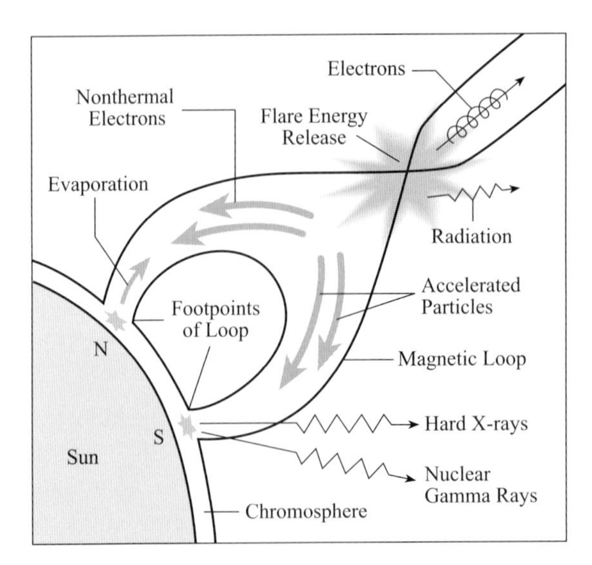

Figure 6.10. Solar flare model. A solar flare is powered by magnetic energy released from a magnetic interaction site above the top of a coronal loop. Electrons are accelerated to high speed during a solar flare, generating a burst of radio energy as well as impulsive loop-top hard x-ray emission. Some of these nonthermal electrons are channeled down the loop and strike the chromosphere at nearly the speed of light, emitting hard x-rays by electron–ion *bremsstrahlung* at the loop footpoints. When beams of accelerated protons enter the dense, lower atmosphere, they cause nuclear reactions that result in gamma-ray spectral lines and energetic neutrons. Material in the chromosphere is heated very quickly and rises into the coronal loop, accompanied by a slow, gradual increase in soft x-ray radiation. This upwelling of heated material is called *chromospheric evaporation* and it occurs in the decay phase of the flare.

coronal mass ejection moves through space at a speed of about 400 km s^{-1}, carrying a kinetic energy of about 10^{24} J, which is comparable to the explosive energy of a large solar flare. When directed at the Earth, a coronal mass ejection arrives at the planet about four days after being ejected from the Sun.

The physical size of the mass ejections dwarfs that of solar flares and even the active regions in which flares occur. However, like solar flares, the rate of occurrence of coronal mass ejections varies in step with the 11-year cycle of solar-magnetic activity, ballooning out of the corona several times a day during activity maximum. Large coronal mass ejections can occur with or without a solar flare, but they both appear to be powered by the abrupt release of the corona's magnetic energy, with threatening effects for the Earth and nearby space.

6.4 Space Weather

Earth's Protective Magnetosphere

Our planet is immersed within the hot, electrically charged solar wind that blows out from the Sun in all directions and never stops, carrying with it a magnetic field rooted in the Sun. Solar flares and coronal mass ejections produce powerful gusts in the Sun's winds, producing space weather – the cosmic equivalent of a terrestrial blizzard or hurricane.

Figure 6.11. Coronal mass ejection. A huge coronal mass ejection is seen in this coronagraph image, taken on December 5, 2003 with the Large Angle Spectrometric Coronagraph (LASCO) on the SOHO spacecraft. The black area corresponds to the occulting disk of the coronagraph that blocks intense sunlight and permits the corona to be seen. An image of the singly ionized helium, denoted He II, emission of the Sun, taken at about the same time, has been scaled appropriately and superimposed at the center of the LASCO image. The full-disk helium image was taken at a wavelength of 30.4 nanometers, corresponding to a temperature of about 60,000 K, using the Extreme-ultraviolet Imaging Telescope (EIT) aboard SOHO. (Courtesy of the SOHO LASCO and EIT consortia. SOHO is a project of international cooperation between ESA and NASA.) (See color plate.)

Fortunately, we are protected from the full force of this relentless, stormy gale by the Earth's magnetic field.

In 1600, William Gilbert, physician to Queen Elizabeth I of England, authored a treatise in Latin, with the grand title *De Magnete, Magneticisque Corporibus, et de Magno Magnete Tellure,* which is translated into English as *Concerning Magnetism, Magnetic Bodies, and the Great Magnet Earth*. In this work, which is still available in an English version, Gilbert showed that the Earth is itself a great magnet, which explains the orientation of compass needles. It is as if there were a colossal bar magnet at the center of the Earth.

At the equator, the two ends of a compass needle point north or south, toward the Earth's magnetic poles. At each magnetic pole, the needle stands upright, pointing into or out of the ground. In between, at intermediate latitudes, the compass needle points north or south with a downward dip of one end but not vertically as at a pole. Because the geographic poles are located near the magnetic poles, a compass needle is aligned in the north–south direction. We usually put an arrow on the north end of the needle; therefore, an arrowed compass points north.

We can describe the Earth's magnetism by invisible magnetic field lines, which orient compass needles. These lines of magnetic force emerge from the south magnetic pole, loop through nearby space, and reenter at the north magnetic pole. The lines are close together near the magnetic poles, where the magnetic force is strong, and spread out above the Earth's Equator, where the magnetism is weaker than at the poles. We cannot see the invisible magnetic field lines, but compass needles point along them, and other instruments can be used to measure their strength.

The magnetic field strength at the Earth's magnetic equator is 0.00003 tesla. Measurements of the surface magnetic fields of the Earth show stronger fields near the poles, where the magnetic field lines congregate, at roughly twice the strength of the field at the equator. The magnetic strength anywhere on the Earth is several times weaker than a toy magnet, but the comparison is somewhat misleading because our planet is a huge magnet.

Fortunately for life on the Earth, the terrestrial magnetic fields influence the space near our planet. Although these fields decrease in strength as the inverse cube of the distance, they remain strong enough to divert most of the solar wind around the Earth at a distance far above the atmosphere, thereby protecting humans on the ground from possibly lethal solar particles.

When any charged particle encounters a magnetic field, it must change direction, moving away from or around the magnetism. When the protons and electrons in the gusts or steady flow of the solar wind encounter the Earth's magnetic fields, they are deflected around it, like a rock in a stream or a windshield deflecting air around an automobile.

The Earth's magnetic fields hollow out a protective cavity in the solar wind, which is called the *magnetosphere*. It is that region surrounding any planet in which its magnetic field dominates the behavior of electrically charged particles, such as electrons, protons, and other ions.

The dipolar (i.e., two poles) magnetic configuration applies near the surface of the Earth, but, farther out, the magnetic field is distorted by the Sun's perpetual wind. Although it is exceedingly tenuous, the solar wind is powerful enough to mold the outer edges of the Earth's magnetosphere into a changing asymmetric shape, like a teardrop falling toward the Sun (Fig. 6.12).

The solar wind usually bends around the Earth's magnetic field at a distance from the Earth's center of about 10 times the Earth's radius on the dayside that faces the Sun. Here, the solar wind pushes the Earth's magnetism in, compressing its outer magnetic boundary and forming a shock wave, shaped like waves that pile up ahead of the bow of a moving ship and resembling the flow of air around a supersonic aircraft. After forming this bow shock, the solar wind is deflected around the Earth, pulling the terrestrial magnetic field into a long magnetotail on the night side. Thus, the Earth's magnetosphere is not precisely spherical. It has a bow shock facing the Sun and a long magnetotail in the opposite direction. The term *magnetosphere* therefore does not refer to form or shape but instead implies a sphere of influence.

The Earth's magnetic shield is so perfect that only 0.1 percent of the mass of the solar wind that hits it manages to penetrate inside. Yet, even that small fraction of wind particles has a profound influence on the Earth's nearby environment in space; they create an invisible world of energetic particles and electric currents that flow, swirl, and encircle the Earth.

Trapped Particles

One of the first scientific discoveries of the Space Age was the finding, by James A. Van Allen and his students, of high-energy electrons and protons that girdle the Earth

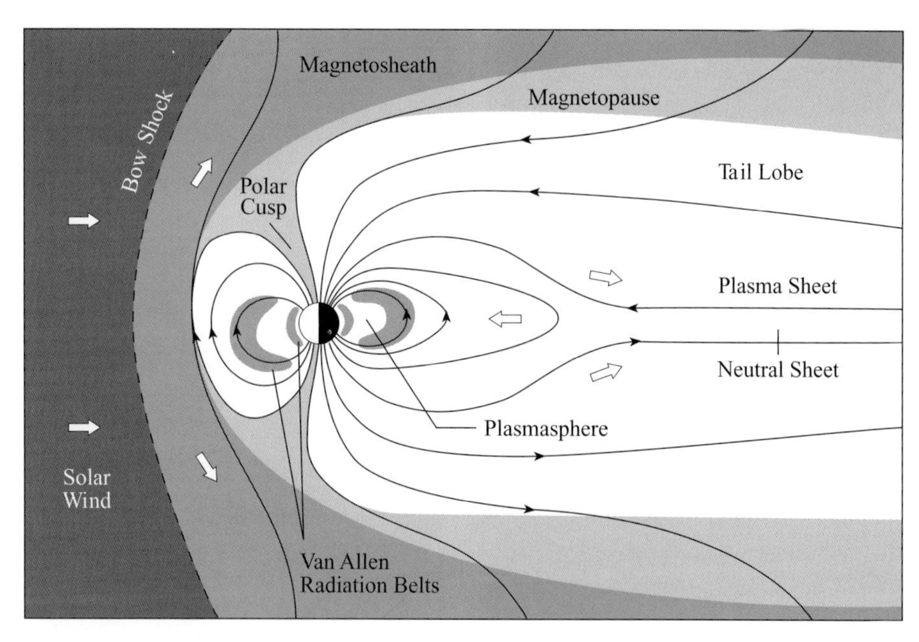

Figure 6.12. Elements of the magnetosphere. The Earth's magnetic field carves out a hollow in the solar wind, creating a protective cavity called the *magnetosphere*. A bow shock forms at about 10 Earth radii on the sunlit side of our planet. The location of the bow shock is highly variable because it is pushed in and out by the gusty solar wind. The magnetopause marks the outer boundary of the magnetosphere, at the place where the solar wind takes control of the motions of charged particles. The solar wind is deflected around the Earth, pulling the terrestrial magnetic field into a long magnetotail on the nightside. Plasma in the solar wind is deflected at the bow shock (*left*), flows along the magnetopause into the magnetic tail (*right*), and then can be injected back toward the Earth within the plasma sheet (*center*). The Earth, its auroras, atmosphere, and ionosphere, and the two Van Allen radiation belts all lie within this magnetic cocoon.

far above the atmosphere. They move within two belts that encircle the Earth's magnetic equator but do not touch it, like a gigantic, invisible, torus-shaped doughnut.

These regions sometimes are called the *inner* and *outer Van Allen radiation belts*. Van Allen used the term radiation belt because the charged particles were then known as corpuscular radiation; the nomenclature does not imply either electromagnetic radiation or radioactivity. The radiation belts lie within the inner magnetosphere at distances of 1.5 and 4.5 Earth radii from the center of the Earth, creating a veritable shooting gallery of high-speed electrons and protons in nearby space.

In 1907, about a half-century before the discovery of radiation belts, the Norwegian geophysicist Carl Størmer showed how electrons and protons could be almost permanently confined and suspended in space by the Earth's dipolar magnetic field. An energetic charged particle moves around the magnetic fields in a spiral path toward one magnetic pole. Its trajectory becomes more tightly coiled in the stronger magnetic fields close to a magnetic pole, where the intense polar fields act like a magnetic mirror, turning the particle around so that it moves back toward the other pole.

Thus, the electrons and protons bounce back and forth between the north and south magnetic poles. It takes about 1 minute for an energetic electron to make one trip between the two polar mirror points. The spiraling electrons also drift eastward, completing one trip around the Earth in about a half hour. There is a similar drift for protons but in the westward direction. The bouncing can continue indefinitely for particles trapped in the Earth's radiation belts, until the particles collide with one another or some external force distorts the magnetic fields.

The problem at the time Størmer developed his theory was that there was no mechanism known to allow electrically charged particles into the dipolar magnetic field. After all, if electrons and protons cannot leave the magnetic cage, how could they get into it in the first place? They can arrive via the solar wind and penetrate the Earth's magnetic defense through a temporary opening in it.

The solar wind carries the Sun's magnetic field with it, and the solar magnetism is draped around the magnetosphere when encountering it. The solar magnetic field can open up the Earth's magnetic field if the two fields are pointing in opposite directions when they touch. With this orientation, they can join one another and become linked, similar to how the opposite poles of two toy magnets stick together. The merging process, known as *magnetic reconnection*, can create an opening in the Earth's magnetic field, forming a portal through which the solar particles can flow.

The solar wind then is plugged into the Earth's "electrical socket," and our planet becomes wired to the Sun along magnetic fields that can stretch all the way back to the solar corona. Tons of high-energy particles then may flow into the magnetosphere along this magnetic highway and through the opening before it closes again. The magnetic reconnection can occur during either a frontal assault near the bow shock and magnetic poles or from the rear in the immense magnetotail.

Earth's Magnetic Storms

If the magnetic fields of a coronal mass ejection and of the Earth are pointing in opposite directions when they meet, the two fields become linked, resulting in intense, geomagnetic storms that cause compass needles to swing widely. The flow of currents associated with these great magnetic storms can interfere with electrical power grids here on the Earth, creating voltage surges on long-distance power lines and overheating or melting the windings of transformers. They can send cities into complete darkness, especially in high-latitude regions where the currents are strongest (e.g., Canada, the northern United States, and Scandinavia). This does not occur often, perhaps once a year; however, the potential consequences are serious enough to employ early warning systems.

Out in Space There Is No Place to Hide

When directed at our planet or at humans in deep space, both solar flares and coronal mass ejections produce dangerous gusts and squalls in the Sun's winds. Here on the ground, we are shielded from many of the effects by the Earth's atmosphere and magnetic fields, but out in space there can be no protection, and both humans and satellites are vulnerable (Fig. 6.13).

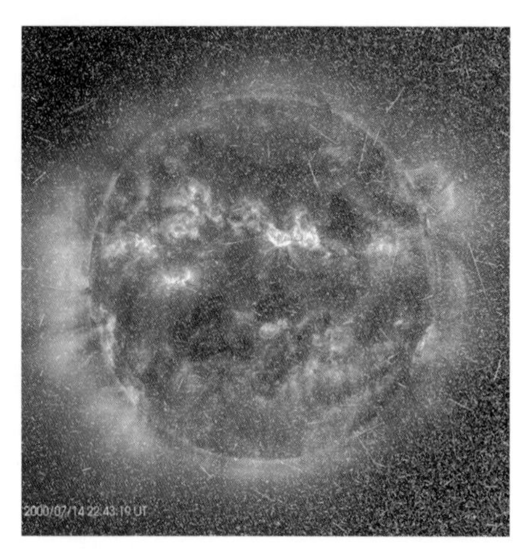

Figure 6.13. Solar flare produces energetic particle storm. A powerful solar flare (*left*), occurring on Bastille Day, July 14, 2000, unleashed high-energy protons that began striking the SOHO spacecraft near the Earth about 8 minutes later, continuing for many hours, as shown in the image taken 12 hours after the flare began (*right*). Both images were taken at a wavelength of 19.5 nanometers, emitted at the Sun by 11 times ionized iron, denoted Fe XII, at a temperature of about 1.5 million K, using the EIT on the SOHO spacecraft. (Courtesy of the SOHO EIT consortium. SOHO is a project of international cooperation between ESA and NASA.) (See color plate.)

Energetic charged particles generated during a solar flare will threaten our planet only if the flare occurs at the right place on the Sun – that is, at one end of the spiral magnetic field that connects the Sun to the Earth. Given the right circumstances, when a flare occurs near the west limb and the solar equator, the magnetic spiral acts as an interplanetary highway for high-speed charged particles that can threaten astronauts or satellites. The spiral magnetic pattern, produced when the solar wind carries the rotating Sun's magnetic field into surrounding space, has been confirmed by tracking the radio emission of charged particles thrown out during solar flares, as well as by spacecraft that have sampled the interplanetary magnetism near the Earth (Fig. 6.14).

Interplanetary shocks generated by coronal mass ejections can accelerate particles to high energy. As suggested by the Cornell astrophysicist Thomas Gold, closed magnetic fields can be ejected from the corona, generating shocks as they move into interplanetary space. The expanding magnetic loops even can remain attached to the Sun at both ends while moving all the way to the Earth.

When a coronal mass ejection travels out into space, it can take the form of a magnetic cloud that moves behind an interplanetary shock (Fig. 6.15). The mass ejection plows into the slower-moving solar wind, like an automobile out of control, driving huge shock waves millions of kilometers ahead of it. The shock waves cross magnetic field lines and accelerate particles as they go, much as ocean waves propel a surfer. The magnetic cloud can remain attached magnetically to the Sun, carrying its looping magnetic fields all the way to the Earth.

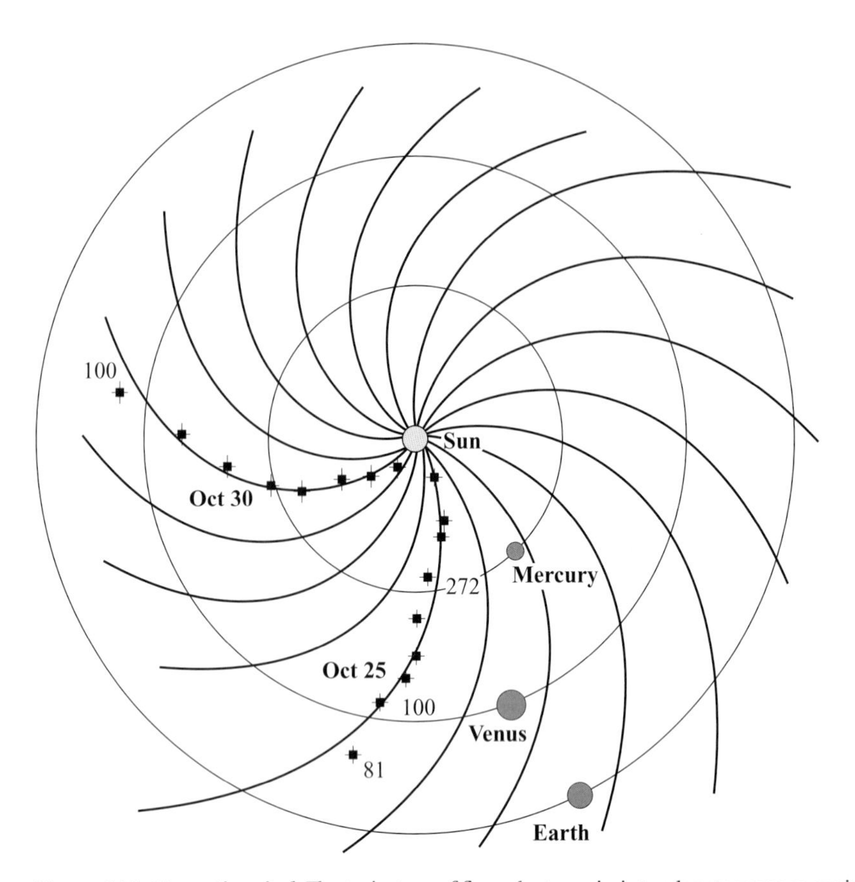

Figure 6.14. Magnetic spiral. The trajectory of flare electrons in interplanetary space as viewed from above the Sun's polar regions using the Ulysses spacecraft. The squares and crosses show Ulysses radio measurements of Type III radio bursts. As the high-speed electrons move out from the Sun into progressively more tenuous plasma, they excite radiation at successively lower plasma frequencies. The numbers denote the observed frequency in kiloHertz (kHz). Because the flaring electrons are forced to follow the interplanetary magnetic field, they do not move in a straight line out from the Sun but instead travel along the spiral pattern of the interplanetary magnetic field, shown by the solid curved lines. The magnetic fields are drawn out into space by the radial solar wind and attached at one end to the rotating Sun. The locations of the orbits of Mercury, Venus, and the Earth are shown as circles. (Courtesy of Michael J. Reiner. Ulysses is a project of international collaboration between ESA and NASA.)

High-Flying Humans at Risk

Because of their higher mass, it is solar-flare protons rather than electrons that provide the greatest threat to humans. Such high-speed protons, called *solar energetic particles*, can endanger the health and even the lives of astronauts when they are in outer space, unprotected by the Earth's magnetic field that deflects charged particles.

High-energy protons from a solar flare or coronal mass ejection easily can pierce a space-suit, causing damage to human cells and tissues and even threatening the life of unprotected astronauts who venture into space to unload spacecraft cargo, construct a space station or walk on the Moon, or Mars. Therefore, solar astronomers keep careful watch over the Sun during space missions to warn of possible activity occurring in just the wrong place or time.

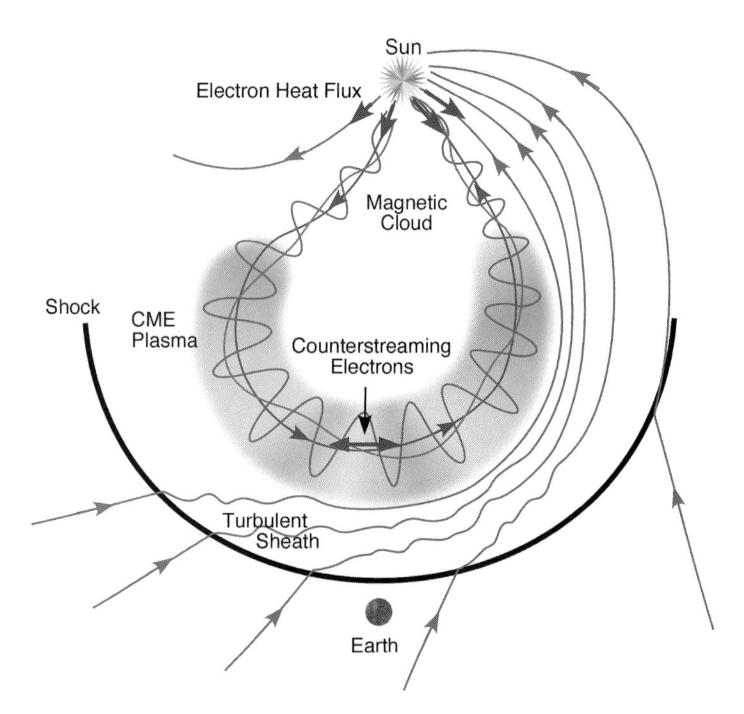

Figure 6.15. Magnetic cloud. When a coronal mass ejection (CME) travels into interplanetary space, it can create a huge magnetic cloud containing beams of electrons that flow in opposite directions within the magnetic loops that are rooted at both ends in the Sun. The magnetic cloud also drives a shock ahead of it. Magnetic clouds are present only in a subset of observed interplanetary coronal mass ejections. (Courtesy of Deborah Eddy and Thomas Zurbuchen.) (See color plate.)

If we go far enough into space, the chemical bonds in our molecules will be broken apart by storms from the Sun, increasing the risk of cancer and errors in genetic information. Therefore, space agencies set limits to an astronaut's exposure to solar energetic particles and radiation while traveling or working unprotected in space. Because of potential genetic damage, astronauts are supposed to have had all of their children before flying in outer space. Probably because of hormones, men are more radiation-resistant than women, and the resistance peaks between the ages of 45 and 50. So, if genetic harm and other health risks are the dominant factors, most astronauts are typically middle-aged men.

Failing to Communicate

Eight minutes after an energetic solar flare, a strong blast of x-rays and extreme ultraviolet radiation reaches the Earth and radically alters the structure of the planet's upper atmosphere, known as the *ionosphere*, altering its ability to reflect radio waves. During even moderately intense flares, long-distance radio communications can be silenced temporarily over the Earth's entire sunlit hemisphere. These radio blackouts are particularly troublesome for the commercial airline industry, which uses radio transmissions for weather, air traffic, and location information. The U.S. Air Force operates a global system of ground-based radio and optical telescopes and taps into the output of national space-borne x-ray telescopes and particle detectors to continuously monitor the Sun for intense flares that might severely disrupt military communications and satellite surveillance.

Although solar flares do not affect short-wavelength microwave signals that pass right through the ionosphere to communication satellites, solar explosions can destroy the satellites.

Satellites in Danger

More than 1,000 commercial, military, and scientific satellites are now in operation, affecting the lives of millions of people, and the performances and lifetime of all of these satellites are affected by Sun-driven space weather. Geosynchronous satellites, which orbit the Earth at the same rate that the planet spins, stay above the same place on the Earth to relay and beam down signals used for cellular phones, global positioning systems, and Internet commerce and data transmission. They can guide missiles or vehicles to their destinations, enable aviation and marine navigation, aid in search-and-rescue missions, and permit nearly instantaneous money exchange or investment choices.

Geosynchronous satellites are endangered by the coronal mass ejections that cause intense geomagnetic storms. These satellites orbit our planet at about 6.6 Earth radii, or about 4,200 km, moving around the Earth once every 24 hours. A coronal mass ejection can compress the Earth's protective magnetic fields from their usual location at about 10 Earth radii above the Equator to below the satellites' synchronous orbits, exposing them to the full brunt of the gusty solar wind and its charged, energized ingredients.

Other satellites revolve around our planet in closer, low-Earth orbits at altitudes of 300 to 500 km, scanning the air, land, and sea for environmental change, weather forecasting, and military reconnaissance. Space weather can increase noticeably the atmospheric friction exerted on these satellites, causing their orbit to decay more quickly than expected. The enhanced extreme ultraviolet and x-ray radiation from solar flares heats the atmosphere and causes it to expand; similar or greater effects are caused by coronal mass ejections. The expansion of the terrestrial atmosphere brings higher gas densities to a given altitude, increasing the friction and drag exerted on a satellite, pulling it to a lower altitude, and sometimes causing ground controllers to lose contact. Space stations, for example, periodically must be boosted in altitude to a higher orbit to avoid a similar fate.

Infrequent, anomalously large eruptions on the Sun can hurl energetic protons toward the Earth and elsewhere in space. The solar protons can enter a spacecraft like a ghost, producing erroneous commands and crippling the microelectronics. Such single-event upsets already have destroyed at least one weather satellite and disabled several communications satellites. However, to put the space-weather threat in perspective, of the thousands deployed, only a few commercial satellites have been lost to storms from the Sun. The U.S. military builds satellites that can withstand the effects of a nuclear bomb exploded in space.

Forecasting Space Weather

Recognizing our vulnerability, astronomers use telescopes on the ground and *in situ* particle detectors or remote-sensing telescopes on satellites to carefully monitor the Sun, and government agencies post forecasts that warn of threatening solar activity. This enables evasive action that can reduce disruption or damage to communications, defense, and weather satellites and electrical power systems on the ground. Once it is known that a Sun storm is imminent, the launch of manned space-flight missions can be postponed and a walk outside a spacecraft or on the Moon or Mars can be delayed. Airplane pilots can be warned of potential radio-communication failures. Operators can power down sensitive electronics on communication and navigation satellites until the danger passes. Utility companies can reduce load in anticipation of trouble on power lines, in that way trading a temporary "brown out" for a potentially disastrous "black out."

Everyone wants to know how strong a storm is and when it is going to hit. Like winter storms on the Earth, some of the effects can be predicted days in advance. A coronal mass ejection, for example, arrives at the Earth one to four days after solar astronomers watch it leave the Sun. Solar flares are another matter; as soon as a solar flare is observed on the Sun, its radiation and fastest particles have already reached the Earth, taking only 8 minutes to travel from the Sun. One promising prediction technique is to observe when the magnetism on the Sun has become twisted into a stressed situation because it then may be about to release a solar flare. Another technique employs helioseismology to look through the Sun and watch active regions develop before they rotate to face the Earth.

7

Comparisons of the Sun with Other Stars

7.1 Where and When Can the Stars Be Seen?

Go outside and look up at night to locate a familiar bright star or a pattern of stars such as the Big Dipper. To make this sighting, we must get our bearings here on the Earth, as well as in the sky, and also know the time. After all, the stars only come out at night.

So, once we know where we are, where the stars are located in the sky, and what time it is, we can become astronomers, from the Greek *astronomos* for "an arranger of stars." Then, as the great Italian poet Dante Alighieri wrote in the last line of his *Inferno*, we can "come forth, to look once more upon the stars."[15]

The Earth has a spherical shape and we can position ourselves on it by using a grid of *great circles* that specify our longitude and latitude. A great circle divides the sphere in half; the name derives from the fact that no greater circles can be drawn on a sphere.

Our planet spins on an axis that is speared through its center and emerges at the two geographic poles, the North and South Poles. A great circle halfway between these poles is called the Equator because it is equally distant from them. Circles of longitude are great circles that pass around the Earth from pole to pole, perpendicular to the Equator. The starting point for measuring longitudes is the great circle passing through the old Royal Observatory in Greenwich, England, and our longitude is measured westward along the Equator from this reference (Fig. 7.1). Our latitude is the angle measured northward (positive) or southward (negative) along a circle of longitude from the Equator to our place on the Earth.

The Earth's northern rotation axis now points close to Polaris, also known as the North Star or the Pole Star, which would lie approximately overhead when viewed from the Earth's geographic North Pole. The latitude of any location in the Earth's Northern Hemisphere is equal, within about 1 degree, to the angular altitude of Polaris. The uncertainty is due to the fact that Polaris is not exactly at the north celestial pole, where the north end of the Earth's rotation axis pierces the night sky.

We can locate Polaris by following the line joining the two stars farthest from the handle of the Big Dipper. When we observe Polaris, we are looking north, which accounts for the phrase "follow the drinking gourd" that was used by southern slaves escaping to northern parts of the United States. Mariners also have used the North Star for navigation, to find

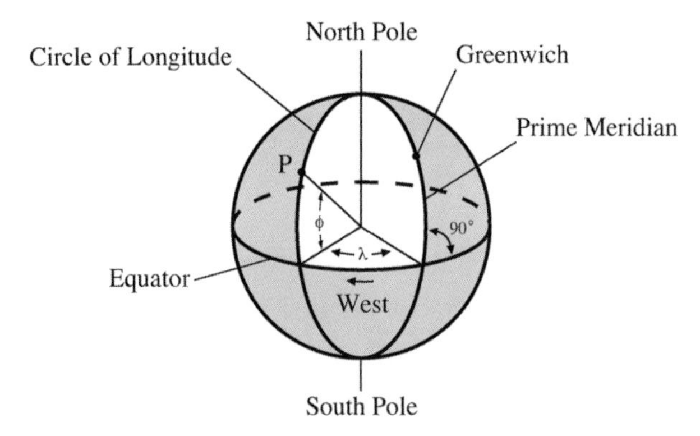

Figure 7.1. Latitude and longitude. Great circles through the North and South Poles of the Earth create circles of longitude. They are perpendicular to the Equator where they intersect it. The circle of longitude that passes through Greenwich, England, is called the Prime Meridian. The longitude of any point, P, is the angle λ, measured westward along the Equator from the intersection of the Prime Meridian with the Equator to the equatorial intersection of the circle of longitude that passes through the point. The latitude is the angle phi, φ, measured northward (positive) or southward (negative) along the circle of longitude from the Equator to the point. In this figure, the point P corresponds to San Francisco.

the direction of north and the latitude of their ship. Nowadays, we might use the Global Positioning System (GPS) to determine where we are. The numbers after the N (north) notation on the GPS receiver indicate latitude; the numbers after W (west) indicate longitude.

All of the stars seem to be placed on their own larger sphere, which is concentric with the Earth. Such a celestial sphere explains why people located at different places on the Earth invariably see only half of all the stellar sky. As our planet spins, day turns into night and these stars glide by. They rise at the horizon on one side of the Earth, slowly move overhead, and eventually set below the horizon on the other side of the planet, only to reappear the next night.

Astronomers define points and circles on the celestial sphere similar to geographers on the Earth. If we extend the Earth's rotation axis in both directions, it intersects the celestial sphere at the north and south celestial poles. They are the pivotal points of the night sky's apparent daily rotation. Whereas most stars sweep by as the Earth rotates, a star aligned with our planet's rotation axis, at the north celestial pole, seems to remain in an unchanging location.

When the plane of the Earth's Equator is extended outward in all directions, it cuts the celestial sphere in half, at the celestial equator. *Right ascension*, denoted α, is a celestial object's longitude, measured eastward along the celestial equator from a specific point called the Vernal Equinox (Fig. 7.2). The right ascension is expressed in hours and minutes of time, with 24 hours in the complete circle of 360 degrees, denoted as 360°. For conversion, 1 hour of time is equivalent to 15 degrees of angle, or $1 \text{ hr} = 15^{\circ}$; 1 second of time is equal to 15 seconds of arc, or $1 \text{ s} = 15''$; and 1 minute of arc equals 4 seconds of time, or $1' = 4\text{s}$.

Just as latitude is a measure of one's distance from the Equator of the Earth, *declination*, denoted δ, is a celestial object's angular distance from the celestial equator. It is positive in

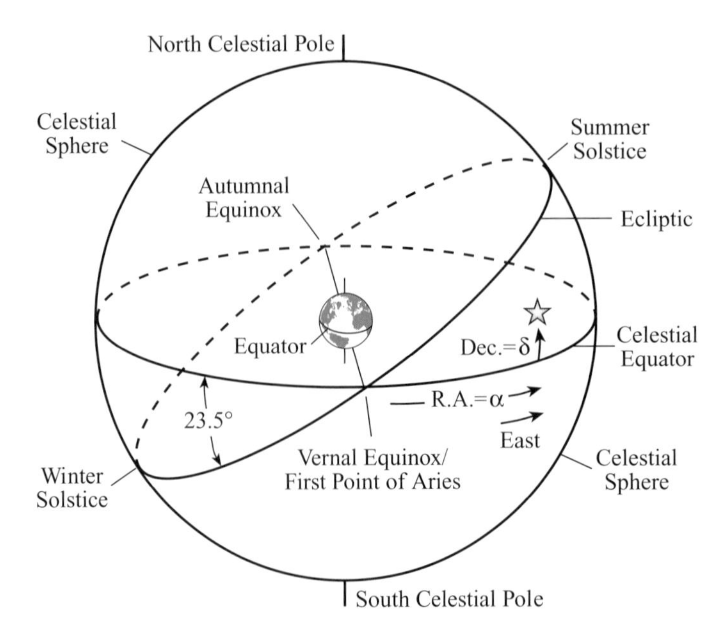

Figure 7.2. Celestial coordinates. Stars, galaxies, and other cosmic objects are placed on an imaginary celestial sphere. The celestial equator divides the sphere into northern and southern halves, and the ecliptic is the annual path of the Sun on the celestial sphere. The celestial equator intersects the ecliptic at the Vernal Equinox and the Autumnal Equinox. Every cosmic object has two celestial coordinates. They are the right ascension, designated by the angle alpha, α, or by R.A., and the declination, denoted by the angle delta, δ, or Dec. Right ascension is measured eastward along the celestial equator from the Vernal Equinox to the foot of the great circle that passes through the object. Declination is the angular distance from the celestial equator to the object along the great circle that passes through the object, positive to the north and negative to the south. Precession results in a slow motion of the Vernal Equinox, producing a steady change in the celestial coordinates.

the north and negative in the south. Polaris, for example, is located near 90 degrees north declination.

Right ascension, α, and declination, δ, specify the position of any object on the celestial sphere in equatorial coordinates. For centuries, astronomers have used catalogues of the right ascensions and declinations of celestial objects to locate them in the sky.

If we want to observe a star and we know its celestial coordinates, we also need to know what time it passes overhead. Our clocks and watches are set to the time it takes the Sun to rise and set and rise again. This solar day is 24 hours, or 86,400 seconds, long. Astronomers also tell time by the stars; for an object with an observable declination, the time at which it can be observed depends on this star time. This is known as *sidereal time*, meaning related to the stars. The sidereal day is the time it takes for a star – or any other observable celestial object – to proceed from its highest point in the sky one day to its highest point the next day. A star is overhead, crossing the local meridian, when its right ascension is equal to the local sidereal time; however, a terrestrial clock records star time on only one day a year, around September 26. After that, the local sidereal time gains about 2 minutes a day and 2 hours a month on a terrestrial clock. There is another observational caveat: the slow precession in celestial positions (Focus 7.1).

Focus 7.1 The Locations of the Stars Are Slowly Changing

The location of any object on the celestial sphere is displaced gradually as the result of gravitational pull of the Moon and the Sun on the Earth's elongated shape. They send the planet into a wobbling rotation, like a spinning top. This causes a progressive change of the celestial positions, called *precession*. The related change in the location of bright stars was first observed by Hipparchus, a Greek astronomer who lived in the second century BC; the telescope was not invented until 17 centuries after Hipparchus established the positions of bright stars using his eyes.

Because the Moon and the Sun lie in the ecliptic plane, which is inclined by 23.5 degrees (denoted 23.5°) to the plane of the Earth's Equator, they exert a changing gravitational force on the Earth's equatorial bulge. This causes the rotation axis to sweep out a cone in space, completing one circuit in about 26,000 years (Fig. 7.3).

So, the Earth is not placed firmly in space; instead, it wobbles about causing the identity of the Pole Star to gradually change over time-scales of thousands of years. The northern projection of the Earth's rotation axis is currently within about 0.75° of Polaris and will move slowly toward it in the next century. After that, the north celestial pole will move away from Polaris and, in about 12,000 years, the Earth's rotation axis will point to within 5 degrees of the bright star Vega.

The slow conical motion of precession carries the Earth's Equator with it; as the Equator moves, the celestial position of any object slowly changes by about 50 seconds of arc per year. Because of these changes, the *equinox*, or reference date, must be given when specifying the right ascension or declination of any cosmic object.

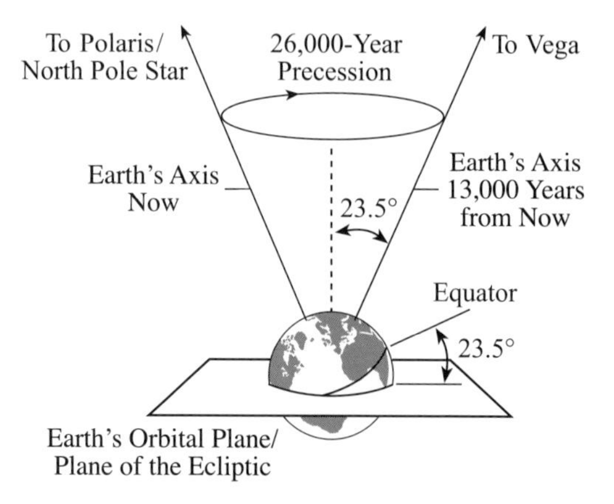

Figure 7.3. Precession. The Earth's rotation axis traces out a circle on the sky once every 26,000 years, sweeping out a cone with an angular radius of about 23.5 degrees. The Greek astronomer Hipparchus (c. 146 BC) discovered this precession in the second century BC. The north celestial pole, which marks the intersection of this rotation axis with the northern half of the celestial sphere, now lies near the bright star Polaris. However, as the result of precession, the rotational axis will point toward another bright north star, Vega, in roughly 13,000 years. This motion of the Earth's rotational axis also causes a slow change in the celestial coordinates of any cosmic object.

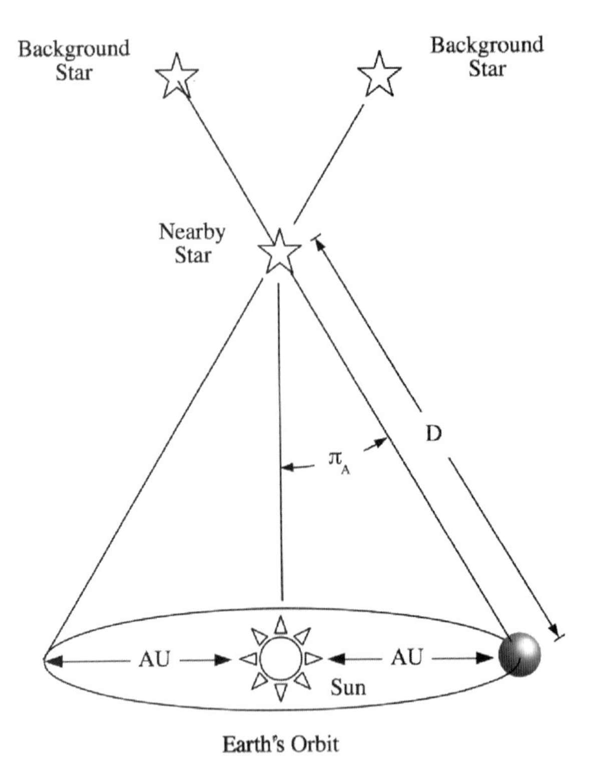

Figure 7.4. Annual parallax. When a distant and nearby star are observed at six-month intervals, from opposite sides of the Earth's orbit around the Sun, astronomers measure the angular displacement between the two stars. It is twice the annual parallax, designated by π_A, which can be used to determine the distance, D, of the nearby star. From trigonometry, $\sin \pi_A = AU/D \approx \pi_A$ for small angles, where 1 AU is the mean distance between the Earth and the Sun. The distance D to the star in units of parsecs is given by $1/\pi_A$, if the parallax angle is measured in seconds of arc. This angle is greatly exaggerated in the figure, for all stars have a parallax of less than 1 second of arc or less than 1/3,600th of a degree. The German astronomer Friedrich Wilhelm Bessel (1784–1846) announced the first reliable measurement of the annual parallax of a star in 1838.

7.2 How Far Away Are the Stars?

We receive so little light from the stars that they seem to hang in the dark night sky like small dim lanterns. Yet, some of them are more luminous than the Sun, and the faint light that reaches the Earth is due to their great distances from us.

To determine the distance of a nearby star (other than the Sun), astronomers measure its angular displacement when viewed from opposite sides of the Earth's orbit, or from a separation of twice the AU. (The AU is the mean distance between the Earth and the Sun, with a value of about 149.6 million kilometers.) The angle is known as the *annual parallax*, from *annual* for the Earth's year-long orbit and the Greek *parallaxis* for the "value of an angle." Once the parallax is combined with the known value of the AU, the star's distance can be established by *triangulation*, the geometry of a triangle.

The stellar measurement involves careful scrutiny of two stars that appear close together in the sky: a bright one relatively nearby and another fainter one much farther away (Fig. 7.4). The annual parallax of the nearer star then can be determined by comparing its position to that of the distant one for a year or more. During the course of the year, the nearby star seems to sway to and fro, in a sort of cosmic minuet that mirrors the Earth's orbital motion. The nearer the star, the larger the annual parallax sways. This angle is usually expressed in seconds of arc, denoted by the symbol ″.

A convenient unit for measuring stellar distance is the *parsec*, abbreviated pc, which is equivalent to 206,000 AU and 31 million billion, or 3.1×10^{16}, meters. Neighboring stars often are separated by about 1 parsec. The distance in parsecs is equal to the reciprocal of the annual *par*allax in *sec*onds of arc; the name derives from the italicized parts of these two words.

The fact that light travels through space at a constant speed provides another convenient way to describe distance. That is, we can specify the distance of an astronomical object in terms of the time it takes light to move through space from the object to the Earth, which is known as the *light-travel* time. Light from the Moon takes 1.5 seconds to reach the Earth, so we say that the Moon is 1.5 light-seconds away from the Earth. Sunlight takes 499 seconds, or 8.3 minutes, to cover the average distance between the Sun and the Earth. The nearest star (other than the Sun) is at a distance of a little more than 4 light-years. To convert between the two commonly used units of stellar distance, 1 parsec is equal to 3.26 light-years.

Many of the brightest stars are hundreds of light-years away, so we can walk outside at night and see stars whose light was emitted before our parents were born. We also can observe stars in our Milky Way that are many millions of light-years away. Starlight may reach us from stars that now have extinguished the internal nuclear fires that made them shine. The most distant objects in the universe are billions of light-years away, and the light we now detect from them was generated that long ago. So, radiation provides a method of looking back into time, to decipher the history of the universe – looking back to the time before the Sun and the Earth were formed, some 4.6 billion years ago.

The first star whose distance was reliably determined by the annual parallax method was 61 Cygni, which is 11.4 light-years and 3.49 pc away from the Earth. Its annual parallax, first measured by the German astronomer Friedrich Wilhelm Bessel in 1838, has a value of 0.286″.

The star with the largest annual parallax (i.e., 0.769″) is Proxima Centauri (Fig. 7.5), making it the closest star to the Earth other than the Sun. The distance to Proxima is 1.30 pc, and light takes only 4.23 years to reach us from this star. It is about 268,000 times more distant than the Sun, indicating that there are vast, seemingly empty spaces between the stars.

Because the Earth's atmosphere usually limits the angular resolution of a ground-based telescope to no more than 0.05″, the annual-parallax method can be used only for the very nearest stars, those that are closer than about 65 light-years, or 20 pc. However, instruments aboard the ESA *HIPPARCOS* satellite, which orbited the Earth above its atmosphere in the 1990s, pinpointed the position of more than 100,000 stars with an astonishing precision of 0.001″ and obtained accurate measurements for the distances of stars up to 3,260 light-years and 1,000 pc away. This explains the spacecraft's name, which is an acronym for *HIgh Precision PARallax COllecting Satellite*; the name also alludes to the ancient Greek astronomer Hipparchus, who recorded accurate star positions more than 2,000 years ago. A successor to this mission is the ESA *GAIA* mission, short for *Global Astrometric Interferometer for Astrophysics*, currently scheduled for launch in March 2013. This mission is intended to measure 1 billion stellar distances to perhaps 36,200 light-years and 10,000 pc.

Figure 7.5. Proxima Centauri, the closest star. The red star centered in this image is the nearest star to the Earth, other than the Sun; it is at a distance of 4.24 light-years from us. Despite its closeness, the star has an absolute luminosity of just 2/1,000ths, or 0.002 times, the Sun's luminosity, and is too faint to be seen by the unaided human eye. Named Proxima Centauri because of its proximity, the star is close enough for its angular diameter to be measured by interferometric techniques, yielding a radius of 1/7th the radius of the Sun. Proxima Centauri orbits the bright double star Alpha Centauri (see Fig. 7.8) at a distance of about 0.24 light-year and an angular separation of 2.2 degrees. All three stars are located in the southern sky and cannot be seen from the Northern Hemisphere of the Earth. (Courtesy of David Malin/UK Schmidt Telescope/DSS/AAO.) (See color plate.)

7.3 How Bright and Luminous Are the Stars?

Apparent Brightness of the Stars

At the end of a day, when the Sun has set, a bright star could be the first to appear, reminiscent of the familiar nursery rhyme:

> Star light, star bright,
> The first star I see tonight;
> I wish, I may, I wish I might,
> Have the wish I wish tonight.[16]

As twilight fades into darkness, a few more bright stars appear, together with many other faint ones. The brilliant stars seen tonight may be among the 10 brightest stars in the sky (Table 7.1).

Because a human eye does not register directly the relative amount of radiation entering it, the Greek astronomer Hipparchus divided the stars that he could see into six groups that better measure their relative brightness, relative to the eyes. This way of measuring brightness is called the *apparent visual magnitude* and is designated by the lowercase letter m or, to be explicit about the visual aspect, by m_V with the subscript V denoting visual. Hipparchus designated the brightest stars, such as Sirius or Rigel, with the first and most important magnitude, $m = 1$; Polaris and most of the stars in the Big Dipper were designated

Table 7.1. *The 10 brightest stars as seen from the Earth*[a]

Star Name	R.A. (2000) h	m	Dec (2000) °	′	m	Spectral Class[b]	D[c] (light-years)	L ($L_\odot$)	Mag	M ($M_\odot$)	R ($R_\odot$)
Sun					−26.74	G2 V	0.000016	1.0	+4.83	1.0	1.0
Sirius	06	45.2	−16	43.0	−1.46	A1 V	8.6	25.4	+1.42	2.02	1.71
Canopus	06	24.0	−52	41.8	−0.72	F0 Ib	310	13,600	−5.53	8.5	65.0
Alpha Centauri[d]	14	39.6	−60	50.0	−0.01	G2 V	4.3	1.52	+4.38	1.10	1.23
Arcturus	14	15.7	+19	11.0	−0.04	KI III	36.7	210	−0.29	1.5	25.7
Vega	18	36.9	+38	47.0	+0.03	A0 V	25.0	37	+0.58	2.14	2.5
Capella	05	16.7	+45	42.2	+0.08	G1 III	41	78	+0.20	2.6	9.2
Rigel	05	14.5	−08	12.1	+0.18	B8 Ia	772.5	66,000	−6.7	17.0	78.0
Procyon	07	39.3	+05	13.5	+0.34	F5 IV	11.46	7.73	+2.65	1.42	2.05
Achernar	01	37.7	−57	14.2	+0.50	B3 V	144	3,311	−2.77	6 to 8	10
Betelgeuse	05	55.2	+07	24.4	+0.42v	M2 Ia	643	140,000	−6.05	18 to 19	≈1,180

[a] The stars are listed in order of increasing apparent visual magnitude, m, or decreasing apparent brightness for the brightest component if it is a binary system. The absolute magnitude is designated as Mag.

[b] The luminosity classes are Ia = supergiant of high luminosity; Ib = supergiant of lower luminosity; II = bright giant; III = normal giant; IV = subgiant; V = main-sequence star, or dwarf star; VI = subdwarf.

[c] The luminosity, L, is in units of the Sun's luminosity $L_\odot = 3.828 \times 10^{26}$ J s^{-1}; the mass, M, is in units of the Sun's mass $M_\odot = 1.989 \times 10^{30}$ kg; and the radius, R, is in units of the Sun's radius $R_\odot = 6.955 \times 10^8$ m.

[d] Alpha Centauri also is known as Rigel Kentaurus.

m = 2; and the faintest stars visible to the unaided eye received the sixth magnitude or m = 6. Thus, in the magnitude system, brighter stars have lower magnitudes, and fainter stars have higher ones.

About two millennia later, in 1856, the British astronomer Sir Norman Pogson noted that the stars of the first magnitude were 100 times as bright as stars of the sixth magnitude and that each magnitude unit is 2.512 times brighter than the next one down, where the number 2.152 is the fifth root of 100, or $100^{1/5}$. He therefore adopted a logarithmic scale that caused the very brightest stars to climb to negative apparent magnitudes. Sirius is m = −1.44, the planets Venus and Jupiter are a little brighter than Sirius, and the Sun is so close and bright that it is m = −26.73 (Table 7.2).

The number of stars increases dramatically with increasing apparent visual magnitude. There are 14 stars brighter than m = 1 and about 4,600 stars brighter than m = 6, which are all of the stars detectable by the unaided eye. There are 335,000 stars brighter than m = 10, 1.5 million stars brighter than m = 12, and 4.8 billion stars brighter than m = 25. A backyard telescope can detect stars of apparent magnitude between 10 and 15; the Hubble Space Telescope can approach apparent magnitude 30. These stars are 4 billion times fainter than the human eye can see without a telescope, and there are 100 billion of them.

Intrinsic Luminosity of the Stars

Luminosity is an intrinsic measure of a star, and it is not related to the star's distance from the observer. It is the amount of energy a star radiates per unit time in units of J s^{-1}, which also is the emitted power in watts, where 1 J s^{-1} = 1 watt. A star's luminosity usually

Table 7.2. *Apparent visual magnitudes, m, of a few bright astronomical objects*

Object Name	m
Sun	−26.73
Full Moon	−12.7
Venus[a]	−4.5
Jupiter[a]	−2.5
Sirius	−1.44
Rigel	0.12
Saturn[a]	0.7
Polaris	1.97

[a] At maximum brightness, when the planet is in the part of its orbit that brings it closest to the Earth.

is compared to the luminosity of the Sun, designated $L_{\odot}$, with a value of $L_{\odot} = 3.828 \times 10^{26}$ J s^{-1}. There are stars that are 1 million times more luminous than the Sun (Table 7.3) and other stars that are 1 million times less luminous than the Sun. The exceptionally luminous beacons are rare, and the most common stars are not even as luminous as the Sun; they are so dim that telescopes are required to see them. The most luminous stars also are among the most massive, largest, and hottest stars, and the progressive decrease in stellar luminosity usually corresponds to a decrease in stellar mass and radius (see Table 7.3).

The luminosity of a star can also be expressed as an *absolute magnitude*, which by definition is the apparent magnitude a star would have it were 32.6 light-years or 10 parsecs away. A smaller absolute magnitude corresponds to a greater luminosity.

The apparent brightness of stars can be combined with measurements of their distances to determine their luminosity. These extrapolations indicate that our eyes provide an incomplete view of the celestial realm. Some of the stars that appear bright to the eye are relatively nearby and no more luminous than the Sun, but many are distant stars that can pour forth hundreds of thousands of times more energy every second than that of the Sun. Thus, the exceptional brightness of the brightest stars, as seen from the Earth, can be due to either the immense power of the radiation they emit or to their relative closeness, compared with other stars.

The brightest star in the night sky, for example, is Sirius, whose name is derived from the Ancient Greek *Seirios* for "glowing" or "scorcher." However, Sirius is only 26 times more luminous than the Sun, and its brilliance is due to its relative proximity, just 8.6 light-years away. In contrast, the third and tenth brightest stars, Rigel and Betelgeuse, are supergiant stars located hundreds of light-years away and shining with a luminosity of tens of thousands of times that of the Sun.

Table 7.3. *The range in stellar luminosity*[a]

Star Name	Luminosity ($L_\odot$)	Mass ($M_\odot$)	Radius ($R_\odot$)	Temperature (K)
R 136a1[b]	8,700,000	265	35.4	53,000
LBV 1806~20	5,000,000	200	150	36,000
Pistol Star	1,700,000	150	340	20,000
Betelgeuse	200,000	19	1,180	3,500
Rigel	85,000	17	78	11,000
Polaris	2,000	7.5	30	3,200
Aldebaren	425	1.7	44	4,010
Vega	37	2.1	2.5	9,600
Sirius A	25.4	2.0	1.7	9,940
Alpha Centauri A	1.5	1.1	1.2	5,790
Sun	1.0	1.0	1.0	5,780
Sirius B[c]	0.026	0.978	0.0084	25,200
Gliese 229B[d]	0.000006	0.03 to 0.05	0.1	950

[a] The luminosity is in units of the Sun's luminosity $L_\odot = 3.828 \times 10^{26}$ J s^{-1}; the mass is in units of the Sun's mass $M_\odot = 1.989 \times 10^{30}$ kg; the radius is in units of the Sun's radius $R_\odot = 6.955 \times 10^{8}$ m; and the temperature is the effective temperature of the visible stellar disk in degrees kelvin, denoted as K.

[b] In the Large Magellanic Cloud, a nearby irregular galaxy and satellite of the Milky Way.

[c] Sirius B is an Earth-sized white dwarf star, which has a mass about equal to that of the Sun but has depleted its thermonuclear fuel.

[d] Gliese 229B is a substellar brown dwarf object, whose mass is below the lower limit, at about 0.08 solar mass, to sustain hydrogen fusion. This brown dwarf object has a radius about equal to that of Jupiter, which is very close to 1/10th the Sun's radius but with a mass of 30 to 50 times the mass of Jupiter, or about 0.03 to 0.05 times the mass of the Sun.

7.4 How Hot Are the Stars?

The Hottest and Coolest Stars

An understanding of the physical properties of a star requires knowledge of its temperature as well as its luminosity. The effective temperature of the Sun's visible disk, the photosphere, is inferred from the solar radius and luminosity, with a value of 5,780 K; however, we do not have direct knowledge of the radius of most stars. They are too far away and too small in angular size for a telescope to resolve them.

Fortunately, there are two methods to infer a star's temperature even when we do not know its size and luminosity. We can estimate the temperature from the color of the star or infer its temperature from the relative intensities of absorption lines observed in its spectrum. These are the effective temperatures of the stellar photospheres. The hottest stars that we can see have temperatures of more than 100 times those of the coolest stars, with a range between 50,000 and 2,000 K.

These are the stars we look at, but our eyes do not see all of the radiation that a star produces. At extreme hot or cold temperatures, a star can become visibly dim, even invisible, because most of its radiation is produced outside the visible part of the radiation spectrum and often is absorbed in the Earth's atmosphere before reaching the ground.

The hottest star, with a temperature of more than 100,000 K, is so hot that we cannot even see it. A very hot star emits most of its light at ultraviolet wavelengths that are absorbed in the atmosphere. These hottest stars also are exceptionally luminous and massive.

The coolest star-like objects emit most of their radiation at infrared wavelengths, also absorbed in the Earth's atmosphere and outside our range of vision. There are the substellar, brown dwarf objects, for example, that do not have enough mass to begin nuclear fusion of hydrogen in their core. These stellar disks emit heat associated with their formation or by burning deuterium that was already present in them. The brown dwarf objects often are no warmer than the kitchen oven and sometimes even colder than room temperature, or below 300 K. Molecules of water and methane sometimes are found in their photosphere.

Therefore, what we see depends on how we look at it, among the stars as well as the universe. Nevertheless, it is the visible colors, spectra, and temperatures that dominate classical studies of the stars, and we focus on them.

The Colors of Stars

There are reddish stars like Betelgeuse and Antares, yellowish stars like the Sun and Capella, and whitish stars like Vega and Sirius. These colors provide a rough indication of the temperature of a star's photosphere. As the temperature rises, the colors change from red – near 3,000 K, to yellow – around 6,000 K, to white – at about 10,000 K.

A star often has a certain color because most of its radiation is emitted at the wavelengths corresponding to that color. The wavelength of maximum starlight intensity varies inversely with temperature. A blue star, for example, is hotter than a red star. The coloring of a star, however, is very subtle, depending on the relative amount of light seen in different colors.

Blue-colored stars, for example, are not just bluer than red stars. For a star of the same radius, a blue star is more luminous than a red one. Exceptionally hot stars emit most of their radiation at invisible ultraviolet wavelengths, and such a star is even more luminous. There is enhanced radiation intensity at adjacent wavelengths, and this ultraviolet spillover produces more blue light than expected for a cooler star. In this case, the temperature of the star is much hotter than that inferred from blue light alone.

Astronomers therefore decided to quantify color by comparing the apparent magnitudes measured in different wavelength bands. They are denoted U, B, and V for ultraviolet, blue, and visual bands. The difference between the amount of light received at one color and the amount at another is known as the *color index*, which is usually measured by the difference between blue, designated B, and visual, denoted V, with a color index denoted by B–V. It provides a reasonable estimate of photosphere temperature by using the ratio of luminosities at two wavelengths, which is better than a temperature estimated from observations at only one wavelength.

However, in addition to ultraviolet spillover into the blue colors, interstellar dust reddens starlight as it travels through space to arrive at the Earth, and the amount of reddening increases with a star's distance. Thus, the observed colors may not reliably reflect the emitted colors. A star's spectral lines provide a more accurate indication of the temperature of a star's photosphere.

Table 7.4. *The spectral classification of stars*[a]

Class	Dominant Lines	Color	Color Index	Effective Temperature	Examples
O	He II	Blue	−0.3	28,000–50,000	χ Per, ε Ori
B	He I	Blue-White	−0.2	9,900–28,000	Rigel, Spica
A	H	White	0.0	7,400–9,900	Vega, Sirius
F	Metals; H	Yellow-White	0.3	6,000–7,400	Procyon
G	Ca II; Metals	Yellow	0.7	4,900–6,000	Sun, α Cen A
K	Ca II; Ca I	Orange	1.2	3,500–4,900	Arcturus
M	TiO; Ca I	Orange-Red	1.4	2,000–3,500	Betelgeuse

[a] H denotes hydrogen, He denotes helium, Ca denotes calcium, and TiO is a molecule. The Roman numeral I denotes an electrically neutral, un-ionized atom; the number II describes an ionized atom missing one electron; and the temperatures are in degrees kelvin, denoted as K.

The Spectral Sequence

More than a century ago, astronomers noticed that stars of different colors exhibit different spectral lines. Strong absorption lines of hydrogen, for example, dominate the spectra of white stars like Vega and Sirius, whereas some blue stars have noticeable helium-absorption lines. Yellow stars like the Sun have strong absorption lines of calcium and heavier elements, called metals, in their spectra.

The different spectral lines that are emitted among stars depend on the physical conditions in the visible disk – the photosphere – and therefore the level of ionization of the emitting atoms. Stars that display spectral lines of highly ionized elements must be relatively hot because high temperatures are required to ionize atoms. These hot stars have relatively weak hydrogen lines because nearly all of the hydrogen is ionized and all of its electrons have been set free from their atomic bonds, no longer emitting or absorbing radiation. In other words, stars that display hydrogen lines have moderate photosphere temperatures. Those exhibiting molecular lines have even cooler temperatures because molecules break apart into their component atoms when the temperature increases.

A system of stellar classification based on spectra was developed in the early twentieth century and is still in use today. Working under the direction of Edward C. Pickering, astronomers at the Harvard College Observatory examined the spectra of hundreds of thousands of stars. The astronomers were mainly women who had studied physics or astronomy at nearby women's colleges, including Wellesley and Radcliffe. Harvard did not educate women at that time and did not permit women on its faculty.

One of these faithful, stalwart workers was Annie Jump Cannon, who classified the spectra of roughly 400,000 stars in her lifetime. She distinguished the stars on the basis of the absorption lines in their spectra and arranged most of them in a smooth and continuous spectral sequence. The hottest stars, with the bluest colors, were designated as spectral type O, followed in order of declining photosphere temperature by spectral types B, A, F, G, K, and M (Table 7.4).

Cannon further refined each spectral class by adding numbers from 0 to 9, running from hot to cold; the larger the number the cooler the star in that class. For example, the hottest

Table 7.5. *Some well-known large stars*[a]

Star Name	Radius ($R_{\odot}$)	Luminosity ($L_{\odot}$)	Mass ($M_{\odot}$)	Temperature (K)
Supergiant Stars				
VY Canis Majoris	$\approx$ 2,000	$\approx$ 450,000	$\approx$ 40	$\approx$ 3,000
VV Cephei A	$\approx$ 1,900	$\approx$ 300,000	$\approx$ 30	$\approx$ 3,300
Mu Cephei	1,650	60,000	15	3,690
Betelgeuse	1,180	140,000	19	3,500
Antares	800	65,000	15	3,500
Red Giant Stars				
Mira A	400	9,000	1.2	3,000
R Doradus	370	6,500	$\approx$ 1.0	2,740
Aldebaren	44.2	425	1.7	4,010
Polaris	30	2,200	7.5	7,200
Arcturus	25.7	210	1.1	4,300
Pollux	8.0	32	1.86	4,865

[a] The radius, R, is in units of the Sun's radius $R_{\odot} = 6.955 \times 10^8$ m; the luminosity, L, is in units of the Sun's luminosity $L_{\odot} = 3.828 \times 10^{26}$ J s^{-1}; the mass, M, is in units of the Sun's mass $M_{\odot} = 1.989 \times 10^{30}$ kg; and the temperature, T, is the effective temperature of the stellar disk in degrees kelvin, abbreviated as K.

F star is designated as F0 and the coolest as F9, followed by G0. In this system, our Sun is classified as G2.

7.5 How Big Are the Stars?

Large stars come in two varieties: the giants and the supergiants. A relatively common type of big star is the red giant star, which is about 100 times bigger than the Sun; the other exceedingly rare kind, the supergiant, is about 1,000 times larger than the Sun. The benchmark size is the radius of the Sun, denoted $R_{\odot}$, with a value of $R_{\odot} = 6.955 \times 10^8$ m. The red giants can be found almost anywhere in the night sky, whereas the supergiants are sparsely scattered within the Milky Way. Only one in a million stars is likely to be a supergiant. Well-known examples of both types of large stars are given in Table 7.5.

The biggest stars will become the smallest ones. When the red giants have completed their stellar life, consuming their supply of nuclear fuel, they will collapse into Earth-sized white dwarf stars. When the more massive and luminous supergiants expire, they will leave a city-sized neutron star behind or collapse into oblivion as a stellar black hole.

Supergiants, as the name implies, are simply extreme examples of the giant stars. They are the rare anomaly that stands out because of its size, like some members of a professional football team. We pay attention because supergiants are not normal. They are exceptionally big, massive, luminous, and often bright.

The brightness of supergiants, as seen at the Earth, is one feature that makes them interesting. Some can be seen by the unaided human eye. To supply their high intrinsic luminosity – perhaps 500,000 times that of the Sun – the supergiants must generate energy at a prodigious rate, exhausting their available thermonuclear energy supply in about a half million years. In contrast, the life span of a red giant star is a few million years and that of a Sun-like star is about 10 billion years. Stars that last longer are more likely to be seen, which explains why we observe few supergiants, more red giants, and many stars like the Sun.

The supergiants are so big that the angular diameter of some has been measured but not by an ordinary telescope anywhere on the Earth. Even these behemoths appear as mere points of light when viewed with the largest optical telescope, just as they do when we walk outside and look at them at night.

We can measure the angular size of the largest stars using an interferometer that employs two or more connected mirrors. The radiation waves detected by any two of the mirrors are combined to produce an interference pattern – hence, the name "interference-meter," or *interferometer* for short. If the waves of electromagnetic radiation detected by the two mirrors are in phase when combined, their wave crests combine and strong light is detected. When they are out of phase, one wave crest matches the trough of the other and they cancel one another. In the earliest applications, the two mirrors were separated gradually to produce a set of light and dark bands, or "fringes"; when the fringes disappeared altogether, the star was resolved. The angular diameter of the source, in units of radians, is the ratio of the wavelength to that mirror separation in which the fringes disappear.

The American physicist Albert Michelson teamed up with the American astronomer Francis Pease to measure the size of Betelgeuse with an interferometer in 1921. They mounted two moveable mirrors and two fixed mirrors on a 20-foot (6-meter) steel beam that was placed across the frame of the 100-inch (2.3-meter) Hooker telescope on Mount Wilson. By measuring the mirror separation when the interference fringes disappeared, they concluded that Betelgeuse has an angular diameter of $\theta = 0.047''$, where the symbol $''$ denotes seconds of arc. By way of comparison, if the Sun were placed at the distance of the next nearest star, Proxima Centauri, it would have an angular diameter of approximately $0.007''$.

Modern visible-light interferometry has been used to measure the angular diameters of about 100 stars, including both supergiant stars and relatively nearby red giant stars. Current observations of the angular diameter of the supergiant Betelgeuse, at $0.055''$, indicate that it has a radius of 1,180 solar radii, which is equivalent to 5.48 AU – where 1 AU $= 1.496 \times 10^{11}$ m is the mean distance between the Earth and the Sun. The orbital distance of Jupiter from the Sun is 5.2 AU; therefore, Betelgeuse and other supergiant stars would fill much of our major planetary system. The Hubble Space Telescope (HST) was used to obtain an image of Betelgeuse, obtaining the first direct picture of the visible disk of any star other than the Sun. Interferometric measurements in a recent 15-year period suggest that Betelgeuse may be shrinking, even though its visible brightness showed no significant dimming during the same period – a perplexing result.

When two stars are orbiting a common center of mass and the orbit can be viewed edge on, then a star's radius can be inferred from the duration of its eclipse and the orbital speed; however, accurate measurements of the size of such eclipsing binary stars are not very numerous. One example is the supergiant star VV Cephei A, whose blue companion star can be observed passing behind it. The eclipsed star disappears for about 1.2 years. When combined with the star's speed, this long eclipse duration gives an approximate radius for

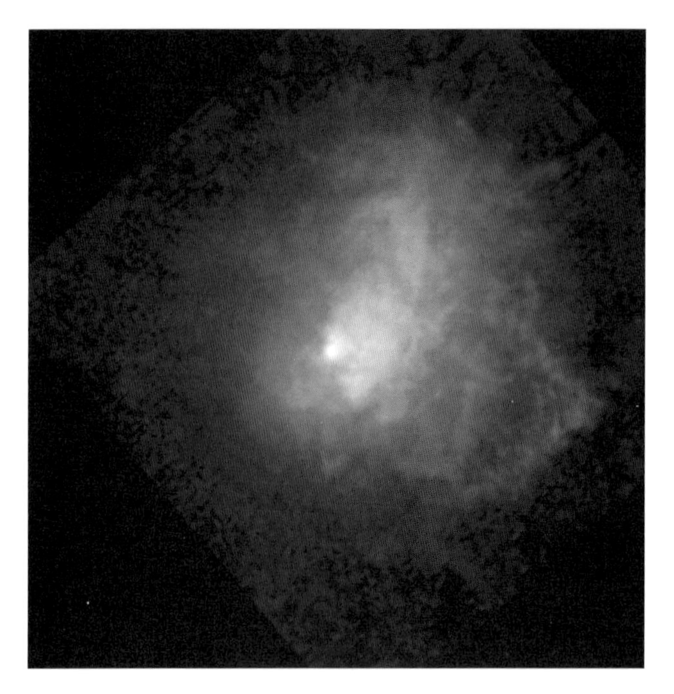

Figure 7.6. Massive star's outbursts. The exceptionally massive and luminous supergiant star VY Canis Majoris is nearing the end of its estimated life of 0.5 million years. As revealed in this image taken from the Hubble Space Telescope, the star's extended, outer atmosphere is being ejected into surrounding space in arcs, loops, filaments, and knots. When these data are combined with spectroscopy using the ground-based Keck telescope, it is found that the numerous features are moving at different speeds and in various directions and, hence, were produced from separate events and at different locations. One of the arcs is moving at a speed of 46,000 km s^{-1}, close to the escape velocity of the star. (Courtesy of NASA/ESA/Roberta Humphreys, University of Minnesota.)

VV Cephei A of 1,900 times that of the Sun, or about 8.8 AU, and somewhat smaller than Saturn's mean distance from the Sun at 9.58 AU. For comparison, the partial eclipse of a companion star by Algol lasts only a few hours, due to Algol's smaller size of about three times that of the Sun.

The supergiant stars are so large that they cannot hold onto their outer atmosphere. They are slowly falling apart and cannot remain self-contained. The supergiants are enveloped by dust and gas blown out by the stars' winds and have no well-defined apparent "edge."

The supergiant VY Canis Majoris, for example, is so big that the stellar gravity at its disk is just 1/100,000th, or 10^{-5}, the gravity at the Sun's photosphere. The average atom in the outer atmosphere of VY Canis Majoris is moving fast enough to overcome the weak gravitational pull and break away from the star. Images taken from the HST reveal arcs, filaments, and knots of material formed by the massive outflows from this supergiant star (Fig. 7.6).

Another supergiant, VV Cephei A, is surrounded by opaque shells of a highly extended atmosphere and is not entirely spherical in shape. It is a member of a binary-star system placed so close together that matter flows from the red supergiant to its companion blue-colored star, of roughly solar size.

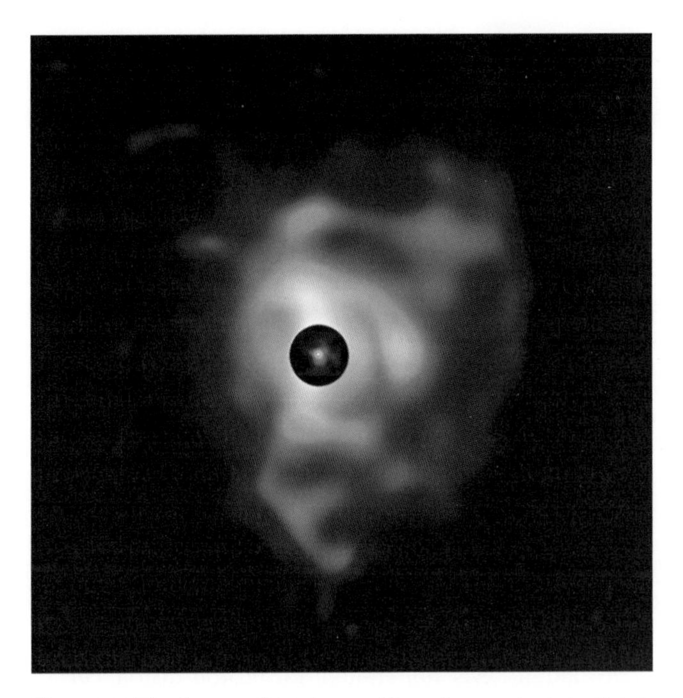

Figure 7.7. The flames of Betelgeuse. The red supergiant star Betelgeuse is slowly shedding its outer atmosphere, producing out-flowing gas that envelops the star and a much bigger nebula of gas and dust that surrounds it. The small circle in the middle of this composite image denotes the edge of the supergiant's optically visible disk; it has a diameter of about 4.5 AU, where 1 AU is the mean distance between the Earth and the Sun. The black disk masks the bright central radiation of the star to detect the infrared radiation of the outer plumes. They stretch to about 400 AU, or 60 million million, 6×10^{13}, meters from the supergiant Betelgeuse. (Courtesy of ESO/VISIR/VLT/Pierre Kervella.) (See color plate.)

A little more than a decade after the pioneering HST observations, the Very Large Telescope (VLT) at the Paranal Observatory in Chile was used to resolve not only the face of Betelgeuse but also surrounding gas and dust, which extend out to 400 AU from the star and indicate that it is shedding gas as it nears the end of its life (Fig. 7.7).

For most stars, the radius is determined from the luminosity and temperature using the Stefan–Boltzmann law, which states that a star's luminosity increases with the square of its radius and the fourth power of its disk temperature. This is the temperature of a thermal (i.e., blackbody) gas emitting the observed luminosity, which is close to the temperature of the visible stellar disk, known as the photosphere. The radius is that of the photosphere, which is the level at which the stellar gases become opaque at visible wavelengths. However, appearances can be deceiving. All stars are gaseous, without a distinct edge or surface, and they are all enveloped in tenuous, transparent atmospheres through which we look.

7.6 How Massive Are the Stars?

The mass of a star usually is expressed in units of the mass of the Sun, $M_{\odot} = 1.989 \times 10^{30}$ kg, which is determined from Kepler's third law and the orbital period and distance of the Earth, or from the length of the year and the AU. The range in the mass of most stars is relatively small, between about 0.1 $M_\odot$ and 100 $M_\odot$. The Sun is on the lower side of the stellar mass range, as are most stars. The most massive stars are relatively rare due to their relatively short lifetimes.

Although there is not much variation between the masses of the stars, a little difference can become important. The mass of a star controls nearly everything, from a star's luminosity to its effective disk temperature and the length of its life and its ultimate fate. A small increase in a star's mass, for example, implies a big increase in its luminosity. Stars of lower mass have less weight pressing down on their core, so their core is cooler, the rate of their thermonuclear reactions is slower, and the stars are dimmer. The life span of stars also depends on their mass. The more massive a star is, the shorter its life span. A star of greater mass is more luminous, burns its nuclear fuel at a greater rate, and depletes its available energy in a shorter time. It is as if bigger fires are more luminous and last a shorter time.

At about 0.08 $M_\odot$, we reach the lower limit for a gaseous body to become a star. Its central regions are too rarefied and too cool to sustain the hydrogen-burning reactions that energize a Sun-like star and make it shine. Some of these nonstellar objects, known as brown dwarfs, can glow for a brief time as the result of heat generated during their formation by gravitational contraction. Although never hot enough for proton fusion, certain brown dwarf stars can shine for a time by burning deuterium that was present in the star at the time of its birth. The low-mass brown dwarfs eventually cool, compressing their near-stellar mass into the size of planets and disappearing from view.

As the mass increases, so does the central temperature. Nevertheless, the temperature inside a star cannot become too high. It can become so hot that it is blown apart from inside; this self-destruction sets an upper bound to the mass of any star at 120 $M_\odot$. As demonstrated by the English astronomer Arthur Eddington, it is the internal radiation pressure of such an exceptionally hot star that blows away its outer atmosphere.

The mass of a star can be determined by observing the motion of another body revolving around it. The mass of the Sun, for example, is established from the Earth's orbital period and distance. Although planets have been observed revolving about other nearby stars, the majority of stars are too distant for the observations of planetary-motion effects.

Fortunately, many stars are not content to be alone and are accompanied through their stellar life by a companion. A direct measurement of stellar mass can be obtained from observations of the relative motion of two stars in such binary-, or double-, star systems. The components of binary stars are in mutual orbit around one another, revolving about a common center of mass. If the orbital period and the distance separating the two stars are measured, for example, the sum of their masses can be determined from Kepler's third law.

Some members of binary-star systems are tens of thousands of AU apart, whereas others touch one another. Moreover, the binary stars come in at least four varieties. There are visual binaries the two components of which both can be resolved with a telescope and separately observed (Fig. 7.8); however, they are separated so widely that their orbital period about a common center of mass is greater than a human lifetime.

The periodic motion of just one component of an astrometric binary is observed, whereas its companion is too faint to be seen. An eclipsing binary is a pair of stars the orbital plane of which contains the Earth's line of sight, so we periodically observe the stars when they pass

Figure 7.8. Alpha Centauri. Two of the most brilliant stars in the southern sky appear as a single star, named Alpha Centauri, to the unaided eye, but they can be resolved into two stars with the aid of binoculars or a small 5-cm (2-inch) telescope. The yellowish Alpha Centauri A (*lower left*), also known as Rigil Kentaurus, and the blue Alpha Centauri B (*upper right*) are locked together in a gravitational embrace, orbiting each other every 80 years. The two components of this binary-star system can approach one another within 11.2 AU and may recede as far as 35.6 AU, where the mean distance between the Earth and the Sun is 1 AU. Both stars have a mass and a luminosity that are comparable to those of the Sun. They appear bright because they are very nearby, at a distance of just 4.37 light-years. A third and faint companion, Proxima Centauri (see Fig. 7.5), is located at about 15,000 AU, or 2.2 degrees, from the two bright stars. At a distance of 4.24 light-years from the Earth, Proxima Centauri is the closest star other than the Sun. (Courtesy of ESO/Yuri Beletsky.) (See color plate.)

in front of or behind one another (Fig. 7.9). A famous example of an eclipsing binary system is the two brightest stars in the Algol system: They have an orbital period of 2.87 days and a combined mass of about 4.5 solar masses. The two stars are so close to one another that the more massive and bigger component has entered the gravitational sphere of influence of the other, transferring mass to it.

Algol also is an example of the spectroscopic type of binary stars. In spectroscopic binaries, the stars usually are not resolved and their separations cannot be measured; however, oscillations in the line-of-sight velocities are inferred from Doppler shifts of spectral lines. These velocities reveal the orbital motion of the unresolved components and can be used to infer the individual masses of the two stars.

When the measured masses of stars are combined with observations of the stellar luminosity, we find that luminosity increases rapidly with increasing mass. The reason for this increase is the hotter temperature at the center of a high-mass star when compared to that of

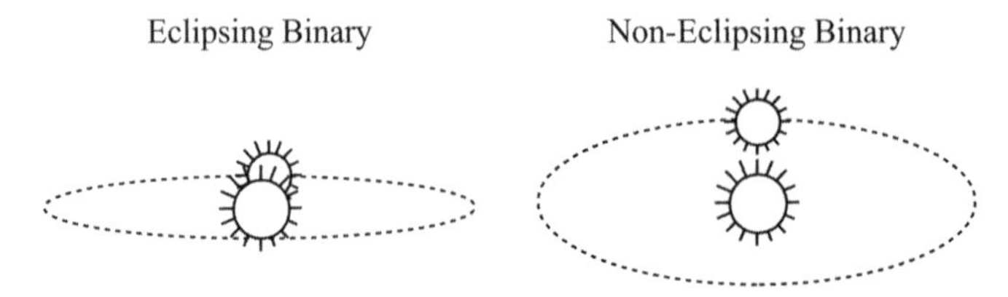

Figure 7.9. Double stars. Two close stars are joined in a gravitational embrace, orbiting each other and forming a binary-star system (*left* and *right*). The orbital period and linear star separation can be used to determine the sum of their masses. If the orbital plane of the two companions is sufficiently inclined and within the line of sight, the star system becomes an eclipsing binary (*left*), in which one star is observed to pass behind the other and vice versa; this can provide additional information about the stars.

a low-mass star. The rate of nuclear reactions is greater at the higher temperature; therefore, the luminosity of the massive star is greater. The luminosity of most stars increases in rough proportion to the fourth power of the mass (Fig. 7.10).

As might be expected, bigger stars are more massive, and there are fewer stars of high mass than those with low mass. The distribution of stars relative to mass is known as the *initial mass function*, with the term *initial* meaning the mass with which the stars were formed before their subsequent evolution. The mass function for stars more massive than the Sun was first derived by the Austrian-born American astronomer Edwin Salpeter. It indicates that the number of stars falls off roughly as the inverse square of the mass and that the star-formation process results in many more stars of low mass than high mass. When compared to the number of stars with a mass equal to that of the Sun, denoted $M_\odot$, there are roughly 100 times more stars with 1/10th of that mass, at 0.1 $M_\odot$, and about 1/100th fewer stars with a mass of 10 solar masses, or 10 $M_\odot$. Stars of higher mass also are bigger, whereas those of low mass are relatively small; the small stars outnumber the large.

7.7 Motions of the Stars

Are the Stars Moving?

To the ancients, the stars seemed firmly rooted in the night sky, cemented into place on the celestial sphere. That is why they always appeared at the same place in the night sky, forever maintaining fixed positions relative to one another and never changing their apparent separations. These unchanging locations enabled the identification of long-lived patterns among groups of stars located close together in the sky and known as the *constellations*.

Looking up when outside at night, we might notice that the stars rise, move slowly across the dark sky, and then disappear from view. This slow movement of the stars is due to the rotating Earth, spinning beneath the celestial sphere with its fixed stars that do not move. Despite eons of stellar observations in antiquity, there was not a shred of evidence to contradict the ancient Greek belief that the stars are motionless.

Yet, if the stars were motionless, their mutual gravitation eventually would pull them together into a single mass. Without motion, there would be nothing to keep the stars apart,

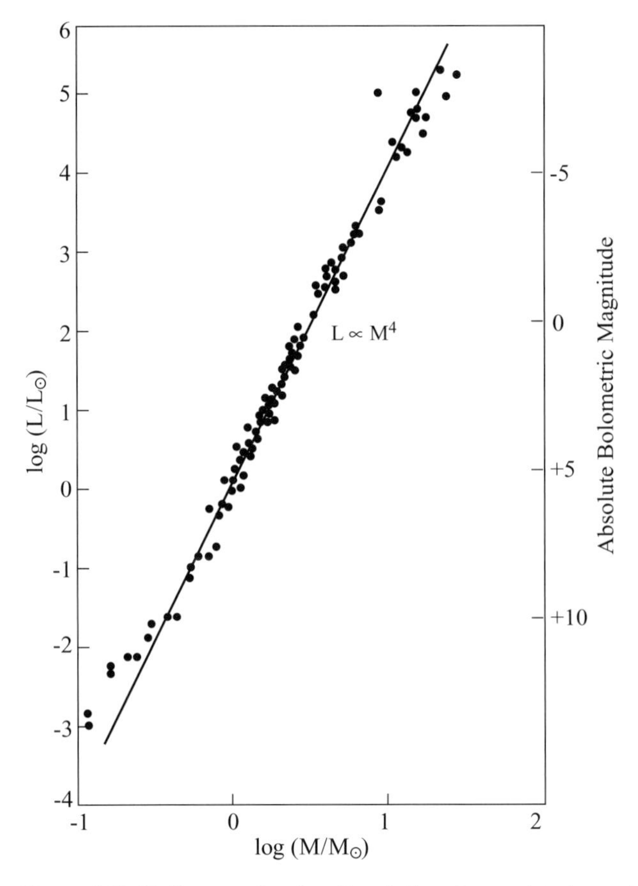

Figure 7.10. Stellar mass-luminosity relation. An empirical mass-luminosity relation for main-sequence stars of absolute luminosity, L, in units of the solar luminosity, $L_{\odot}$; and mass, M, in units of the Sun's mass, $M_{\odot}$. The straight line corresponds to luminosity that is proportional to the fourth power of the mass. The English astronomer Arthur Eddington (1882–1944) proposed a theoretical explanation for this relation in 1924.

and they could not be suspended in space. So, the apparently serene universe – which is frozen into a nighttime glimpse of the stars – is an illusion. There is no star that is completely at rest, and the stars must be moving ever so slightly from their apparent places in the night sky.

Moreover, the speeds of the moving stars are not modest. Observations indicate that the stars are moving around at speeds of about 10 km s^{-1} relative to their stellar neighbors. The Sun, for example, is currently traveling at a speed of about 20 km s^{-1}, or 20,000 m s^{-1}, relative to other nearby stars. This is about 1,000 times faster than an automobile moves on a highway.

Stars also move together at larger speeds in directed motions. Both the Sun and nearby stars, for example, are whirling about the remote center of the Milky Way at a speed of 220 km s^{-1}. If these stars traveled at faster speeds, they would move off into space, even out of the Milky Way; if they were moving at slower speeds, they would be pulled by gravitation into a million-solar-mass black hole at the center of the Milky Way.

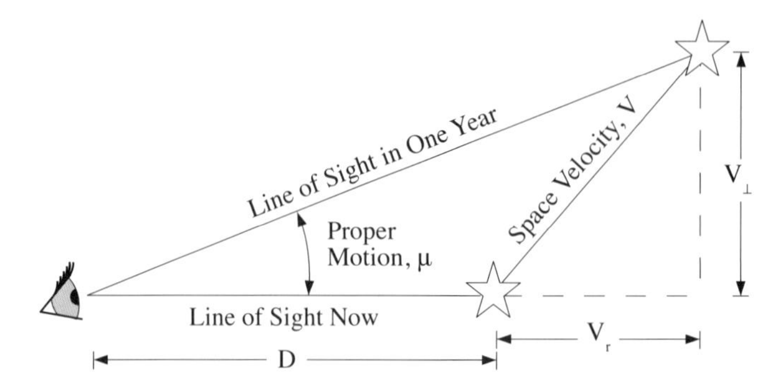

Figure 7.11 A star moves. The space velocity, V, of a star relative to an observer can be resolved into two mutually perpendicular components: (1) the radial velocity, V_r, directed along the line of sight; and (2) the tangential velocity, $V_\perp$, which is perpendicular or transverse to the line of sight. From the Pythagorean theorem $V^2 = V_r^2 + V_\perp^2$. Over a given interval of time, shown here as one year, the star will move through a proper-motion angle μ, which depends on $V_\perp$ and the star's distance, D, from the observer. In this figure, the proper motion $\mu = V_\perp/D$ is exaggerated greatly by more than 10,000 for even the closest star. At a distance of only 6 light-years, Barnard's star has the largest known proper motion of 10.3 seconds of arc per year.

Because they are so far away, the stars seem to be moving gently and slowly through space, like leaves falling from a tree or an airplane traveling far above to a distant country. The moving stars therefore only gradually change their apparent separation and grouping; nevertheless, this means that the constellations are temporary and eventually will disappear, and new ones inevitably will form. At some remote future time, the Big Dipper will tip out its imaginary contents. We will never notice the changes, which occur over eons, not during a human lifetime.

The stars move but what started the motion in the first place is not always obvious. It might be related to their origin or to subsequent events, such as a nearby exploding star or a close encounter with another star. A more poetic answer was provided by Dante Alighieri in the last line of his *Divine Comedy*: Led by Beatrice to the brilliant, endless radiance of paradise, he saw "the Love which moves the Sun and other stars."[17]

Components of Stellar Velocity

Stars seem to be moving here, there, and everywhere, so it is not easy to figure out where they are going. However, a star's motion manifests in two ways, depending on the method used to observe it, and these two components of velocity can be combined to give the direction of motion (Fig. 7.11). The "sideways" velocity component is directed perpendicular or transverse to the line of sight; the other component, the *radial velocity*, is the velocity moving toward or away along the line of sight to the star. When these two velocity components are known, we can determine the speed and direction of a star in three dimensions.

A star's motion across the line of sight produces an angular change in position, called *proper motion*, which depends on both the star's distance and the perpendicular speed. The radial velocity is observed through the Doppler effect, which measures how the star's spectral lines appear to shorten or lengthen in wavelength depending on the relative velocity of the star and observer, and whether the motion is toward or away from the observer. When a star is moving directly away, then there is no perpendicular component to its motion; if a star is moving directly across our line of sight, then the radial component of the star's motion is reduced to zero.

It is difficult to judge a star's speed if it is headed straight toward or away from us, just as it is difficult to determine how fast a distant automobile is moving on a highway. However, if a star crosses at right angles to our line of sight, we should see a change in its position. To detect that change, astronomers needed to look at nearby stars where the angular change in position is greatest.

Assuming that all stars move through space at roughly the same speed, those closest to the Earth should display the largest shift in position during a given period of time. A nearby bird flying overhead similarly travels rapidly across a great angle, whereas a high-altitude bird moving at the same speed barely creeps across the sky. This is how a hunter estimates the distance of a duck – by its angular speed.

Given enough time, the displacement of a nearby star's celestial position can be detected. In 1717, for example, the English astronomer Edmond Halley first noticed the change when he compared the positions of extremely bright stars (e.g., Sirius and Arcturus), with those measured by the Greek astronomer Hipparchus around 150 BC and recorded in Ptolemy's reproduction of Hipparchus' catalogue. Halley's comparison indicated that at least three stars had changed position and moved. So, it took more than 1,800 years before anyone noticed that a star could move. Nowadays, with vastly improved technology and observations from spacecraft above our distorting atmosphere, the motions of many tens of thousands of stars are known with great accuracy.

Proper Motion

The stellar motion that Halley detected is an angular change in a star's position over time, due to its velocity transverse or perpendicular to the line of sight. The angular rate of change is known as *proper motion*, which is intrinsic to the star and belongs to it, in contrast to any improper motion that might be caused by the Earth's movement in space.

Proper motion is not a velocity; it is the angular rate at which a star moves across the sky over years or centuries, and it does not alone determine the speed of motion. To convert a star's proper motion into a velocity or speed, we must know the star's distance; in Halley's time, no one knew the distance of any star other than the Sun.

Radial Velocity

The other component of a star's velocity, the *radial velocity* directed along the line of sight, can be measured using the Doppler shift of a spectral feature in the star's radiation. If the motion is toward the observer, the shift is to shorter wavelengths; when the motion is away, the wavelength becomes longer. The greater the speed along the line of sight, the bigger the wavelength shift.

Observed Proper Motions of Stars

The star with the largest proper motion races across the sky at about 10.4 seconds of arc, denoted as $10.4''$, each year. This is Barnard's star, named after the American astronomer

Table 7.6. *Stars with the highest proper motion*[a]

Star	$\mu_\alpha \cos\delta$ (mas yr^{-1})	μ_δ (mas yr^{-1})	Parallax (mas)	Radial Velocity (km s^{-1})
Barnard's star	−798.71	10,337.77	549.30	−110.6
Kapteyn's star	6,500.34	−5,723.17	255.12	+245.5
Groombridge 1830	4,003.69	−5,814.64	109.22	−98.0
Lacaille 9352	6,766.3	1,327.99	303.89	+9.7
Gliese 1 (GJ 1)	5,633.95	−2,336.69	229.32	+23.6

[a] The proper motions are in units of milliarcseconds per year, or 10^{-3} ″ yr^{-1}, and abbreviated mas yr^{-1}; the + or − sign of the radial velocity indicates motion away or toward the observer, respectively. Astronomers specify the proper motion μ_α in right ascension α and the proper motion μ_δ in declination δ. The magnitude of the total proper motion, μ, is given by the vector addition of its components $\mu^2 = \mu_\delta^2 + \mu_\alpha^2 \cos^2 \delta$, where the *cos* δ factor accounts for the projection of μ_α on the celestial sphere.

Edward E. Barnard, who discovered it in 1916. In our lifetime, it will move by roughly half the angular diameter of the Moon; however, because it is a dim, faint star, a telescope is required to see it. Barnard's star is moving at a speed of about 100 km s^{-1}. Stars with the highest proper motions are listed in Table 7.6.

Most proper motions are exceedingly small and usually measured in seconds of arc per century, or milliarcseconds per year, which means the same thing. However, the effect is cumulative; therefore, successive generations of astronomers can measure proper motion. After 20 centuries, the proper motion of many stars might be 20″, which explains why Halley was able to detect the effect using ancient observations. Due to atmospheric blurring, the angular resolution of the best telescope at the best location on the Earth is only about 0.2″, and we would have to wait more than 20 years to measure a proper motion of this size.

It is much easier to measure proper motion from space, outside the Earth's atmosphere. Instruments aboard the *HIPPARCOS* satellite have pinpointed the positions and established the proper motions of more than 100,000 stars, with an astonishing precision of 0.001″. The mission also obtained stellar distances using the parallax techinque; therefore, perpendicular velocities can be inferred from the proper motions.

Motions in Star Clusters

Gravitation can constrain the paths of stars that are congregated within star clusters (Table 7.7). As many as 1 million stars, for example, are crowded together in a typical *globular star cluster*. The cluster is tightly bound by gravity, which gives it a distended spherical shape as well as a relatively high stellar density toward the center (Figs. 7.12, 7.13, and 7.14). The name of this category of star cluster is derived from the Latin *globules* for "a small sphere." Another type of stellar grouping, known as an *open star cluster*, includes as many as a few thousand stars that were formed at the same time but are bound only loosely to one another

Table 7.7. *Physical properties of star clusters*

Open Star Cluster
N_S = total number of stars = 100 to 1,000
R_C = radius = 3.2 to 32.6 light-years = 1 to 10 pc $\approx$ (3 to 31) $\times$ 10^{16} m
Age = 10^7 to 10^9 yr
Globular Star Cluster
N_S = total number of stars = 10^4 to 10^6
R_C = radius = 32.6 to 326 light-years = 10 to 100 pc $\approx$ (3 to 31) $\times$ 10^{17}m
Age = (10 to 20) $\times$ 10^9 yr = 10 to 20 Gyr

by mutual gravitational attractions (Figs. 7.15 and 7.16). Unlike a globular star cluster, which can be held together by its stars' mutual gravitational pull for tens of billions of years, an open star cluster will disperse within a few million years.

The stars in a globular cluster are moving in all directions, like a swarm of bees, and their collective stellar motions oppose the combined gravitational attraction of all of the stars, preventing them from gathering together and collapsing to the center of the star cluster.

Figure 7.12. Globular star cluster M 80. Hundreds of thousands of ancient stars are held together by their mutual gravitational attraction in this globular star cluster, which is designated M 80 or NGC 6093 and is located about 28,000 light-years from the Earth. Most of the stars displayed in this Hubble Space Telescope image are about as old as the observable universe, with ages of nearly 14 billion years – much older than our Sun, which is 4.6 billion years old. Stellar collisions result in more massive and relatively young "blue straggler" stars in the dense core of the globular cluster. (Courtesy of NASA/AURA/STScI/Hubble Heritage Team.) (See color plate.)

Figure 7.13. Faint stars in a globular cluster. This five-day exposure from an instrument aboard the Hubble Space Telescope includes the faintest detectable stars in the globular star cluster NGC 6397, which is located about 8,500 light-years away from the Earth. Some of these objects are white dwarf stars – the collapsed, burned-out relics of former stars like the Sun. White dwarfs cool down at a predictable rate, which can be used to measure the age of this globular cluster, estimated to be about 12 billion years. The crossed lines radiating from the bright stars are diffraction spikes caused by the struts that support the telescope mirror. (Courtesy of NASA/ESA/Harvey Richer, University of British Columbia.)

In a short elegant discussion in 1916, the great English astronomer Arthur Eddington demonstrated that one could infer the average speed of stellar motion in a star cluster by assuming that the stars behave like the atoms or molecules in a gas. The average kinetic energy of the moving stars is simply equal to half the combined gravitational potential energy of all of the stars in the cluster; this specifies the equilibrium speed. If the stars move on average at a slower speed, they will be pulled gravitationally into each other and the cluster will collapse like a failed soufflé. If the stars move at a faster speed, they eventually will disperse because the cluster cannot hold together for very long. This is what is happening to open star clusters and to star associations that are bound even more loosely. In fact, some stars now are moving out of certain stellar associations at unexpectedly rapid speeds.

Runaway Stars

Some stars have "gone ballistic," racing through space like bullets with an abnormally high velocity relative to the surrounding interstellar medium. These high-speed stars are known as *runaway stars* because they are moving away from their place of birth, like children running away from home. They are former members of very loose star clusters, known as *stellar associations*, which contain massive, luminous, and relatively hot and young

Figure 7.14. Globular star cluster NGC 6934. Several hundred thousand stars swarm around the center of the globular star cluster NGC 6934, which lies at a distance of about 50,000 light-years from the Earth. These ancient stars are estimated to be about 10 billion years old. This sharp image, obtained from the Hubble Space Telescope, is about 3.5 minutes of arc and 50 light-years across. (Courtesy of NASA/ESA.)

Figure 7.15. Open star cluster NGC 265. A brilliant cluster of bright blue-colored stars is located in the Small Magellanic Cloud, about 200,000 light-years away and about 65 light-years across. This Hubble Space Telescope image subtends an angle of about 70 seconds of arc. (Courtesy of NASA/ESA/E. Olszewski, University of Arizona.) (See color plate.)

Figure 7.16. Pleiades star cluster. These luminous blue-colored stars are members of the open star cluster known as the Pleiades, or Seven Sisters, as well as M 45. It can be seen by the unaided human eye without binoculars or a telescope. The Pleiades contains about 3,000 stars and is only 440 light-years away and about 13 light-years across. Interstellar dust reflects the blue light of the hot stars, creating the hazy nebulous appearance. The telescope causes the prominent, cross-shaped diffraction spikes. Astronomers estimate that the young Pleiades stars were formed just 100 million years ago and that the cluster will survive for another 250 million years before dispersing. This color-composite image was taken in 1986 with a 48-inch Schmidt telescope as part of the second Palomar Observatory Sky Survey. (Courtesy of NASA/ESA/AURA/Caltech.)

stars that are designated as O and B stars. Most of these very massive O and B stars congregate together in OB associations.

As first noticed by the Armenian astronomer and statesman Viktor Ambartsumian, in the mid-twentieth century, stars in OB associations are expanding away from one another and from a common origin, dispersing and disintegrating but still moving together in a roughly spherical shape due to their relatively young age. They are not expected to stay together for longer than a few tens of millions of years.

The runaway stars are moving with faster speeds than the other stars in the OB associations but with proper motions that often point away from the stellar association to which they once belonged. In 1961, the Dutch astronomer Adriaan Blaauw called attention to the fact that the runaway stars are very massive stars whose high space velocities are comparable to the orbital velocities expected for massive binary-star systems. He proposed that the runaway stars are escaped members of former binary-star systems. Because massive stars burn their nuclear fuel faster and have a shorter lifetime than normal stars, one member of such a binary system will quickly exhaust its thermonuclear reserves and explode as a supernova, thereby releasing the other member as a high-velocity star.

The HST has captured striking images of runaway stars plowing through regions of dense interstellar gas and creating brilliant bow-shock structures, trailing tails of glowing gas (Fig. 7.17). These features are formed when the stars' powerful stellar winds slam into the surrounding gas, somewhat like a speeding boat plowing through water on a lake. The bullet-nosed shocks indicate that these runaway stars are traveling at speeds between 50 and 100 km s^{-1} relative to the dense gas through which they are moving. This is five or ten times

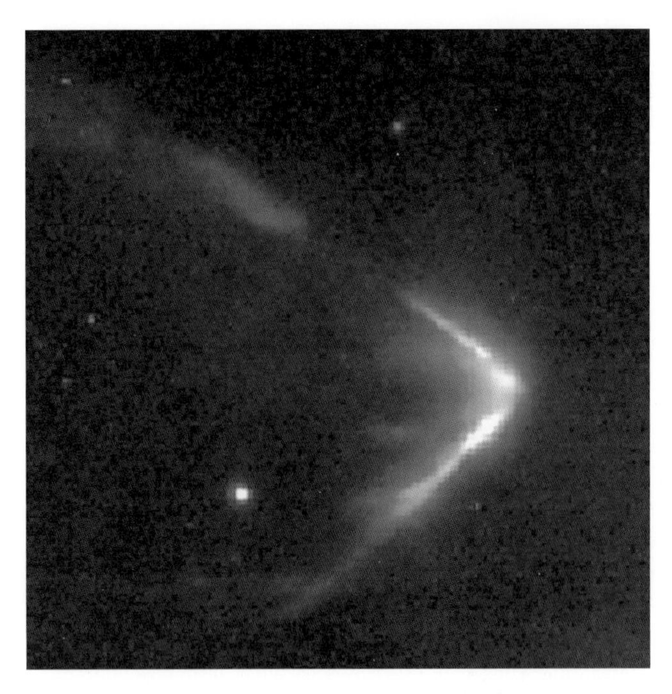

Figure 7.17. Runaway star. A high-speed star slams into dense interstellar gas, creating a bow shock that may be a million million kilometers wide. The star is thought to be relatively young, only millions of years old. Moving at a speed of about 100 km s^{-1}, it has journeyed 160 light-years since its birth, most likely in a loosely bound stellar association. (Courtesy of NASA/ESA/R. Sahai/JPL.)

faster than the expansion speeds of the stellar associations or the average speeds of stellar motions relative to nearby stars or the local interstellar medium.

Stellar Rotation

In addition to moving through space, a star also spins or rotates about its axis. Every known cosmic object rotates, with rotation periods that tend to decrease with size (Table 7.8). This rotation often can be traced back to an origin from a more distended object of slower spin but sometimes is related to a glancing collision in the past.

Sunspot observations have long shown that the Sun rotates with a speed of about 2 km s^{-1} at its equator, with a slower rotation speed nearer its poles. The Sun spins about its axis with a period of about 25 days at the equator, which corresponds to a rotation speed of about 2 km s^{-1} or roughly 7,200 km per hour. But since the Earth is orbiting the Sun in the same direction that the Sun rotates, the solar rotation period observed from the Earth is about 27 days. The shorter rotation period is the star's true rotation period, and is technically known as the sidereal rotation period, from fixed star to fixed star; the longer period required for a fixed feature to rotate back to the same position as viewed from Earth, is called the synodic rotation period.

The rotation of stars other than the Sun is inferred from the Doppler broadening of their spectral lines; however, astronomers did not realize at first that some stars rotate fast enough for such measurements to be meaningful.

Table 7.8. *Rotation periods and rotation velocities of some planets and stars*

Object	Rotation Period	Radius[a] (m)	Rotation Velocity[a] ($m\ s^{-1}$)
Earth	0.99727 Earth days[b]	6.378×10^6	4.651×10^2
Earth's Moon	27.322 Earth days[c]	1.738×10^6	4.627
Mercury	58.6462 Earth days	2.440×10^6	3.026
Venus	–243.018 Earth days	6.052×10^6	1.81
Jupiter	9.9249 hours	7.149×10^7	1.26×10^4
Saturn	10.6562 hours	6.027×10^7	9.87×10^3
Sun (equator)	25.67 Earth days	6.955×10^8	1.97×10^3
Vega	12.5 hours	1.933×10^9	2.74×10^5
White Dwarf Star[d]	186.5 seconds	6.378×10^6	2.02×10^5
Crab Pulsar	0.033 seconds	10^4	1.90×10^6

[a] The equatorial radius is given when the object has a known oblate shape; in this situation, the equatorial rotation velocity is provided.

[b] One Earth day is defined as the time for our planet to revolve once with respect to the Sun; such a day is 24 hours, or 86,400 seconds, long. The Earth's rotation period with respect to stars, or sidereal time, runs 4 minutes slower each day, lasting 23 hours 56 minutes 04 seconds, or 8.6164×10^4 seconds.

[c] The sidereal rotation period of the Earth's Moon, from fixed star to fixed star, is 27.322 Earth days. The time from new Moon to new Moon, known as the synodic Month, is 29.53 Earth days.

[d] The radius of a white dwarf star is assumed to be equal to that of the Earth. Its rotation period is inferred from the rotation period of the Sun and conservation of angular momentum, in which the period scales as the inverse square of the radius.

The Doppler effect of atoms moving in the hot stellar atmosphere once was thought to be substantially greater than that of stellar rotation. The turning point came in a seminal paper published in 1929 by two Russian astronomers, Grigory Ambramovich Shajn and Otto Struve, who showed that relatively young stars rotate faster than older ones and therefore exhibit exceptionally broad spectral lines. They concluded that some young stars have equatorial rotational velocities ranging up to 100 km s^{-1}.

Stars with rapid rotation are massive, hot, and young, and some of the brightest stars in the night sky. They include Achernar, Alpha Arae, Pleione, and Vega, with respective equatorial rotational velocities of about 300, 470, 329, and 274 km s^{-1}. These stars are rotating so rapidly that their equators bulge outward, giving them a flattened shape. Achernar, for example, is thought to have an equatorial diameter that is about 50 percent greater than the distance between its poles.

Less massive, cooler, and older stars like the Sun rotate with much slower speeds of 10 km s^{-1} or less. These stars most likely were formed with fast rotations like the more massive stars but have slowed down as they aged. Stellar magnetic fields coupled to the surrounding interstellar material act as magnetic brakes over long time intervals. The Sun, for example, probably originated 4.6 billion years ago with a rotation velocity of about 100 km s^{-1}; its magnetism helps to explain why it is now rotating at about 2 km s^{-1}.

8

The Lives of Stars

8.1 Main-Sequence and Giant Stars

The Hertzsprung–Russell Diagram

Once the luminosity of stars was obtained from their brightness and measurements of their distance, astronomers were able to show that most stars exhibit a systematic decrease in luminosity as one progresses through the spectral sequence O, B, A, F, G, K, M. (These spectral types are described in Section 7.4.) This progression is exactly what we would expect because the spectral sequence also denotes a scale of decreasing stellar temperatures, and the luminosity of a radiating body depends strongly on temperature.

The luminosity drop is illustrated in the famous Hertzsprung–Russell (H–R) diagram of luminosity or absolute magnitude plotted against the spectral class or effective temperature. The diagram's name derives from the Danish astronomer Ejnar Hertzsprung, who first plotted such diagrams for the Pleiades and Hyades star clusters, and the American astronomer Henry Norris Russell, who next published early versions of this diagram for both cluster and noncluster stars (Fig. 8.1).

Most stars, including the Sun, lie on the main sequence that extends diagonally from the upper left to the lower right, or from the high-luminosity, high-temperature blue stars to the low-luminosity, low-temperature red stars. The stars on the main sequence are the most common type in the Milky Way, constituting about 90 percent of its stars.

The Stefan–Boltzmann law describes the general characteristics of the H–R diagram. This expression indicates that for a fixed radius, the luminosity of a star increases with the fourth power of the temperature; therefore, hotter stars are more luminous. That is exactly what happens along the main sequence, for although the radius varies by a relatively small amount along the main sequence, the luminosity variation is due mainly to a change in temperature.

The observations also showed a different, unanticipated effect that gives the H–R diagram a peculiar shape. Some of the stars retain a high luminosity at decreasing temperature, in a band that extends to the upper right of the H–R diagram (see Fig. 8.1). This could be explained if the luminous cool stars were larger in radius than the less luminous ones, with the increase in size offsetting the drop in temperature. The Stefan–Boltzmann law indicates that for a fixed temperature, the luminosity of a star increases with the square of the radius.

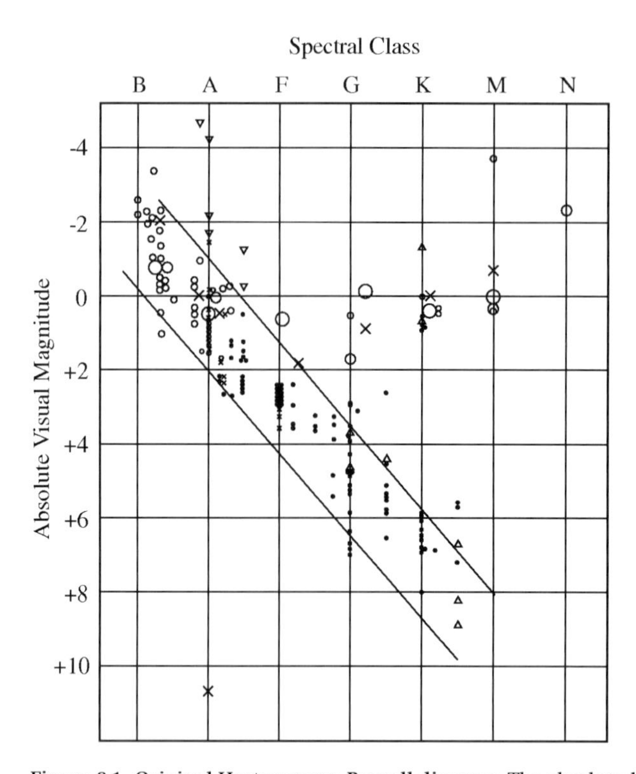

Figure 8.1. Original Hertzsprung–Russell diagram. The absolute luminosity, in magnitude units (*vertical axis*) plotted in 1914 by Henry Norris Russell (1877–1957) as a function of spectral class (*top horizontal axis*) for four moving star clusters: the Hyades (*black dots*), the Ursa Major group (*small crosses*), the large group in Scorpius (*small open circles*), and the 61 Cygni group (*triangles*). The large circles and crosses represent points calculated from the mean parallaxes and magnitudes of other groups of stars. The two diagonal lines mark the boundaries of Ejnar Hertzsprung's (1873–1967) observations of the Pleiades and Hyades open star clusters in 1911; this now is known as the main sequence along which most stars, including the Sun, are located. The giant stars are located at the upper right. In his publication, Russell included a similar diagram for individual bright stars the distances of which had been established from stellar parallax measurements. It closely resembled the diagram shown here with an exceptional point in the lower left-hand corner, which is included here with an "x" mark. This star is the faint companion of a double-star system Omicron^2 Eridani, or 40 Eridani, now known to be a white dwarf star. (Adapted from Henry Norris Russell, "Relations Between the Spectra and Other Characteristics of Stars," *Popular Astronomy* 22, 275–294 [1914].)

Russell realized that he had found another type of star, which he named *giants* for their large size. He also dubbed the more numerous main-sequence stars *dwarfs* because they are smaller than the giants, but the designation is confusing. There is no observable difference between the size and luminosity of the hottest dwarf and most giant stars, and the white dwarf stars are not even on the main sequence. In this book, therefore, we retain the designation *giant stars* but use the term *main-sequence stars* for the other stars.

There are relatively few giant stars when compared to the number of stars on the main sequence. This is because stars spend the majority of their lifetime on the main sequence, and the giant stars belong to a subsequent and shorter-lived part of a star's evolution.

Nearly a century of increasingly accurate and extensive observations confirmed the initial characteristics of the H–R diagram (Fig. 8.2). To assist physical interpretations, the

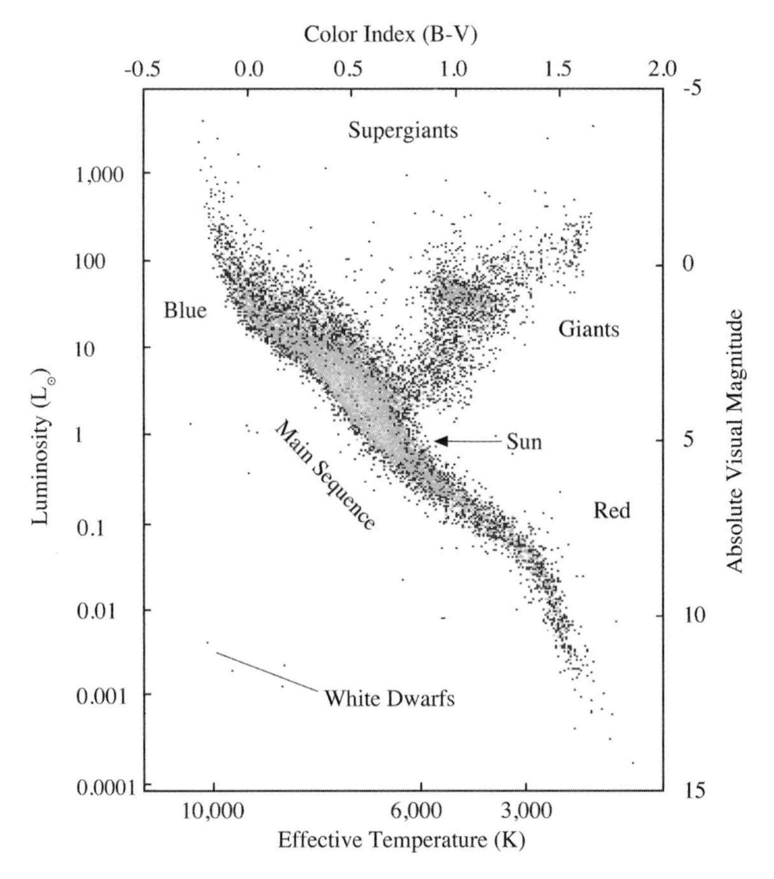

Figure 8.2. Hertzsprung–Russell diagram for nearby stars. A plot of the luminosity (*left vertical axis*) in units of the Sun's absolute luminosity, denoted $L_\odot$, against the effective temperature of the star's disk in degrees kelvin, designated K (*bottom horizontal axis*) for 22,000 stars in the catalogue of the *HIPPARCOS* satellite. This plot is known as the Hertzsprung–Russell (H–R) diagram. The absolute visual magnitude (*right vertical axis*) and color index, B–V (*top horizontal axis*) are also designated. Most stars, including our Sun, lie along the main sequence, which extends from the high-temperature blue-white stars at the top left to the low-temperature red stars at the bottom right. The Sun is a main-sequence star with an absolute visual magnitude $M_V = 4.8$ and color index B–V = 0.68. The radiation from all main-sequence stars is sustained by hydrogen-burning reactions in their core. Stars of about the Sun's mass evolve into helium-burning red giant stars, located in the upper-right side of the diagram. Very rare bright giant stars and extremely scarce and luminous supergiants are found above the giant stars and along the top of the diagram. Faint and initially hot white dwarf stars are located in the lower left side. Due to their low luminosity, these endpoints of stellar evolution are relatively difficult to observe. (Datapoints courtesy of the ESA/*HIPPARCOS* mission.)

luminosities are displayed along the left vertical axis, and the color index or, equivalently, the effective temperature of the stellar disk is on the bottom horizontal axis.

The Luminosity Class

There was an unresolved uncertainty in the H–R diagram, which created a dilemma for specifying the physical characteristics of a star. A star could be small or large as well as hot or cold. A red cool star, for example, might be either much more luminous than the Sun

Table 8.1. *The Morgan–Keenan, abbreviated M–K, luminosity classes*

Ia	Bright supergiants
Ib	Supergiants
II	Bright giants
III	Giants
IV	Subgiants
V	Main-sequence stars (or dwarfs)
VI	Subdwarfs (or SD)
D (or VII)	White dwarfs

or much fainter. Once the spectral type establishes the temperature, the star is on either the luminous giant or the dimmer main-sequence part of the H–R diagram. To resolve this ambiguity, astronomers found a way of classifying stars by their luminosity in addition to their spectral type.

Pioneering investigations by the American astronomer Walter Adams and the German astronomer Arnold Kohlschütter found that the relative intensities of certain neighboring spectral lines could be used to determine the luminosities of both main-sequence and giant stars.

In the mid-twentieth century, William W. Morgan and Philip C. Keenan of the Yerkes Observatory in Chicago introduced the M–K system in which the most luminous and largest stars have the lowest numbers, given in Roman numerals (Table 8.1). In the M–K system, the Roman numeral III designates giant stars and V denotes the main-sequence stars; the Class IV of subgiants is located between them. The most luminous Class I stars are the supergiants, shown near the rarely occupied, upper edge of the H–R diagram. Both the spectral type and the M–K luminosity class can be specified in the H–R diagram (Fig. 8.3).

Because the spectral type O, B, F, G, K, or M depends solely on the physical properties of a star's outer atmosphere – the photosphere – it is not sufficient to determine the star's internal properties and evolutionary status. To solve this problem, both the spectral type and luminosity class are provided in a two-dimensional scheme; for example, the Sun is designated as G2V.

If we know a star's luminosity class, we can find its luminosity, or absolute magnitude, and a star's distance can be inferred from the apparent magnitude and luminosity. This is known as the *spectroscopic distance*, or *spectroscopic parallax*.

Life on the Main Sequence

Russell thought that his diagram showed stars caught at different stages in their life and that most stars began life as hot, blue-white stars and ended their life as cool red ones, moving from upper left to lower right along the main sequence. That is reasonable and it even seems inevitable. If something starts out hot, it is going to cool with time like some human relationships.

Astronomers of the time also speculated that the giants are the youngest stars now in the process of gravitational collapse, causing them to become smaller and hotter and to move

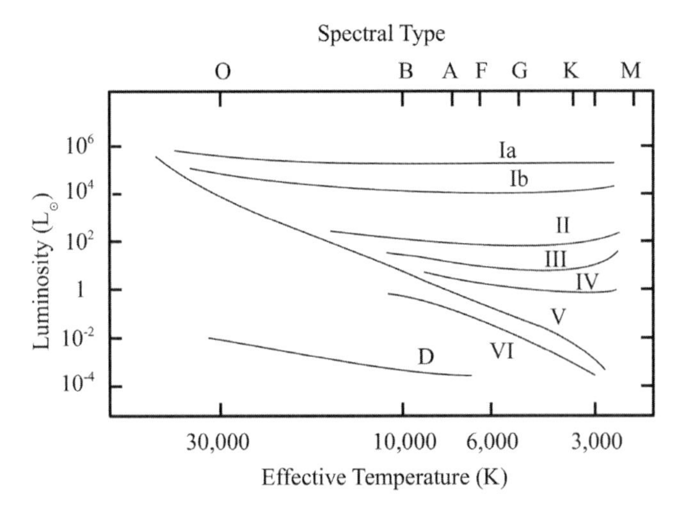

Figure 8.3. Spectral type and luminosity class in the H–R diagram. When both a star's spectral type (*top horizontal axis*) and luminosity class (*Roman numerals*) are known, the star's luminosity (*left vertical axis*) in units of the Sun's luminosity, denoted $L_\odot$, and effective disk temperature (*bottom horizontal axis*) can be obtained. The spectral types are shown at the coolest temperature for each type. A Roman numeral V designates the main-sequence stars: subgiants by IV, giants by III, bright giants by II, and supergiants by Ia and Ib. VI or SD denotes the subdwarfs, and D or VII designates the white dwarf stars.

leftward along the upper part of the H–R diagram. Therefore, there were supposed to be stars with rising temperatures (the collapsing giants) and stars with falling temperatures (the main sequence) with the maximum-temperature, A-type stars in between. They were thought to be the youngest "early-type" stars on the main sequence.

Today, the H–R diagram remains a primary tool for tracing the path of stellar evolution, but the routes are more complex than Russell imagined. Once scientists understood the ways that nuclear fusion makes a star shine, the early speculations proved to be dead wrong. The main sequence is not a singular evolutionary pathway, as once thought; it is simply a portrait of the sky at one moment, depicting different stars of varying mass. The giant stars represent later rather than earlier stages in a star's life cycle. As it turns out, a star begins its bright shining life on the main sequence.

Like our Sun, other stars on the main sequence generate energy by converting hydrogen into helium; as long as it shines in this way, a star's position on the main sequence does not change substantially. It simply slowly becomes more luminous and moves slightly to the upper right of the H–R diagram. Moreover, the different positions along the main sequence are closely related to only one property of the stars – their mass – and not to their different evolutionary status. The mass sets the central temperature and nuclear-fusion rate at which the outward pressure and the inward gravitational force remain in balance.

The stellar masses decrease downward from upper left to lower right on the main sequence (Fig. 8.4). The high-mass stars are more luminous than the low-mass stars because the central temperatures of the former are higher, to support the greater mass, and their nuclear-reaction rates are faster, producing radiation of much greater luminosity. The hot, luminous

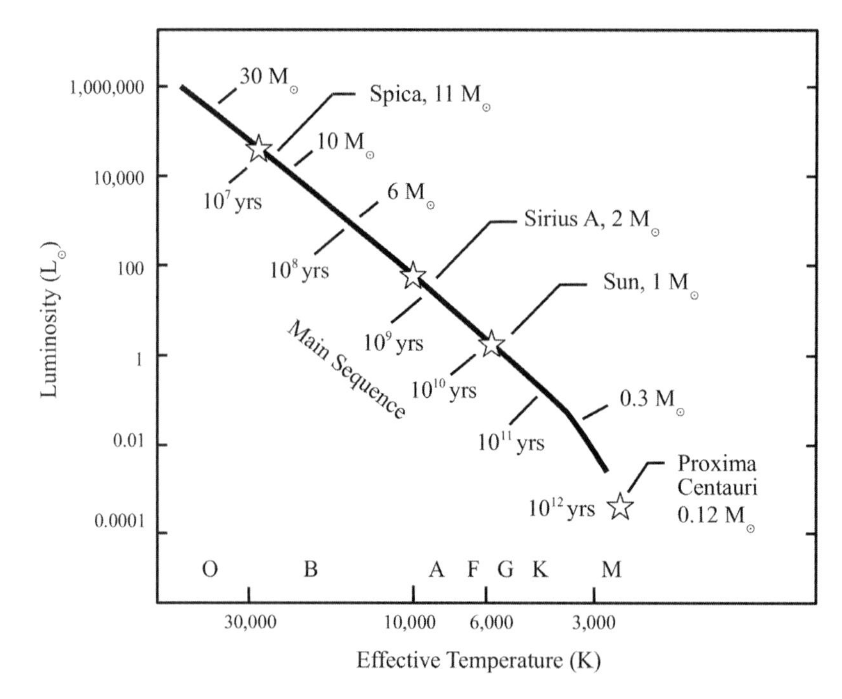

Figure 8.4. Stellar mass and lifetime on the main sequence. The relation between a star's luminosity (*left vertical axis*), in units of the Sun's luminosity, denoted $L_\odot$, and the star's effective disk temperature (*bottom horizontal axis*) in degrees kelvin, designated K, for the main-sequence stars in the Hertzsprung–Russell diagram. The stellar masses given along the main-sequence curve are in units of the Sun's mass, denoted $M_\odot$. Stars of higher mass are hotter and more luminous. All of these stars shine by hydrogen burning with a lifetime that also is denoted along the main-sequence curve. More massive stars burn their hydrogen fuel at a faster rate and have a shorter lifetime.

O stars can have masses as high as 150 times that of the Sun, whereas the cool, dim, main-sequence M stars might have as little as 0.08 solar mass.

All of these main-sequence stars are shining by converting hydrogen into helium. The effective disk temperature, mass, luminosity, radius, and lifetime of main-sequence stars of different spectral types are listed in Table 8.2.

Because a star begins shining with a limited supply of hydrogen, it can remain on the main sequence for only a limited lifetime; that is, the time it takes to deplete all of the hydrogen fuel in its hot core. Although more massive stars certainly contain more hydrogen in their larger core, they are much hotter inside and fuse this hydrogen into helium at a faster rate, resulting in a shorter life. The main-sequence lifetimes range from a few million to 100 billion years from spectral type O5 to M5.

Very massive stars have remarkably short lifetimes, cosmically speaking, which helps explain why we see relatively few massive stars. That is, because the nuclear energy is being radiated away at a faster rate for the more luminous and massive stars, their lifetime on the main sequence is shorter. Most of the very massive stars that ever were born have already extinguished their nuclear fires. In fact, it is so difficult to find exceptionally massive O-type stars that they are not even visible on many H–R diagrams. Stars of lower mass, which account for most of the stars on the main sequence, spend almost all of their life there. Some

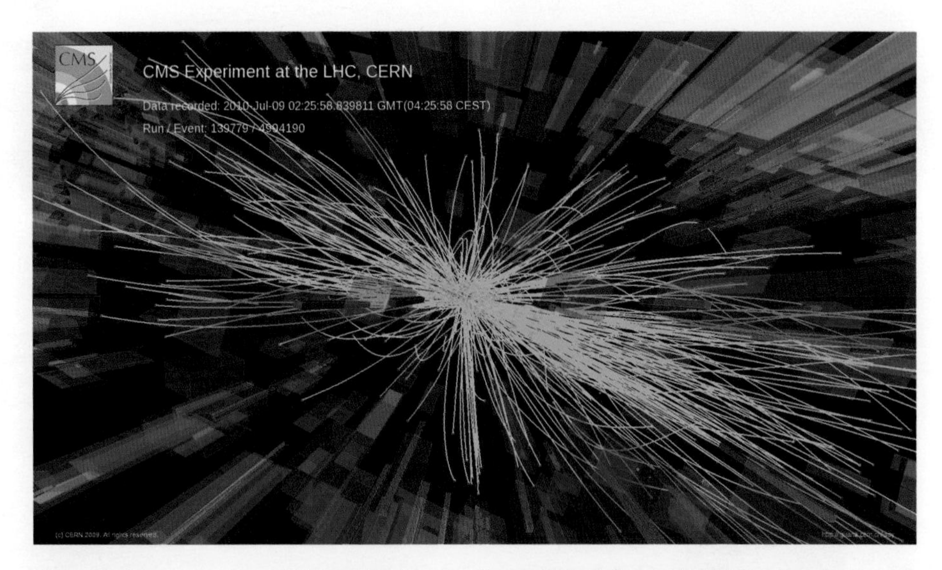

Plate 4.5. Proton-proton collision. Two beams of protons have been whirled in opposite directions to nearly the speed of light, each with an energy of 7 trillion electron volts, or 7 TeV $= 7 \times 10^{12}$ eV $= 1.12 \times 10^{-6}$ J, and directed into collision with one another at CERN's Large Hadron Collider (LHC). This image displays the tracks of more than 100 charged particles as they fly away from the point of proton collision. By studying the collision-particle debris, including correlations among them, scientists hope to gain an improved knowledge of how subatomic particles interact at extremely high energies, including the hot, dense conditions only a small fraction of a second after the "big bang." At the point of proton impact, temperatures of more than 1 million million, or 10^{12}, K are generated, exceeding 100,000 times the temperature at the center of the Sun. Experiments with this instrument have provided evidence for a new, previously unseen particle, named the Higgs boson, in the mass-energy range of 10^{11} eV. CERN is a French acronym for the Conseil Européen pour la Recherche Nucléaire (the European Organization for Nuclear Research). The Compact Muon Solenoid (CMS) particle detector created this image. (Courtesy of CERN.)

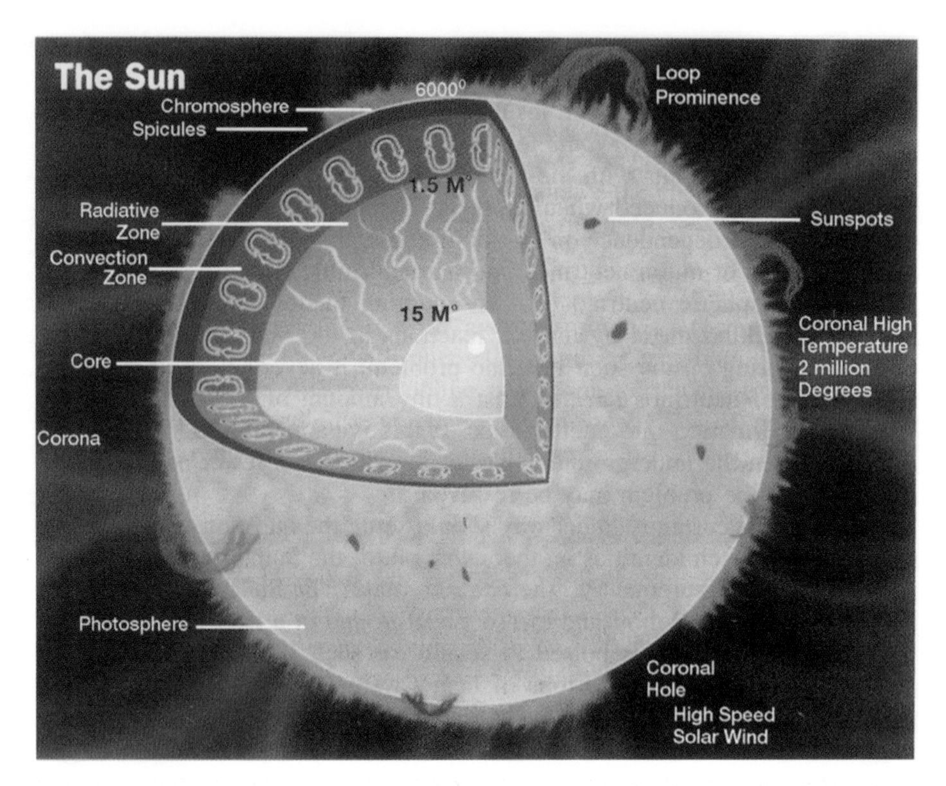

Plate 5.5. Anatomy of the Sun. The Sun is an incandescent ball of ionized gas powered by the fusion of hydrogen in its core. As shown in this interior cross section, energy produced by nuclear fusion is transported outward, first by countless absorptions and emissions within the radiative zone and then by convection. The visible disk of the Sun, called the *photosphere*, contains dark sunspots, which are Earth-sized regions of intense magnetic fields. A transparent atmosphere envelops the photosphere, including the low-lying chromosphere with its jet-like spicules and the 1-million-degree corona that contains holes with open magnetic fields, the source of the high-speed solar wind. Loops of closed magnetic fields constrain and suspend the hot million-degree gas within coronal loops and cooler material in prominences.

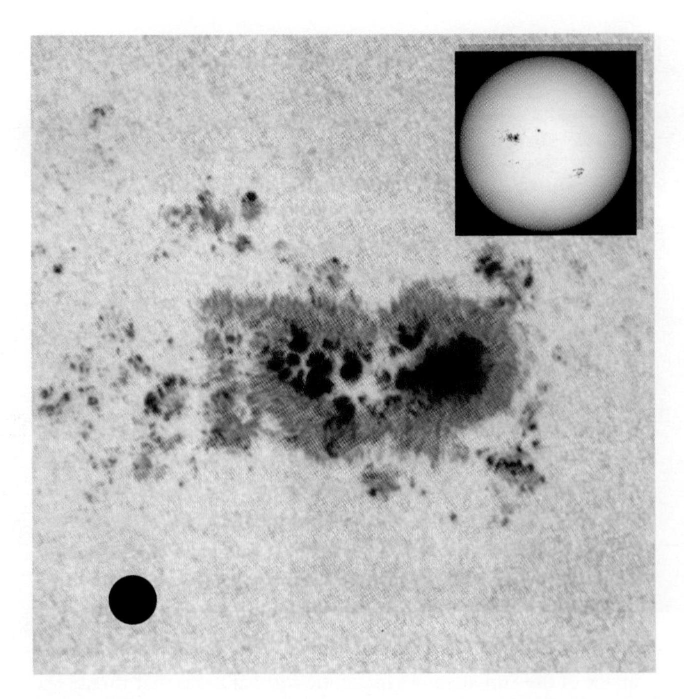

Plate 6.2. Sunspot group. Intense magnetic fields emerge from the interior of the Sun through the Sun's visible disk, the photosphere, producing groups of sunspots. The sunspots appear dark because they are slightly cooler than the surrounding photosphere gas. This composite image was taken in white light; that is, all of the colors combined. The enlarged image shows the biggest sunspot group, which is about 12 times larger than the Earth, the size of which is denoted by the black spot *(lower left)*. (Courtesy of *SOHO*/ESA/NASA.)

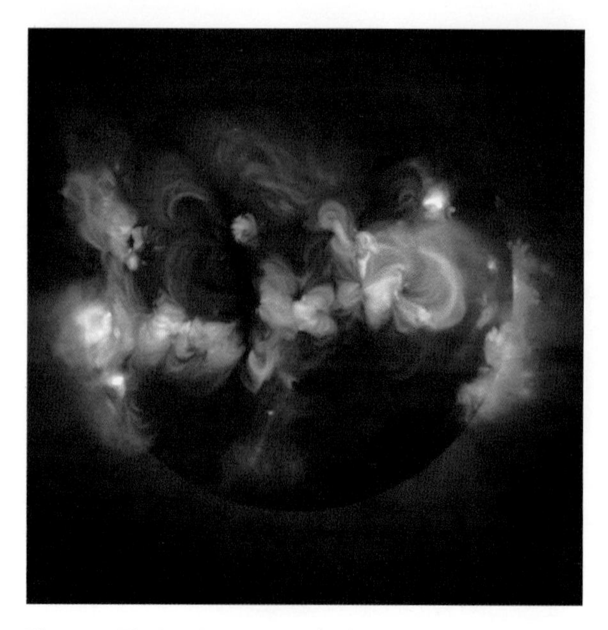

Plate 6.8. The Sun in x-rays. Ionized gases at a temperature of a few million K produce the bright glow seen in this x-ray image of the Sun. It shows magnetic coronal loops that thread the corona and hold the hot gases in place. The brightest features are called active regions and correspond to the sites of the most intense magnetic field strength. This image of the Sun's corona was recorded by the Soft X-Ray Telescope (SXT) aboard the Japanese *Yohkoh* satellite on February 1, 1992, near a maximum of the 11-year cycle of solar magnetic activity. Subsequent SXT images, taken about five years later near activity minimum, show a remarkable dimming of the corona when the active regions associated with sunspots have almost disappeared, and the Sun's magnetic field has changed from a complex structure to a simpler configuration. (Courtesy of NASA/ISAS/LMSAL/NAO Japan, University of Tokyo.)

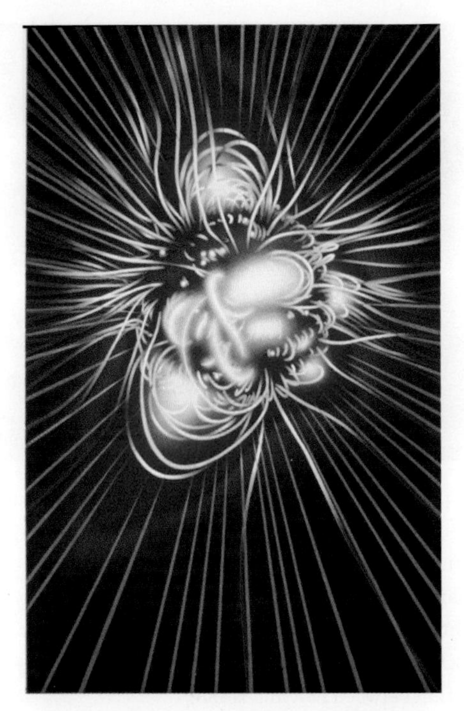

Plate 6.9. Magnetic fields near and far. In the low solar corona, strong magnetic fields are tied to the Sun at both ends, trapping hot, dense electrified gas within magnetized loops. Far from the Sun, the magnetic fields are too weak to constrain the outward pressure of the hot gas, and the loops are opened up, allowing electrically charged particles to escape. They form the solar wind that carries the magnetic fields away from the Sun. (Courtesy of *Newton Magazine*, the Kyoikusha Company.)

Plate 6.11. Coronal mass ejection. A huge coronal mass ejection is seen in this coronagraph image, taken on December 5, 2003 with the Large Angle Spectrometric Coronagraph (LASCO) on the SOHO spacecraft. The black area corresponds to the occulting disk of the coronagraph that blocks intense sunlight and permits the corona to be seen. An image of the singly ionized helium, denoted He II, emission of the Sun, taken at about the same time, has been scaled appropriately and superimposed at the center of the LASCO image. The full-disk helium image was taken at a wavelength of 30.4 nanometers, corresponding to a temperature of about 60,000 K, using the Extreme-ultraviolet Imaging Telescope (EIT) aboard SOHO. (Courtesy of the SOHO LASCO and EIT consortia. SOHO is a project of international cooperation between ESA and NASA.)

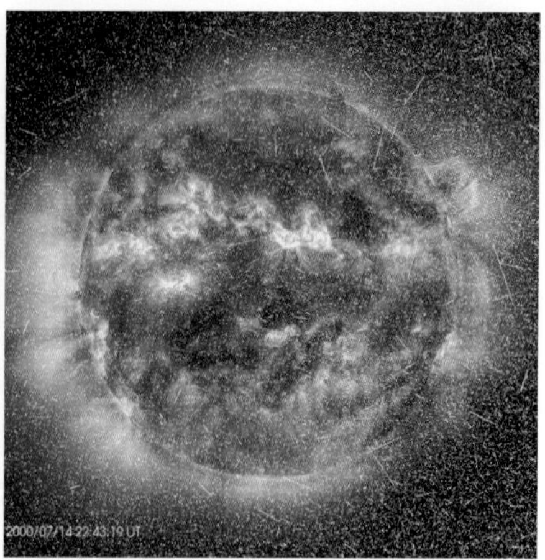

Plate 6.13. Solar flare produces energetic particle storm. A powerful solar flare (*left*), occurring on Bastille Day, July 14, 2000, unleashed high-energy protons that began striking the SOHO spacecraft near the Earth about 8 minutes later, continuing for many hours, as shown in the image taken 12 hours after the flare began (*right*). Both images were taken at a wavelength of 19.5 nanometers, emitted at the Sun by 11 times ionized iron, denoted Fe XII, at a temperature of about 1.5 million K, using the EIT on the SOHO spacecraft. (Courtesy of the SOHO EIT consortium. SOHO is a project of international cooperation between ESA and NASA.)

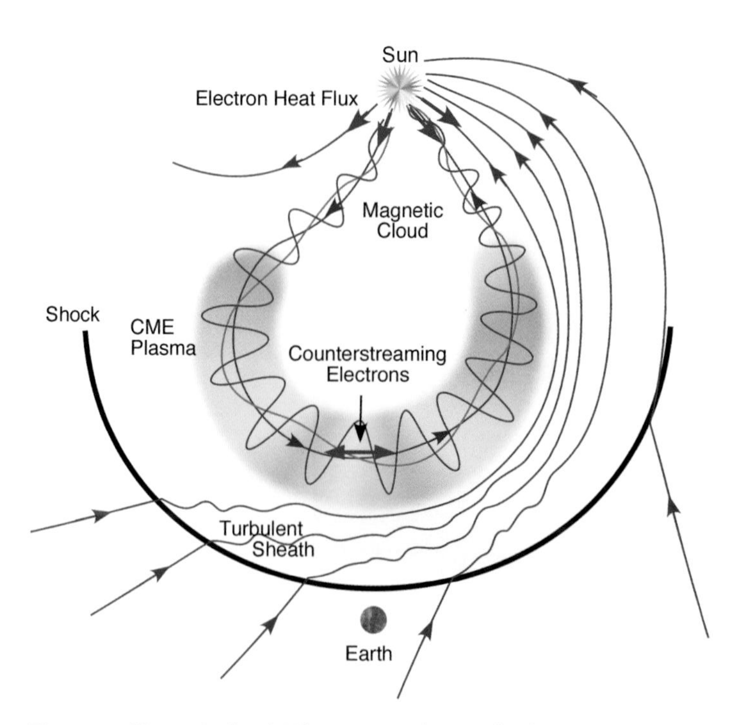

Plate 6.15. Magnetic cloud. When a coronal mass ejection (CME) travels into interplanetary space, it can create a huge magnetic cloud containing beams of electrons that flow in opposite directions within the magnetic loops that are rooted at both ends in the Sun. The magnetic cloud also drives a shock ahead of it. Magnetic clouds are present only in a subset of observed interplanetary coronal mass ejections. (Courtesy of Deborah Eddy and Thomas Zurbuchen.)

Plate 7.5. Proxima Centauri, the closest star. The red star centered in this image is the nearest star to the Earth, other than the Sun; it is at a distance of 4.24 light-years from us. Despite its closeness, the star has an absolute luminosity of just 2/1,000ths, or 0.002 times, the Sun's luminosity, and is too faint to be seen by the unaided human eye. Named Proxima Centauri because of its proximity, the star is close enough for its angular diameter to be measured by interferometric techniques, yielding a radius of 1/7th the radius of the Sun. Proxima Centauri orbits the bright double star Alpha Centauri (see Fig. 7.8) at a distance of about 0.24 light-year and an angular separation of 2.2 degrees. All three stars are located in the southern sky and cannot be seen from the Northern Hemisphere of the Earth. (Courtesy of David Malin/UK Schmidt Telescope/DSS/AAO.)

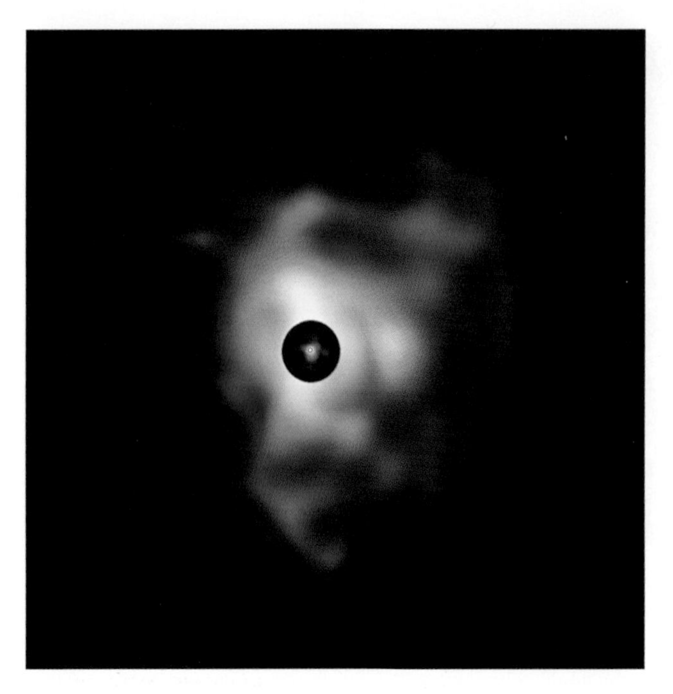

Plate 7.7. The flames of Betelgeuse. The red supergiant star Betelgeuse is slowly shedding its outer atmosphere, producing out-flowing gas that envelops the star and a much bigger nebula of gas and dust that surrounds it. The small circle in the middle of this composite image denotes the edge of the supergiant's optically visible disk; it has a diameter of about 4.5 AU, where 1 AU is the mean distance between the Earth and the Sun. The black disk masks the bright central radiation of the star to detect the infrared radiation of the outer plumes. They stretch to about 400 AU, or 60 million million, 6×10^{13}, meters from the supergiant Betelgeuse. (Courtesy of ESO/VISIR/VLT/Pierre Kervella.)

Plate 7.8. Alpha Centauri. Two of the most brilliant stars in the southern sky appear as a single star, named Alpha Centauri, to the unaided eye, but they can be resolved into two stars with the aid of binoculars or a small 5-cm (2-inch) telescope. The yellowish Alpha Centauri A (*lower left*), also known as Rigil Kentaurus, and the blue Alpha Centauri B (*upper right*) are locked together in a gravitational embrace, orbiting each other every 80 years. The two components of this binary-star system can approach one another within 11.2 AU and may recede as far as 35.6 AU, where the mean distance between the Earth and the Sun is 1 AU. Both stars have a mass and a luminosity that are comparable to those of the Sun. They appear bright because they are very nearby, at a distance of just 4.37 light-years. A third and faint companion, Proxima Centauri (see Fig. 7.5), is located at about 15,000 AU, or 2.2 degrees, from the two bright stars. At a distance of 4.24 light-years from the Earth, Proxima Centauri is the closest star other than the Sun. (Courtesy of ESO/Yuri Beletsky.)

Plate 7.12. Globular star cluster M 80. Hundreds of thousands of ancient stars are held together by their mutual gravitational attraction in this globular star cluster, which is designated M 80 or NGC 6093 and is located about 28,000 light-years from the Earth. Most of the stars displayed in this Hubble Space Telescope image are about as old as the observable universe, with ages of nearly 14 billion years – much older than our Sun, which is 4.6 billion years old. Stellar collisions result in more massive and relatively young "blue straggler" stars in the dense core of the globular cluster. (Courtesy of NASA/AURA/STScI/Hubble Heritage Team.)

Plate 7.15. Open star cluster NGC 265. A brilliant cluster of bright blue-colored stars is located in the Small Magellanic Cloud, about 200,000 light-years away and about 65 light-years across. This Hubble Space Telescope image subtends an angle of about 70 seconds of arc. (Courtesy of NASA/ESA/E. Olszewski, University of Arizona.)

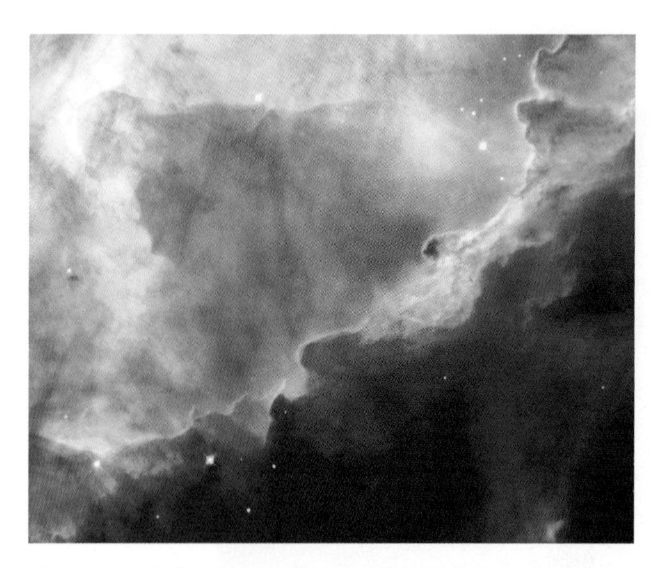

Plate 9.2. Turbulent interstellar gas. Glowing hydrogen gas in a small region of the Omega Nebula, which also is designated as M 17. The wave-like patterns of gas have been sculpted and illuminated by intense ultraviolet radiation from young, massive stars, which lie outside the picture to the upper left. The ultraviolet radiation heats the surface of otherwise cold, dark clouds of interstellar hydrogen. (A Hubble Space Telescope image, courtesy of NASA/ESA/J. Hester, Arizona State University.)

Plate 9.6. Dark clouds in the Carina Nebula. Molecular clouds in the Carina Nebula contain so much gas and dust that they are opaque to optically visible light, forming dark and dense structures where new stars may be born. Energetic stellar winds and intense radiation from nearby massive stars are sculpting the outer edges of the dark clouds. This image is a composite of observations taken in light emitted by hydrogen and oxygen atoms using the Hubble Space Telescope. The Carina Nebula, also designated NGC 3372, is about 7,500 light-years away from the Earth and spans more than 300 light-years. (Courtesy of NASA/ESA/Hubble Heritage Project/STScI/AURA.)

Plate 10.3. Mountains of creation. The infrared heat radiation of hundreds of embryonic stars (*white/yellow*) and windblown, star-forming clouds (*red*), detected from the Spitzer Space Telescope. The intense radiation and winds of a nearby massive star, located just above the image frame, probably triggered the star formation and sculpted the cool gas and dust into towering pillars. (Courtesy of NASA/JPL-Caltech/Harvard-Smithsonian CfA/ESA/STScI.)

Plate 10.4. Star-forming region. This multiple-wavelength portrayal combines infrared *(red)*, visible light *(green)*, and x-ray *(blue)* images of the bright, star-forming region designated NGC 346. It is located in the Small Magellanic Cloud that orbits our Milky Way Galaxy at a distance of about 210,000 light-years. Both wind-triggered and radiation-induced star formations are revealed, primarily by the infrared emission of the cold dust *(red)*, detected from the Spitzer Space Telescope. Young stars enshrouded by dust appear as red spots with white centers. The pressure of intense radiation from massive stars in the central regions of NGC 346 pushed against nearby gas, causing it to expand, and created shock waves that compressed nearby dust and gas into small new stars. Red-orange filaments surrounding the center of the image show where this process occurred. The supernova explosion of a very massive star apparently triggered the formation of even younger stars, seen as a pinkish concentration at the top of the image. Strong winds from this exploding star pushed dust and gas together about 50,000 years ago, compressing it into new stars. The x-rays *(blue)*, observed from ESA's XMM-Newton orbiting telescope, reveal very warm gas. The visible light *(green)* radiation was detected using the European Southern Observatory's New Technology Telescope. (Courtesy of NASA/JPL-Caltech/ESA/MPIA.)

Plate 10.5. The North America Nebula in infrared. Clusters of relatively young stars, about 1 million years old, are found throughout this infrared image of the North America Nebula, also designated as NGC 7000. The infrared detectors aboard the Spitzer Space Telescope penetrated the dark clouds seen in optically visible light, viewing young stars in many stages of formation, including gas and dust cocoons, disks, and jets. The North America Nebula is about 1,500 light-years from the Earth and spans about 50 light-years. (Courtesy of NASA/JPL-Caltech.)

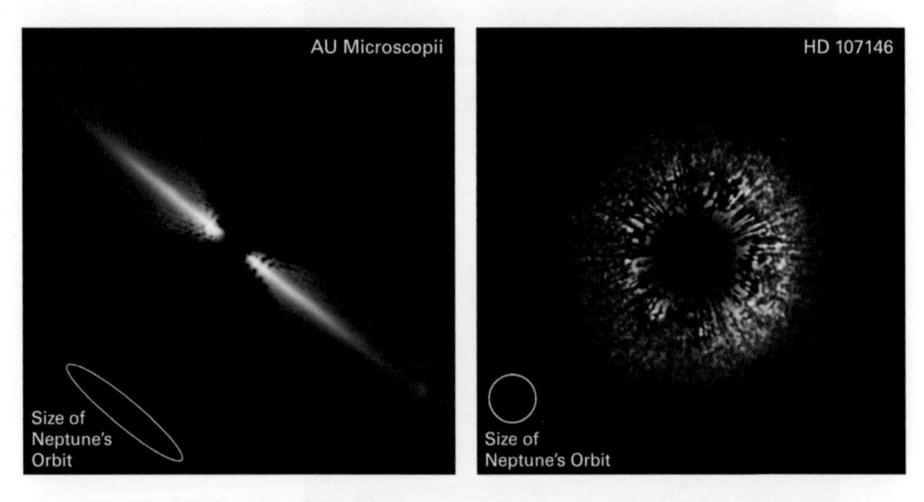

Plate 10.8. Dusty disks around Sun-like stars. Instruments aboard the Hubble Space Telescope have obtained these images of the visible starlight reflected from thick disks of dust around two young stars that still may be in the process of forming planets. Viewed nearly face-on, the debris disk surrounding the Sun-like star known as HD 107146 (*right*) has an empty center large enough to contain the orbits of the planets in our solar system. Seen edge-on, the dust disk around the reddish dwarf star known as AU Microscopii (*left*) has a similar cleared-out space in the middle. HD 107146 is 88 light-years away and is thought to be between 50 million and 250 million years old, whereas AU Microscopii is located 32 light-years away and is estimated to be just 12 million years old. (Courtesy of NASA/ESA/STScI/JPL/David Ardila – JHU [*right*], and John Krist – STScI/JPL [*left*].)

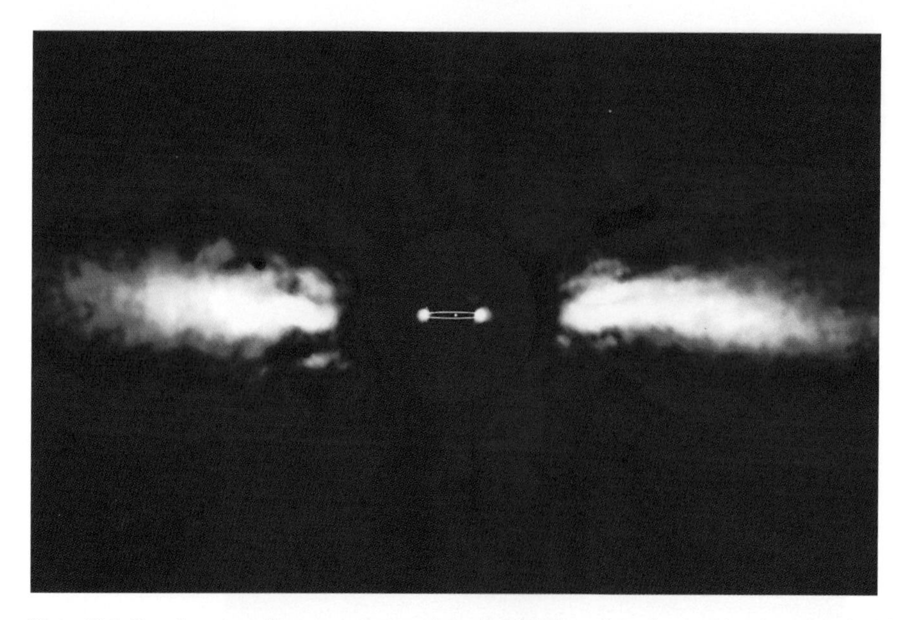

Plate 10.9. Exoplanet on the move. An exoplanet's orbital motion, denoted by the central white elliptical line, was imaged from an adaptive optics instrument attached to the VLT in Chile. The small white spot at the center shows the location of the host star, Beta Pictoris. Observations in 2003 are located at the left side of the planet's orbital ellipse and those in 2009 are on the right side. The larger dust disk surrounding the host star also is shown by the large flattened blue image at the left and the right. (Courtesy of ESO/A. M. Lagrange.)

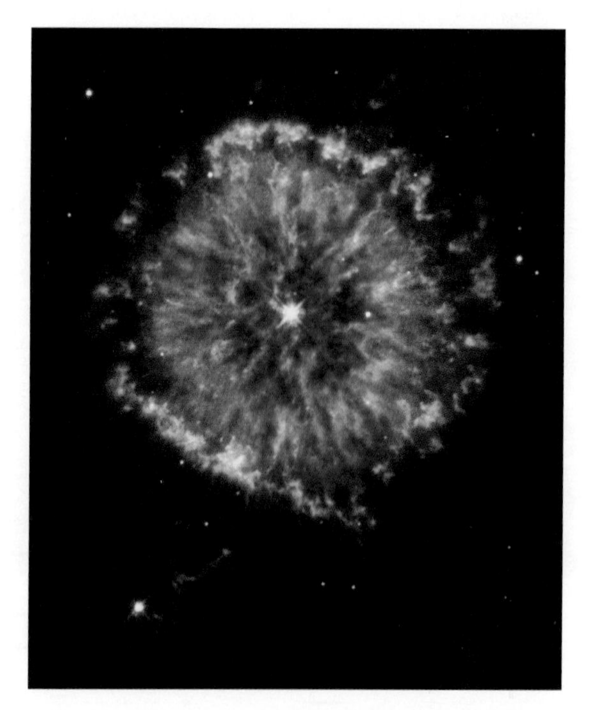

Plate 11.1 Planetary nebula. When a Sun-like star uses up its nuclear fuel, the star's center collapses into an Earth-sized white dwarf star and its outer gas layers are ejected into space. Such a planetary nebula is named after its round shape, which resembles a planet as seen visually in small telescopes, and is unrelated to planets. The shells of gas in the planetary nebula NGC 6751, shown here, were ejected several thousand years ago. The hot stellar core, exposed by the expulsion of the material surrounding it, has a disk temperature of about 140,000 K. Its intense ultraviolet radiation causes the ejected gas to fluoresce as a planetary nebula. (A Hubble Space Telescope image courtesy of the NASA/STScI/AURA/Hubble Heritage Team.)

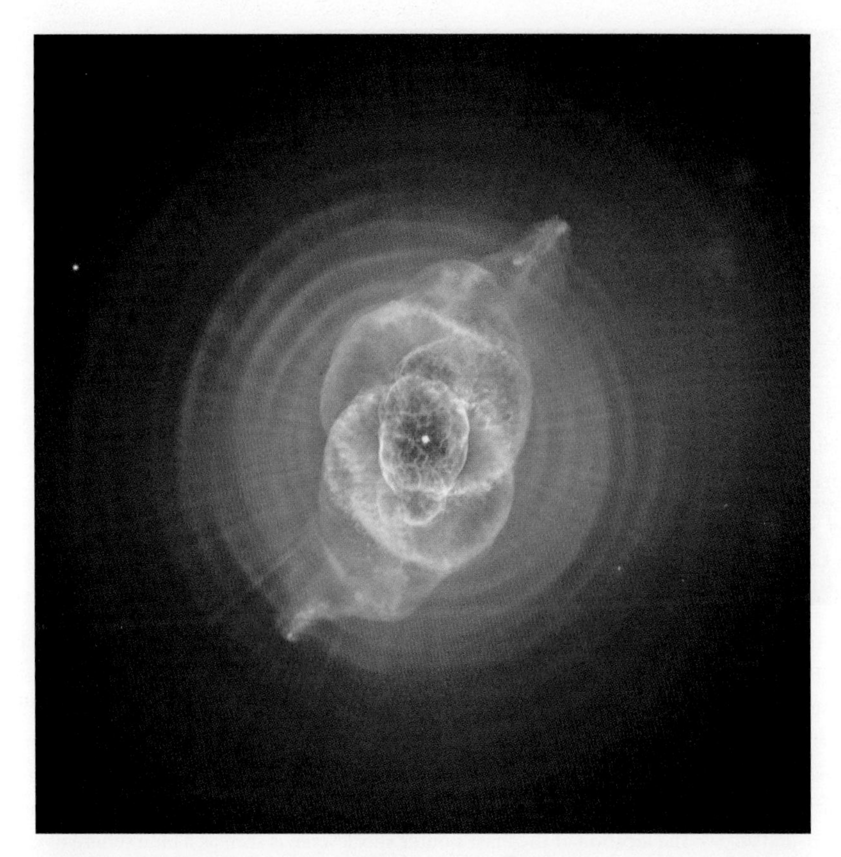

Plate 11.4 The Cat's Eye Nebula. A detailed view of the planetary nebula known as the Cat's Eye Nebula and formally designated NGC 6543. It reveals concentric rings, jets of high-speed gas, and shock-induced knots of gas. At least 11 large concentric shells mark the dense edges of spherical bubbles of gas and dust that have been ejected from a dying Sun-like star in regular, explosive pulses at 1,500-year intervals. The formation of more complex inner structures is not well understood. (A Hubble Space Telescope image courtesy of NASA/ESA/HEIC/the Hubble Heritage Team, STScI/AURA.)

Plate 11.10 Tycho supernova remnant. The expanding remains from a Type Ia supernova that occurred in 1572. It is named for the Danish astronomer Tycho Brahe (1546–1601), who recorded observations of its brightness in that year. The circular supernova remnant is located at a distance of about 13,000 light-years and is about 20 light-years across. It is bounded by an expanding shock wave and consists of ejected material moving away from the explosion and the interstellar material it sweeps up and heats along the way. The explosion left a hot cloud of expanding debris (*green* and *yellow*). The location of the blast's outer shock wave is seen as a blue sphere of very energetic electrons. Newly synthesized dust in the ejected material and heated preexisting dust from the area radiate at infrared wavelengths (*red*). Foreground and background stars in the image are white. This image is a composite of an x-ray image *(blue, green,* and *yellow)* taken from the Chandra X-Ray Observatory; an infrared image (*red*) taken from the Spitzer Space Telescope; and a visible light image (*white*) taken with the 3.5-m (138-inch) Calar Alto telescope located in southern Spain. (Courtesy of MPIA/NASA/Calar Alto Observatory.)

Plate 11.13. The Crab Nebula supernova remnant. The optically visible light of the Crab Nebula, designated as M 1 and NGC 1952, consists of two distinct parts: (1) a system of expanding filaments forms an outer envelope in which emission lines occur at well-defined wavelengths; and (2) an inner amorphous region that emits continuum radiation at all wavelengths. A Type II supernova explosion observed nearly 1,000 years ago, in 1054, ejected the filaments. The expanding filaments shine mostly in the light of hydrogen (*orange*) but also include the light of neutral oxygen (*blue*), singly ionized sulfur (*green*), and doubly ionized oxygen (*red*). The blue-white continuum glow concentrated in the inner parts of the nebula is the nonthermal radiation of high-speed electrons spiraling in magnetic fields. This continuum emission is powered by a spinning neutron star, the southwesternmost (*bottom right*) of the two central stars. The neutron star is the crushed, ultradense core of the exploded star. It also is a radio pulsar that acts like a lighthouse spinning 30 times a second. (Courtesy of NASA/ESA/J. Hester and A. Loll, Arizona State University.)

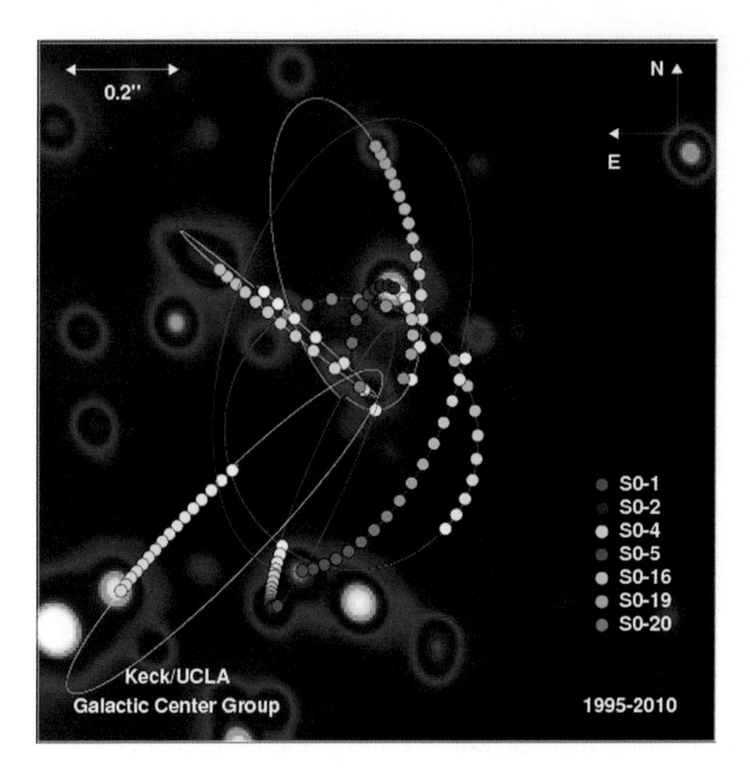

Plate 12.6. Super-massive black hole at the center of the Milky Way. Stars slowly revolve about an unseen center whose gravity controls their motions. This diagram portrays the orbits of infrared stars near the center of our Milky Way. The annual positions of seven stars were determined during a 15-year period (*colored dots*), determining their curved trajectories and inferring the mass of the invisible center. These orbits indicate that a super-massive black hole is located at the center of the Milky Way; it has a mass of 4.1 million times the mass of the Sun. The display covers the central 1.0×1.0 seconds of arc, which at a distance of about 27,700 light-years corresponds to a width of about 0.13 light-year or 1,140 light-hours. (Courtesy of Andrea M. Ghez/UCLA galactic center group.)

Plate 12.8. The Andromeda Nebula. The nearest spiral galaxy, the Andromeda Nebula, also known as M 31 and NGC 224, is located at a distance of about 2.6 million light-years, so its light takes about 2.6 million years to reach us. Both the Andromeda Nebula and our Galaxy are spiral galaxies with total masses of about 1 million million, or 10^{12}, solar masses, and roughly 100 billion, or 10^{11}, optically visible stars. The several distinct stars surrounding the diffuse light from Andromeda are stars within our own Galaxy; these stars lie well in front of Andromeda. Two smaller galaxies also are shown in this image: M 32, also designated NGC 221, shown at the edge of the Andromeda Nebula; and NGC 205, located somewhat farther away. These are elliptical systems at about the same distance as M 31 but with only about 1/100th of its mass. (Courtesy of Karl-Schwarzschild Observatorium, Tautenburg.)

Plate 12.9. Spiral galaxy. This natural-color Hubble Space Telescope image shows the spiral galaxy NGC 4911 in the Coma cluster of galaxies, which lies about 320 million light-years away from the Earth. Central clouds of interstellar gas and dust are silhouetted against glowing, young star clusters and clouds of hydrogen gas. (Courtesy of NASA/ESA/Hubble Heritage Team/STScI/AURA.)

Plate 12.12. Inside the Coma cluster of galaxies. More than 1,000 identified galaxies are located within the Coma cluster, also designated Abell 1656, which has a mean distance of 321 million light-years and is more than 20 million light-years in diameter. Each of these galaxies contains hundreds of billions of stars. Most of the galaxies that inhabit the central portion of this cluster are giant elliptical galaxies. Several spiral galaxies are found farther out from the center, such as the one shown in the upper left of this mosaic of images taken from the Hubble Space Telescope. Nearly every object in this picture is a galaxy. It is a section of the cluster that is several million light-years across, and it is located about one third of the way out from the center. (Courtesy of NASA/ESA/Hubble Heritage/STScI/AURA.)

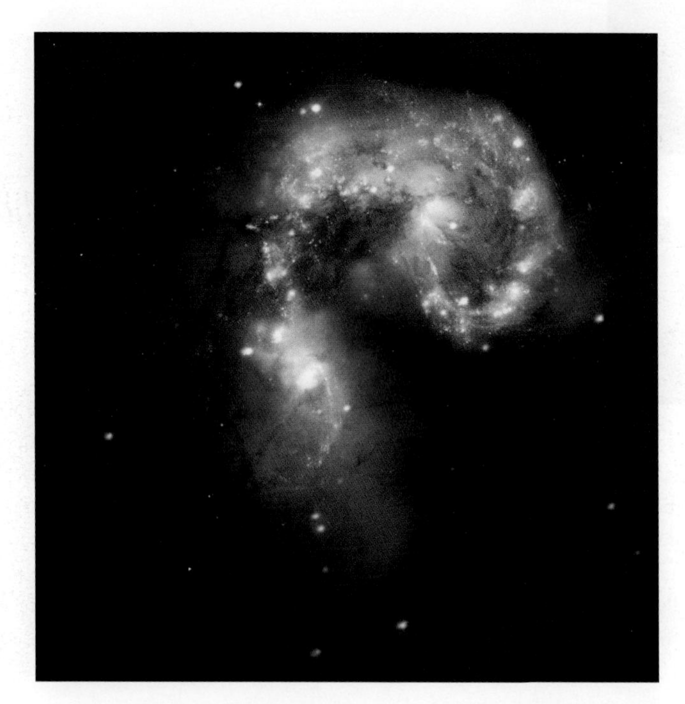

Plate 12.16. Colliding galaxies. Gravitational interaction of the antennae galaxies, catalogued as NGC 4038 and NGC 4039, produces long arms of young stars in their wake. The colliding galaxies are located about 62 million light-years from the Earth and have been merging for the past 800 million years. As the two galaxies continue to churn together, clouds of interstellar gas and dust are shocked and compressed, triggering the birth of new stars. This composite image is from the Chandra X-Ray Observatory (*blue*), the Hubble Space Telescope (*gold* and *brown*), and the Spitzer Space Telescope (*red*). The blue x-rays show huge clouds of hot interstellar gas, the red data show infrared radiation from warm dust clouds that have been heated by newborn stars, and the gold and brown data reveal both star-forming regions and older stars. (Courtesy of NASA/ESA/SAO/CXC/JPL-Caltech/STScI.)

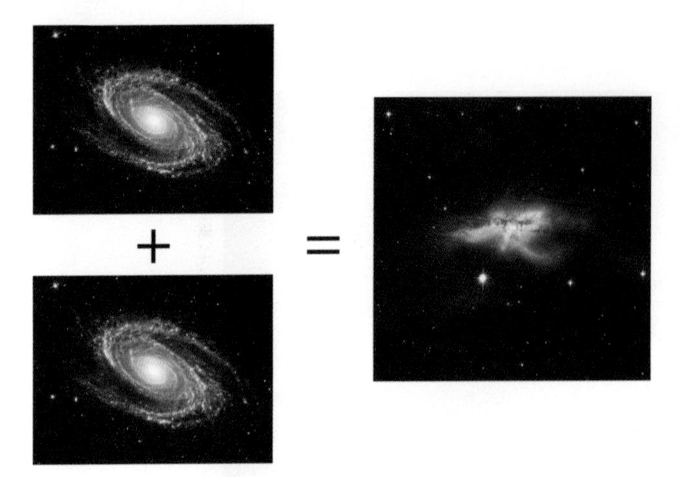

Plate 13.4. Merging galaxies. Two spiral galaxies (*left*), each represented by M 81, can merge to form an elliptical galaxy. A pair of colliding galaxies (*right*), designated NGC 6240, illustrates such a merger just before becoming a single larger galaxy. The prolonged violent collision drastically altered the shape of both galaxies and created large amounts of heat and infrared radiation. All of these images were taken by the infrared camera aboard the Spitzer Space Telescope, with the inclusion of optically visible data for M 81 taken from the Hubble Space Telescope. (Courtesy of NASA/JPL-Caltech/University of Arizona/CfA/NOAO/AURA/NSF [*left*] and NASA/JPL-Caltech/STScI-ESA [*right*].)

Plate 13.10. Supermassive black hole. This barred spiral galaxy, designated NGC 1097, is located about 44 million light-years from the Earth. It is a Seyfert galaxy and a moderate example of an active galactic nucleus (AGN) with jets shooting out of its core. A super-massive black hole, with about 100 million times the mass of the Sun, is located at the center of the galaxy. As shown in this infrared view of NGC 1097, taken from the Spitzer Space Telescope, a star-forming ring surrounds its center. As gas and dust spiral into the central black hole, they cause the ring to light up with hundreds of new stars. The galaxy's spiral arms and the swirling spokes between them show dust heated by other newborn stars as well as older stars. (Courtesy of NASA/JPL-Caltech.)

Table 8.2. *The main-sequence stars*[a]

Spectral Type	Effective Temperature (K)	Mass ($M_S/M_\odot$)	Luminosity ($L_S/L_\odot$)	Radius ($R_S/R_\odot$)	Lifetime (years)
O5	44,500	60	7.9×10^5	12	3.7×10^6
B0	30,000	17.5	5.2×10^4	7.4	1.1×10^7
B5	15,400	5.9	8.3×10^2	3.9	6.5×10^7
A0	9,520	2.9	5.4×10	2.4	2.9×10^8
F0	7,200	1.6	6.5	1.5	1.5×10^9
G0	6,030	1.05	1.5	1.1	5.1×10^9
K0	5,250	0.79	0.42	0.85	1.4×10^{10}
M0	3,850	0.51	0.077	0.60	4.8×10^{10}
M5	3,240	0.21	0.011	0.27	1.4×10^{11}

[a] The mass, M_S, is in units of the Sun's mass $M_\odot = 1.989 \times 10^{30}$ kg; the absolute luminosity, L_S, is in units of the Sun's absolute luminosity, $L_\odot = 3.854 \times 10^{26}$ J s^{-1}; and the radius, R_S, is in units of the Sun's radius, $R_\odot = 6.922 \times 10^8$ m. The lifetime is the amount of time required to exhaust the nuclear hydrogen fuel that supplies the energy of stars on the main sequence.

arrived on the main sequence near the beginning of the observable universe and are still there.

How do we determine the length of time that a main-sequence star continues to shine by converting protons into helium nuclei? These nuclear-fusion reactions are limited to the hot, dense stellar core. Outside of the core, where the overlying weight and compression are less, the gas is cooler and thinner and nuclear fusion cannot exist. For instance, the energy-generating core of the Sun extends about one quarter of the distance from the center to the visible solar disk. When all of the hydrogen within the core has been converted into helium, a star has exhausted its nuclear fuel supply and can no longer reside on the main sequence.

More than a half-century ago, Mario Schönberg and Subrahmanyan Chandrasekhar considered stellar models in which hydrogen is burned inside a star's core, or in a thin shell between the burned-out core and the overlying material. They found that it was impossible to construct models in which more than 12 percent of the mass of the star is included in the exhausted core. This meant that the lifetime of a star on the main sequence is limited to the time it takes to convert 12 percent of its hydrogen into helium.

Stars of intermediate mass, comparable to that of the Sun, will shine for 10 billion to 20 billion years. The Sun, for example, formed about 4.6 billion years ago; in another 7 billion years it is expected to end its main-sequence life. Very massive and luminous stars survive only a few million years; those that we now can see were not shining when the dinosaurs ruled the Earth. Had they been shining then, more than 65 million years ago, they would have burned out by now.

So, the position of a star on the main sequence depends on its mass – the most massive stars being the most luminous and the more massive a star, the shorter it lives and the sooner it evolves off of the main sequence. Ninety percent of all main-sequence stars have a mass below 0.8 solar mass, and they have not yet had time to perish. They have been on the

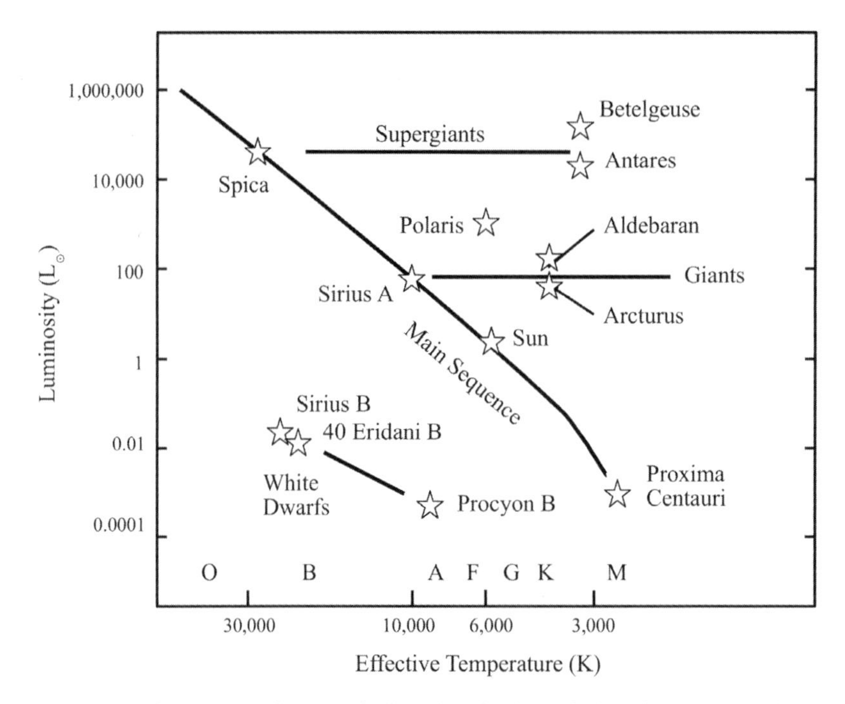

Figure 8.5. Giants, supergiants, and white dwarfs. The majority of stars occupy the main sequence in the H–R diagram. Stars with a mass comparable to that of the Sun will evolve into helium-burning giant stars, illustrated by Aldebaran and Arcturus in this diagram (*middle right*). These red giants are somewhat cooler than the Sun but about 100 times more luminous. The luminosity (*left vertical axis*) is in units of the Sun's luminosity, denoted $L_{\odot}$. More massive stars, which have a shorter lifetime on the main sequence, evolve into supergiant stars that are between 10,000 and 1 million times as luminous as the Sun. Antares and Betelgeuse illustrate the supergiant stars on this diagram (*top right*). After depleting all of their helium fuel, the less-luminous giant stars evolve into white dwarf stars of very low luminosity and initially hot disk temperatures. They are illustrated in the bottom left of this diagram by Sirius B, 40 Eridani B, and Procyon B.

main sequence ever since they were born, thereby providing us with no information about stellar evolution. In contrast, some of the more massive stars, which were born long ago, have had enough time to burn up their available hydrogen fuel and advance to the next stage of stellar life. Thus, to understand stellar evolution, we must examine the upper part of the main sequence in the H–R diagram, which applies to the more massive, shorter-lived stars.

The Red Giants and Supergiants

After the low-mass, main-sequence stars, the most common type of star is the red giant, found in the upper right side of the H–R diagram (Fig. 8.5). These low-temperature stars are not exceptionally massive. They have an intermediate mass of roughly 1 to 10 times that of the Sun and are in a late state of stellar evolution from somewhat hotter main-sequence stars. Although cooler than the Sun, the red giants are about 100 times more luminous due to their much larger size, about 50 times the radius of the Sun. Prominent, bright-red giants include Aldebaran and Arcturus.

Because they are so luminous, we can see red giant stars that are relatively distant without using a telescope. However, they also are much less common than main-sequence stars because relatively few stars have entered this later stage of life. The red giants last only a few million years, which is a brief existence compared to the billions of years that stars of roughly solar mass spend on the main sequence.

The giant stars are enormously distended stars with a low mean mass density and a high luminosity. If we assume that the inner temperatures of giant stars are high enough to generate a gas pressure sufficient to balance gravitation, then their luminosity would greatly exceed that which is actually observed. This enigma was resolved almost a century ago when the great English astronomer Arthur Eddington showed that radiation pressure must stand with gravitation and gas pressure as the third major factor in maintaining the equilibrium of a star.

Although the outward pressure caused by the motion of gas particles, or the gas pressure, indeed does support the Sun and most other stars against the inward force of their immense gravity, it is insufficient for the much larger giant stars. They are supported by radiation pressure, which increases with the fourth power of the temperature. In contrast, gas pressure is simply proportional to the temperature; so, if we sufficiently increase the central temperature, radiation pressure will become much larger than gas pressure.

The supergiants are very massive, evolving from main-sequence stars of 10 to 100 solar masses, and exceptionally large, with a radius of hundreds of times that of the Sun, and 10 to 100 times more luminous than the red giants. Antares and Betelgeuse are supergiant stars.

The supergiants are so exceedingly rare that we can hardly see them as a sparse sprinkling across the top edge of the H–R diagram. They are even less common than the O stars, rarely seen in any given part of the night sky, but so intrinsically luminous that we can see a few without a telescope.

Both the giants and the supergiants are so large that they are slowly blowing away with strong winds that carry their outer atmospheres into surrounding space. For example, the giant star Mira, "The Wonderful," is pulsating like a runner breathing hard at the end of a long race and also losing mass at about one millionth of a solar mass per year. That is more than 1 million times the mass loss rate of the Sun's wind.

Because giants and supergiants are former main-sequence stars that have exhausted their core supply of nuclear hydrogen, they must be undergoing other forms of nuclear fusion. We now consider these various nuclear-burning reactions, including the two methods of burning hydrogen.

8.2 Nuclear Reactions Inside Stars

The Internal Constitution of Stars

Although we cannot see the inside of a star, its internal structure can be explained by a few simple concepts, one of which is a star's equilibrium. Like the Sun, almost every star we see is neither collapsing nor expanding, and it retains the same size throughout most of its long life. At every point inside such a star, the inward pull of its gravity is balanced precisely by the outward push of its internal pressure.

As with the Sun, all of the other main-sequence stars are composed mainly of the lightest element, hydrogen, but it is too hot for whole hydrogen atoms to exist within them. These atoms are fragmented into their subatomic constituents by frequent collisions. The material in these stellar interiors therefore is in the plasma state, composed almost entirely of bare hydrogen nuclei, the protons, and free electrons no longer attached to atoms. Compressed to high density, the protons are still far apart, occupying the vast empty spaces of former atoms, so the plasma behaves like a perfect gas with a pressure that increases with the temperature.

How hot is the center of a main-sequence star and what is the pressure down there? Protons are 1,836 times more massive than electrons, so they dominate the gravitational effects inside a star. The central temperature of a star can be estimated by assuming that a proton at the center is hot enough and moving fast enough to counteract the gravitational compression on the proton from the rest of the star. For the Sun, a central temperature of 15.6 million K establishes this equilibrium between the outward pressure of moving protons and the inward gravitational pull at the Sun's center.

A massive star produces greater compression at its center, and a higher central temperature is required to hold it up. The temperature at the center of a main-sequence star is proportional to its mass, so a star that is 10 times more massive than the Sun is 10 times hotter in its center. At a great enough mass, the star becomes so hot that it is blown apart; this explains why there are no known stars with a mass greater than about 120 times the mass of the Sun.

Where does a star's heat come from? The energy released by nuclear fusion in the stellar core heats the gas and generates its pressure. That is, nuclear reactions that transform a light element into a heavier one liberate subatomic energy that sustains the high temperatures within a star. This energy also makes its way out of the star to provide its luminosity and keep it shining.

Thus, two other fundamental concepts in understanding a star's interior are (1) the way energy is generated by nuclear reactions near its center; and (2) the methods in which the radiation produced by these reactions works its way out to the observed stellar disk, its photosphere. The energy generation depends on the nuclear fuel as well as the mass density and temperature in a star's core. The radiation-energy transfer depends on a star's internal opacity to radiation, which prevents some of the radiation from escaping.

After arrival on the main sequence, which is designated the *zero age*, the internal structure of a star can be determined by only four equations, which describe the equilibrium, energy transport, conservation of mass, and conservation of energy within the star. The crucial equations can be solved without any knowledge of the properties of the star before arrival on the main sequence.

At any given time, stars of the same composition have radii, luminosities, effective temperatures, and mean densities that are determined solely by the star's mass. The German astronomer Heinrich Vogt demonstrated this concept in 1926, and the American astronomer Henry Norris Russell derived it independently the following year; therefore, it is known as the Vogt–Russell theorem. It implies that a star of a given mass, age, and chemical composition occupies a unique position, related to the star's evolutionary history, on the H–R diagram. The mass, age, and composition are all we need to know to understand the life history of a star.

A star will continue shining with a luminosity and temperature determined by its mass, remaining stable and fundamentally unchanged for a long time. The only caveat to this understanding of stellar life is that the core of a star is the only place hot enough for nuclear reactions to occur. The composition of the core slowly changes as the result of these reactions; eventually, there is no more nuclear energy in the stellar core, so it loses its equilibrium.

It appears to be simple, but the theory is complex with detailed applications that are found in advanced texts. We keep it simple in this book, beginning with the fusion of hydrogen into helium in main-sequence stars.

Two Ways to Burn Hydrogen in Main-Sequence Stars

All main-sequence stars generate energy by the thermonuclear fusion of hydrogen nuclei, the protons, into helium nuclei. Because hydrogen is "burned up" or consumed to fuel the nuclear fires, we call this process *hydrogen burning*, although it is a chain of nuclear-fusion reactions rather than the combustion of an ordinary fire.

There are two methods of burning, or fusing, hydrogen into the heavier element helium; the dominant mechanism, which produces the most power, depends on the mass of a star. The main source of energy for main-sequence stars with a mass less than 1.5 times the Sun's mass is the *proton-proton chain* of nuclear reactions, abbreviated as the *P-P chain*. A different sequence of nuclear reactions converts protons into helium nuclei inside main-sequence stars more massive than 1.5 times the mass of the Sun. This is known as the carbon-nitrogen-oxygen (CNO) cycle. As the name suggests, this is a cyclic set of nuclear reactions.

The thermonuclear process responsible for the energy production in any star is not limited to a single nuclear transformation but rather consists of a sequence of linked transformations that together form a nuclear chain reaction. The P-P chain was linear, with only one direction (like a one-way street), whereas the CNO cycle occurs in a closed circular chain (like driving around the block). For both types of hydrogen burning, however, the bottom line is always the same: Four protons combine to make one helium nucleus, thereby releasing energy.

In the lower right-hand, low-mass side of the main sequence, where the great majority of stars are located, nuclear energy is generated as a result of the P-P chain that makes the Sun shine. In 1939, the German-born American physicist Hans Bethe delineated this nuclear transformation of four protons into one helium nucleus.

Because the P-P chain was discussed in detail in Chapter 5, we now limit the discussion to the general result (Focus 8.1), which in shorthand reaction notation is as follows:

$$4\,^1H \rightarrow {}^4He + 6\gamma + 2\nu_e,$$

where nuclei on the left side of an arrow fuse to make the nuclei and other particles or radiation on the right side of the arrow. The hydrogen nuclei, denoted ^1H, combine to make the helium nucleus, designated as ^4He; radiation, denoted by γ; and electron neutrinos, designated as ν_e.

In two papers that appeared in the *Physikalische Zeitschrift* in 1937 and 1938, the German physicist Baron Carl Friedrich von Weizsäcker examined the thermonuclear reactions that might occur in stars. In the first paper, he reasoned that the merger of two protons must have started hydrogen burning in the Sun. In the second paper, Weizsäcker proposed that elements that are heavier than hydrogen already were created before the formation of stars,

Focus 8.1 The Proton-Proton Chain

In the P-P chain of nuclear reactions, four protons, each designated by ^1H or p, combine to form a helium nucleus, denoted by ^4He; releasing powerful gamma-ray radiation, designated γ; positrons with the symbol e^+; electron neutrinos, denoted as ν_e; and an energy of about 4×10^{-12} J per reaction chain.

The chain of reactions begins with the merger of two protons, each designated by ^1H, in the following reaction:

$$^1H + {}^1H \rightarrow {}^2D + e^+ + \nu_e.$$

This is followed by:

$$^2D + {}^1H \rightarrow {}^3He + \gamma,$$

with a last step:

$$^3He + {}^3He \rightarrow {}^4He + {}^1H + {}^1H.$$

In the process, additional gamma rays are released when positrons combine with electrons, denoted e^-, and during pair annihilation, denoted by $e^+ + e^- \rightarrow \gamma + \gamma$. The net result of the P-P chain is the following:

$$4{}^1H \rightarrow {}^4He + 6\gamma + 2\nu_e.$$

as we know them now. He no longer was limited to reactions that began with the lightest element, hydrogen, and this led him to the important discovery of the cyclic CNO chain of reactions in which carbon acts as a catalyst for the synthesis of helium from hydrogen, with the release of radiation and neutrinos (Focus 8.2). The overall result for each CNO reaction chain is as follows:

$$^{12}C + 4{}^1H \rightarrow {}^{12}C + {}^4He + 5\gamma + 2\nu_e,$$

where the carbon nucleus, denoted ^{12}C, forever is being regenerated and acts like a catalyst. That is, the CNO cycle is a circular reaction chain in which carbon is destroyed and then re-created and therefore is available for the sequence of reactions to occur repeatedly.

Carl von Weizsäcker belonged to an interesting family. His brother Richard von Weizsäcker studied philosophy and history at Balliol College, Oxford, and returned to Germany to serve in the army during World War II (1939–1945). He subsequently was elected President of the Federal Republic of Germany, serving in this office from 1984 to 1994. Carl and Richard's father, Ernst von Weizsäcker, was a German diplomat who joined the Nazi Party and the SS in 1938. In 1947, he was sentenced to seven years in prison at the Nuremberg trials for crimes against humanity; some defenders, including Winston Churchill, argued that the sentence was unjust and he was released as part of a general amnesty in 1950.

The circular CNO reaction chain is induced by high temperatures in the cores of massive main-sequence stars and becomes self-sustaining by the catalytic action of carbon. This cycle also could begin at the intermediate stages with nitrogen or oxygen, so the entire cycle is called the CNO cycle.

Focus 8.2 The CNO Cycle

The cyclic CNO chain of reactions starts when a carbon nucleus, ^{12}C, fuses with a proton, ^{1}H, to produce a nucleus of nitrogen, ^{13}N, and gamma-ray radiation, γ. The nitrogen decays to form a nucleus of heavier carbon, ^{13}C; a positron, e^{+}; and an electron neutrino, ν_e. These beginning nuclear-fusion reactions are as follows:

$$^{12}C + {}^{1}H \rightarrow {}^{13}N + \gamma$$

and

$$^{13}N \rightarrow {}^{13}C + e^{+} + \nu_e.$$

The cycle then continues when the heavy carbon combines with a proton to form heavier nitrogen, ^{14}N, which then fuses with another proton to form oxygen, ^{15}O. The oxygen decays to make ^{15}N, which then combines with a proton to make the original carbon, ^{12}C, together with a helium nucleus, ^{4}He. The nuclear-fusion reactions are as follows:

$$^{13}C + {}^{1}H \rightarrow {}^{14}N + \gamma$$

$$^{14}N + {}^{1}H \rightarrow {}^{15}O + \gamma$$

$$^{15}O \rightarrow {}^{15}N + e^{+} + \nu_e$$

and

$$^{15}N + {}^{1}H \rightarrow {}^{12}C + {}^{4}He.$$

The positrons, e^{+}, annihilate the electrons, e^{-}, to produce energetic gamma radiation by the following reaction:

$$e^{+} + e^{-} \rightarrow 2\gamma.$$

Like the P-P chain, the net result of the CNO cycle is that four protons are fused together to form one helium nucleus, gamma rays, and electron neutrinos. By summing the left and right sides of all of the participating reactions, we obtain the following:

$$^{12}C + 4{}^{1}H \rightarrow {}^{12}C + {}^{4}He + 5\gamma + 2\nu_e.$$

Like the noncircular, linear P-P chain, each CNO reaction chain releases about 4×10^{-12} J of energy.

Bethe knew that the P-P reaction, which could explain the luminous output of the Sun, fell short of the much greater luminosity of the hotter and more massive stars. Unlike Weizsäcker, he systematically examined a great number of nuclear reactions that would not operate within stars and eliminated them. He independently found that the CNO cycle would generate about the same energy as the proton-proton process for each nuclear reaction

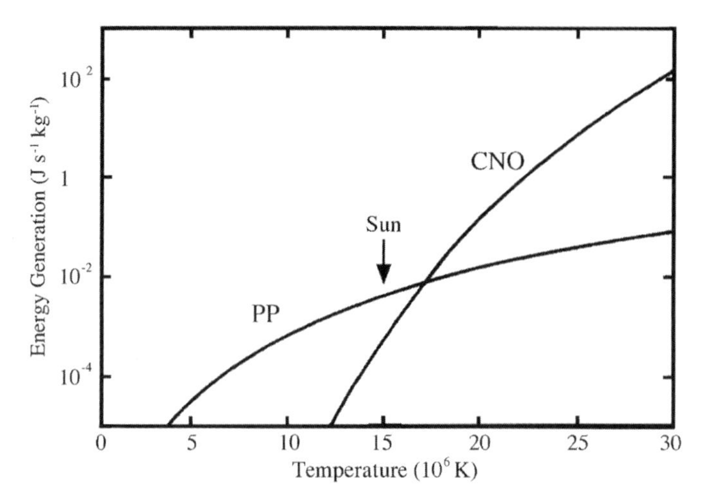

Figure 8.6. Energy generation by two hydrogen-burning processes. The energy output (*left vertical axis*), in units of power per kilogram, or J s^{-1} kg^{-1}, as a function of core temperature (*bottom horizontal axis*) in millions, or 10^6, degrees kelvin, designated K. The proton-proton chain, denoted PP, dominates the hydrogen-burning energy production for the Sun and less massive stars that have lower core temperatures. At the center of the Sun, where the temperature is 15.6×10^6 K, the P-P chain is the dominant nuclear-reaction chain for converting hydrogen nuclei into helium nuclei, with an energy output of 0.016 J s^{-1} kg^{-1}, or 51 million MeV g^{-1} s^{-1} in the units used by nuclear astrophysicists. In more massive main-sequence stars, the central temperature is higher and the CNO cycle of hydrogen burning is the most efficient process. Main-sequence stars of mass less than 1.5 solar masses shine by the P-P chain of nuclear reactions, whereas the main-sequence stars with mass greater that 1.5 solar masses burn hydrogen by the CNO set of nuclear reactions.

chain, but he also showed that the greater rate and temperature dependence of the CNO cycle could account for the high luminosity of the massive stars. Subsequently, this conclusion was placed on a firm basis when William A. "Willy" Fowler and his colleagues at the California Institute of Technology made laboratory measurements of the nuclear-reaction cross sections for every reaction in the chain, in a series of papers spanning several decades.

Both the P-P chain and the CNO cycle operate within main-sequence stars, and they both release about the same amount of energy during each reaction chain; however, the total amount of energy generation differs depending on the mass and central temperature of a star (Fig. 8.6). The CNO cycle contributes little energy in low-mass stars, with low central temperatures, and the P-P chain produces almost all of the energy radiated by main-sequence stars with a mass less than the Sun's mass. However, the CNO cycle generates more energy per unit time in massive stars that have high central temperatures. At high temperatures, the rate of the nuclear reactions is faster, so more reactions occur and more energy is generated every second than at low temperatures. Stars with a mass of about 2 solar masses or above generate almost their entire energy output by the CNO cycle.

In the Sun, only 1.5 percent of its energy is generated by the CNO cycle. This is why the P-P chain explains the Sun so well. However, with increasingly more mass and an increasingly hotter stellar core, the CNO cycle becomes the dominant energy source for a main-sequence star. For a star with a mass of 1.5 times the mass of the Sun, each method of burning hydrogen produces the same amount of total energy and half the luminosity of a star.

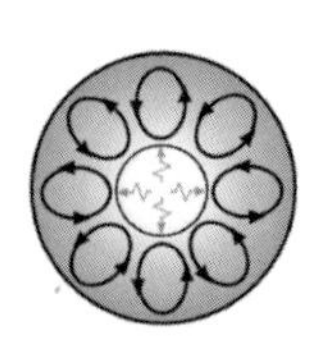

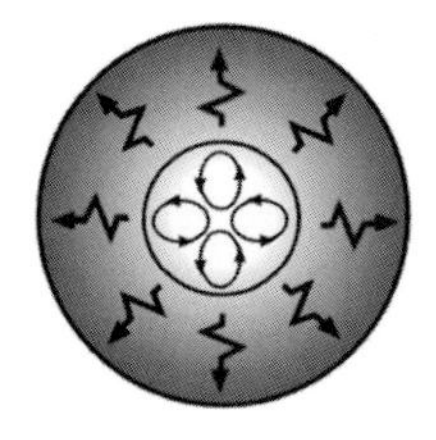

$M < 0.5\ M_\odot$ $M = (0.5 - 1.5)\ M_\odot$ $M > 1.5\ M_\odot$

Figure 8.7. Convection inside stars of different mass. Most stars have convective zones in which energy is transported by the wheeling motion of convection, denoted here by closed curves with arrows for stars of different mass, designated by M, and compared to the Sun's mass, designated $M_\odot$. The symbol < means less than and the symbol > denotes greater than. Low-mass stars, with less than half a solar mass, are fully convective from core to visible disk and therefore of uniform composition. Their low temperatures result in a high opacity to radiation. In intermediate-mass stars, such as the Sun, radiation transport dominates convection in the hot central regions, which are enveloped by a cooler convective region. The visible disks of these stars do not include the nuclear-fusion products from their core but rather retain the same composition as the interstellar medium from which these stars were formed. High-mass stars, with more than 1.5 times the mass of the Sun, have a large radiative zone that is not enveloped by a convective zone. The temperature-sensitive hydrogen-burning reactions of the CNO cycle cause the development of a convective core in these stars.

At the time that Bethe and Weizsäcker were carrying out their pioneering investigations, Eddington's considerations of the internal constitution of stars suggested that the body of the Sun contains about 35 percent hydrogen. This meant that heavier elements could have a dominant role in making our star shine. The CNO cycle, in fact, was once thought to be mainly responsible for solar-energy production. This was the view taken by George Gamow in his influential book, *The Birth and Death of the Sun*, first published in 1940. Given the knowledge available at the time, it was the correct hypothesis. If the relative amounts of hydrogen were less and those of carbon, nitrogen, and oxygen greater, as once was believed, then the CNO cycle would be dominant at the Sun's central temperature.

The role of radiation and convection in transporting energy out of the stellar core also depends on the mass of a main-sequence star (Fig. 8.7). When the stellar mass is comparable to that of the Sun, the energy-generating core is surrounded by a radiative zone and topped by a convective zone. Main-sequence stars with a mass of more than 2 solar masses have a convective core. In these stars, the rate of energy generation by the CNO cycle is sensitive to temperature, so the fusion is highly concentrated in the core. Consequently, there is a high-temperature gradient in the core region, which results in a central convection zone. The outer regions of such a massive star transport energy by radiation with little or no convection.

All stars that begin their relatively long and placid life on the main sequence are composed mainly of hydrogen. These stars are initially uniform balls of plasma with the same composition throughout, like a child with clear eyes and pure, innocent thoughts. As time passes, the central stellar core is changed slowly from hydrogen to helium, so the inside of a main-sequence star eventually becomes different from the outside, as can happen when humans age. Eventually, the core is all used up, exhausting its supply of hydrogen by converting it into helium. The star's nuclear furnace is shut down and all that remains is to leave the main sequence and become hot enough inside to burn helium – the ash of its former hydrogen-burning fires.

Helium Burning in Giant Stars

The fact that giant stars are connected to the main sequence of the H–R diagram suggested that the giants are the next stage of stellar life. However, because giant stars have larger luminosities at lower disk temperatures than main-sequence stars, they seemed to shine by a different and unknown process. The enigma was resolved partially when the Estonian astronomer Ernst Öpik, working at the Armagh Observatory in Northern Ireland, argued that the inside of a giant star can become very hot and dense at the same time that its outer parts become cool and rarefied.

The hydrogen-burning process of a main-sequence star is confined to the central stellar core, which is surrounded by an inert, nonburning envelope in which no nuclear reactions take place. When the core hydrogen is exhausted, the core is forced to contract, for it can no longer support itself under the crush of gravity. The central temperature will rise to about a 100 million K when gravitational forces compress it to a smaller volume and increase the mass density a 1,000 fold, to about 10 million kilograms per cubic meter. The rapid increase in core temperature causes the surrounding hydrogen envelope to expand, producing a vast, cool envelope of low density (Fig. 8.8). These spectacular changes in both the inside and outside of a dying main-sequence star account for the observed characteristics of red giant stars.

Öpik realized that the fantastic increase in central temperatures and densities of giant stars opened up a new source of energy not available to the main-sequence stars, and he proposed that the helium ash produced by hydrogen burning would serve as the nuclear fuel for giant stars. In effect, there wasn't any more food left on the table, so astronomers had to bring in the dessert.

The main difficulty with this scenario was that there is no stable nucleus of atomic weight 5, and this gap seemed to provide an impenetrable barrier for the synthesis of heavier elements from helium of weight 4 using protons of weight 1. A proton could not be attached to a helium nucleus to make the next heavier substance.

It took more than a decade to resolve the difficulty, which was explained almost simultaneously by Öpik and the American astronomer Edwin E. Salpeter. The three helium nuclei must combine to jump over the barrier, like a quarterback hurling himself over the line of scrimmage in a goal-line play. That is, when the core of a star reaches a sufficiently high temperature, helium nuclei can be converted to carbon nuclei by triple collisions of helium nuclei, thus circumventing the mass 5 difficulties.

This release of energy by fusing helium into carbon within a star is known as *helium burning*. It also is called the *triple alpha process* because the helium nucleus is an alpha particle and a triple collision is required to make a nucleus of the carbon atom.

The difficulty in pushing together even two helium nuclei is exacerbated by their electrical charge. Each helium nucleus contains two protons; therefore, the electrical repulsion between two helium nuclei is four times that between two protons. The quantum-mechanical tunneling effect would help, as it does in permitting two protons

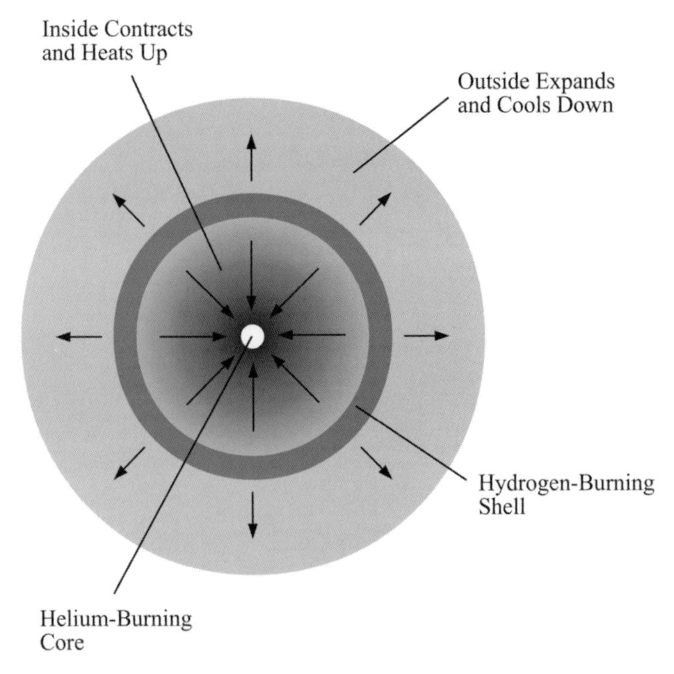

Figure 8.8. Formation of a giant star. When a main-sequence star consumes the hydrogen in its core, the inside of the star contracts and heats up, causing the outside to expand and cool down. Hydrogen burning resumes in a shell that envelops the collapsing core. The center of the star eventually heats up to about 100 million K, which is hot enough to burn helium and stop the core collapse. A giant star then has been created with a luminosity of about 100 times that of the Sun and a radius of approximately 50 times the radius of the Sun.

to fuse together in the Sun. The rise in the central temperature of a giant star is needed to increase the number of helium nuclei moving fast enough to penetrate the larger electrical barrier with the aid of tunneling, which incidentally also explains why helium burning does not occur in low-temperature, main-sequence stars. Only the helium nuclei in the high-velocity tail of the Maxwell speed distribution can merge together in giant stars, which means that most of the helium nuclei are not moving fast enough to merge and that the helium-burning reactions occur relatively slowly.

Under most circumstances, helium burning still would be exceedingly unlikely because it involves the nearly simultaneous collision of not two but rather three helium nuclei. Such a triple collision is favored by two exceptionally large collision cross sections, termed *resonance reactions* in the parlance of nuclear physics. The bigger the cross section for a collision, the more likely it will occur, like the greater possibility of throwing a ball against a house rather than a tree trunk.

The details are somewhat complicated, but suffice it to say that the mass 5 barrier can be overcome. In retrospect, it seems almost miraculous that nature has conspired in this way to make helium burning and the very existence of giant stars occur. Our understanding of the evolution of main-sequence stars into these larger, more luminous counterparts was stimulated by investigations of star clusters.

8.3 Using Star Clusters to Watch How Stars Evolve

Everything is in a perpetual state of transition all across our world and throughout the universe. Like humans, mountains, and even cities, entire stars had a beginning, followed by a long period of growth and inevitable decay, eventually turning into something else. The shining Sun, for example, coalesced from the darkness 4.6 billion years ago and will return to it several billion years hence, becoming a white dwarf star that eventually will fade away. All stars are impermanent beacons that eventually will cease to shine, vanishing like a circle of fire turning to ash.

The atoms moving inside a star are in equilibrium with gravitational forces and seem to be supporting a star forever. However, the balance does not last and the stability eventually is disrupted. A star's nuclear source of energy, which keeps the subatomic particles hot and moving, is depleted and the star's internal equilibrium falls apart.

Once a star is formed, it enters the main sequence, where it produces energy in its core from the fusion of hydrogen into helium. The star then has settled down for a long, rather uneventful life, the longest stop in its life history. The total number of stars and their lifetime both rapidly increase down the main sequence from the high-mass, luminous, short-lived upper end to the low-mass, dimmer, long-lived lower end.

The more massive and luminous stars have a lifetime measured in millions of years, and the hottest stars die first. Some already have extinguished their fire, turned off their light, and descended into the invisible dark where numerous dead stars lie. Other less massive Sun-like stars have a relatively long life of billions of years, but they also inevitably will vanish, shining their substance away.

We can use the H–R diagram of star clusters to map out the stages of stellar transfiguration. As time elapses, the more massive stars evolve into the next phase of stellar life and the main-sequence disappears from the top down. Very massive stars at the upper left of the main-sequence become supergiants; those with intermediate masses comparable to the Sun become red giants.

Stars within a star cluster are all of the same approximate age, within a few million years, dating back to the formation of the cluster. They also began with the same initial composition of material and exhibit a full range of stellar mass. Because the stars in a given cluster are all at the same distance from the Earth, we can obtain direct observations of their relative luminosity without knowing the distance.

A cluster H–R diagram can be used as a clock, dating the age of the cluster and the stars in it by the place of their turnoff from the main sequence to become supergiants or giants. These luminous, high-mass stars are the first to exhaust their hydrogen fuel, leaving the main sequence of the cluster H–R diagram and evolving into supergiants or giants. Stars with a luminosity and temperature greater than the turnoff value all have evolved away from the hydrogen-burning state of stellar life, and the age of the cluster is equal to the main sequence lifetime of stars at this turnoff point. The lower the luminosity and temperature of the turnoff point, the older is the star cluster.

The main-sequence turnoffs of the loosely bound open star clusters like the Pleiades indicate that they are approximately 100 million years old (Fig. 8.9). Such relatively young

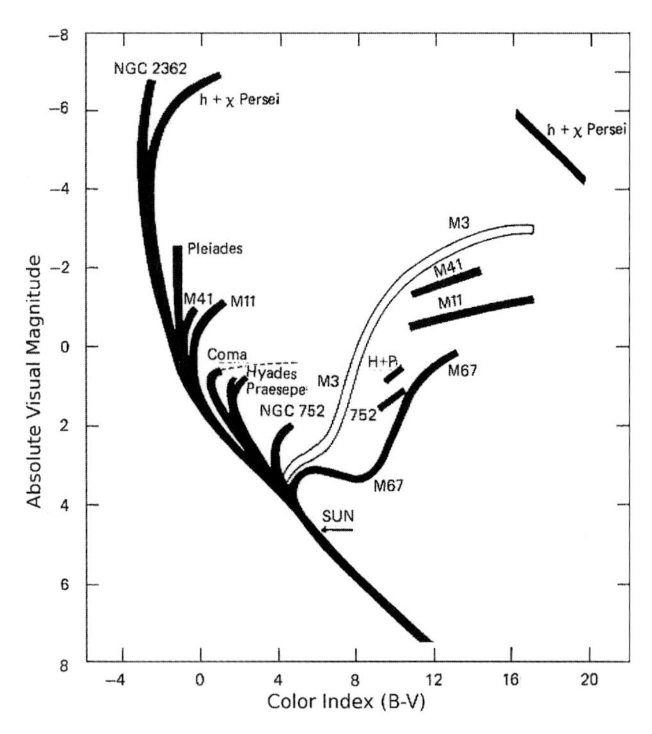

Figure 8.9. Open star clusters in the H–R diagram. The open star clusters are relatively young, and most of their stars have not yet left the main sequence in the H–R diagram. The youngest clusters, such as the Pleiades, retain all but the topmost part of the main sequence. The turnoff point of the Pleiades star cluster from the main sequence indicates an age of roughly 100 million years. The Hyades star cluster, which turns off about halfway down the main sequence, is about 600 million years old. The lowest open cluster in this diagram, M67, is an estimated 5 billion years old, with a main sequence that stops just above the Sun. One globular star cluster, M3, is shown for comparison. Allan Sandage (1926–2010) first published this diagram in 1957 (*Astrophysical Journal* 126, 326 [1957]).

clusters are identified by the membership of O and B stars, which would leave the main sequence in a relatively short time.

The main-sequence of the H–R diagram for globular star clusters does not contain these hot, luminous stars (Fig. 8.10), and these star clusters have ages between 10 billion and 14 billion years – or about two to three times the age of the Sun. This indicates that the masses of the stars at the main-sequence turnoff point of globular clusters are less than the Sun. Due to their great age and numerous stars, the H–R diagrams of these dense stellar concentrations help us watch how stars evolve to the later stages of stellar life.

Such investigations involve theoretical calculations of precisely how long a main-sequence star's central fuel supply can last and models of what happens when its fuel is used up. Martin Schwarzschild, the son of German astronomer Karl Schwarzschild, was one of the first to examine this phase of stellar evolution. After emigrating to the United States, Martin Schwarzschild used theoretical models and primitive computers, developed by his Princeton colleague John von Neumann, to chart the evolutionary trajectory of a star and compare it to

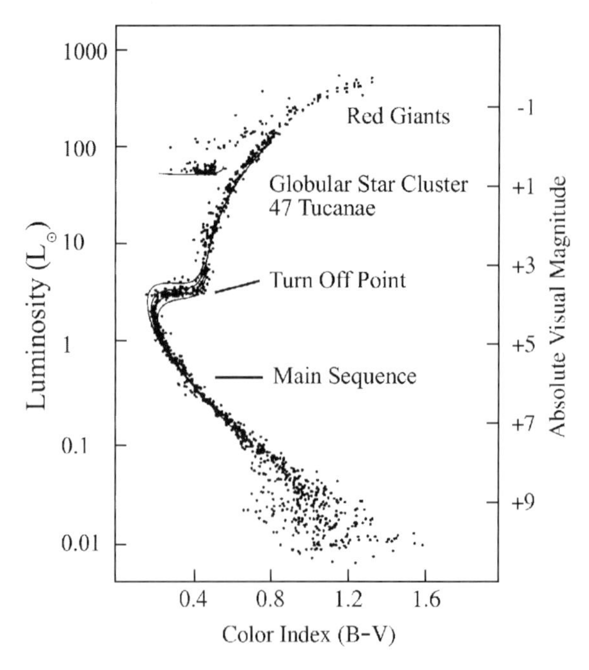

Figure 8.10. How old is a globular star cluster? A plot of the luminosity (*left vertical axis*), in units of the Sun's luminosity $L_{\odot}$, against the color index, B−V (*bottom horizontal axis*), for the stars in the southern globular star cluster 47 Tucanae, also designated NGC 104. The absolute visual magnitudes, M_V, of the stars also are shown (*right vertical axis*). Although low-mass, relatively faint stars are still on the main sequence (*diagonal line from middle left to bottom right*), the massive, bright stars in the cluster have left the main sequence and are evolving into giant stars (*top right*). Theoretical tracks, called *isochrones*, show the evolutionary distributions at different ages of 10 billion, 12 billion, 14 billion, and 16 billion years from top to bottom and left to right. The best fit to the observed data corresponds to an age between 12 billion and 14 billion years for this star cluster. (Courtesy of James E. Hesser.)

the various kinks, bends, and gaps of missing stars on the observed H−R diagrams of globular star clusters. His theoretical evolutionary models were facilitated by the fact that all of the stars in globular star clusters have the same initial chemical composition.

Martin teamed with Allan Sandage, then a graduate student at the California Institute of Technology, whose H−R diagrams included faint stars that connected the main sequence to the red-giant branch (Fig. 8.11); this provided a sound observational basis for stellar aging.

When the hydrogen-burning fires are quenched in a star of roughly solar mass, the shrinking stellar core heats up and causes the star as a whole to swell into a bloated red giant. The gravitational energy released by the inert, nonburning core then is spread over a much larger area, resulting in a lower disk temperature and a shift of the visible starlight into the red part of the spectrum. This accounts for the red-giant branch in the H−R diagram of globular star clusters.

The rising heat from the collapsing core ignites hydrogen burning in an internal shell that envelops a core of inactive helium. As realized by Schwarzschild and the English astronomer Fred Hoyle, the giant core is compressed into a degenerate state and eventually heats up to a temperature of about 100 million K, when core helium burning begins. Once the star

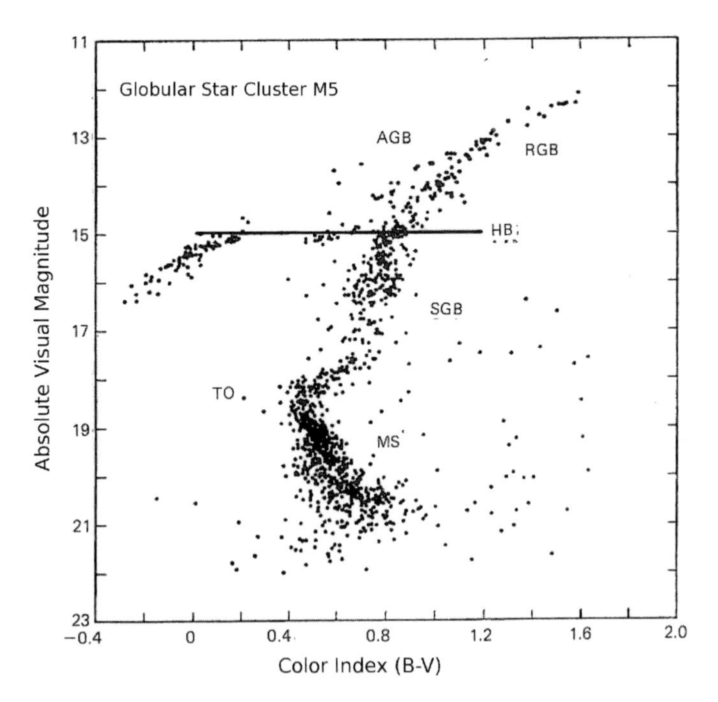

Figure 8.11. Globular star cluster in the H–R diagram. The Hertzsprung–Russell diagram for the globular star cluster M 5, where the absolute visual magnitude (*left vertical axis*) is plotted as a function of color index (*bottom horizontal axis*). It is very different from the H–R diagrams for open star clusters shown in Figure 8.9. The high-mass stars in this globular star cluster have left the main sequences at a relatively low turnoff point, denoted TO, indicating a greater age than the open star clusters. This diagram illustrates the evolutionary tracks of these stars into the red-giant branch, designated RGB (*top right*), as well as other evolutionary stages such as the subgiant branch, denoted SGB; the asymptotic giant branch, designated AGB; and the horizontal branch, denoted HB, that extends to the left. The gap of missing stars in the horizontal branch for the globular star cluster M 5 shows the instability strip of pulsating stars, known as RR Lyrae stars. Halton Arp (1927–) published the data shown in this diagram in 1962 (*Astrophysical Journal* 135, 311 [1962]).

is fusing helium into carbon within its core, it enters the horizontal branch of the cluster H–R diagram and technically no longer is considered a red giant. As then demonstrated by Schwarzschild and Richard Härm, the helium begins to burn abruptly, releasing a flash of intense energy that rejuvenates a star's luminous output. This helium flash initiates the core helium fusion into carbon, which can last about 100 million years.

When the core helium is exhausted, the carbon and oxygen core collapses, helium is burned in a surrounding shell, and the star briefly rises to giant status a second time, along the asymptotic branch of the H–R diagram. A period of instability then has begun, where the star can pulsate, but it is approaching the end of its life.

In contrast to their long, calm life on the main-sequence, the red giants lead a relatively brief life of frenzied activity. Due to the high central temperatures needed for helium burning, their nuclear-fusion rate is higher and their life expectancy shorter than for hydrogen burning in a star of the same mass. The red giants come and go, heating up and cooling down, and crossing back and forth across the H–R diagram. At different phases in their evolution, they might flash, blow off their outer atmospheres, and pulsate in and out. An even briefer stellar life, with more violent winds, is expected for the more massive main-sequence stars that become supergiants, and they help explain where most of the heavier elements originate.

8.4 Where Did the Chemical Elements Come From?

Advanced Nuclear Burning Stages in Massive Supergiant Stars

A supergiant star has a much greater mass, interior compression, central temperature, and luminosity than a giant star, which is why we use the appellation "super." The supergiants pass through the same early stages of stellar life as the giants but at a faster rate. Unlike their counterparts of lesser mass, these massive stars, with a mass above 10 solar masses, quickly consume the hydrogen in their cores and transform into colossal supergiants that burn the next available nuclear fuel: helium. A star with a mass of 25 times that of the Sun, for example, will complete hydrogen burning by the CNO cycle in about 7 million years and helium burning in a mere 660,000 years. In contrast, the Sun will take roughly 10 billion years to complete hydrogen burning by the P-P chain and another 100 million years to consume the helium within its core.

When the core helium is consumed, the evolutionary paths of high-mass and moderate-mass stars diverge. The cores of supergiant stars are so massive that they can contract and heat up enough to burn carbon, thereby stopping the core collapse and shining with renewed vigor. Carbon nuclei have an electrical charge equivalent to six protons; therefore, a formidable electrical repulsion separates them and collisions at great speed are required for its penetration. This can happen when the temperature rises to about a billion K.

In contrast, the giant stars are not sufficiently massive to burn anything heavier than helium because they never get hot enough inside. In technical terms, degeneracy pressure halts the contraction of the inert, nonburning carbon core in a giant star before it can become hot enough for carbon fusion. These stars have nowhere to go, so they expel their outer layers into surrounding space and collapse inside to an Earth-sized white dwarf star that eventually fades away.

A supergiant is so massive that it can enter progressively more advanced nuclear-burning stages. The core helium is converted into carbon, and some of the newly formed carbon nuclei can fuse with helium nuclei to make oxygen. When the helium is gone, the core contracts until it becomes hot enough to burn carbon into neon, which temporarily restabilizes the core. Each time the core depletes the elements that it is fusing, or "burning," it shrinks and heats up until it becomes hot enough for fusion reactions of the nuclear ash, continuing up the chain of successively heavier abundant elements. At the same time, nuclear fusion of the earlier fuel continues in overlapping shells at lower temperatures. Layer upon layer of nuclear burning shells are created deep down inside a supergiant.

The aging of a supergiant star accelerates rapidly, as it often does for humans when they get old. The supergiant consumes its internal fuel sources at ever-increasing central temperatures and rates (Table 8.3). Due to the higher temperatures needed for these nuclear reactions to occur, they also proceed at a much more rapid rate than hydrogen burning, and the thermonuclear lifetime of the supergiant stars therefore is much shorter than those of even the giant stars. This is one reason that so few supergiant stars are observed.

Table 8.3. *Nuclear-fusion processes in a supergiant star of 25 solar masses*[a]

Core Fusion Process	Central Temperature (K)	Central Density (kg m^{-3})	Duration (years)
Hydrogen burning (H → He)	3.7×10^7	3.8×10^3	7,300,000
Helium burning (He → C and O)	1.8×10^8	6.2×10^5	660,000
Carbon burning (C → Ne)	7.2×10^8	6.4×10^8	165
Neon burning (Ne → Mg and Si)	1.4×10^9	3.7×10^9	1.2
Oxygen burning (O → Si)	1.8×10^9	1.3×10^{10}	0.5
Silicon burning (Si → Fe)	3.4×10^9	1.1×10^{11}	0.004

[a] Adapted from Weaver, Zimmerman, and Woosley, *Astrophysical Journal* 225, 1021 (1978).

In the terminal stages of a supergiant's life, the inner core is converting silicon into iron and onion-like overlying layers are burning lighter elements such as oxygen, neon, carbon, and helium. However, when the iron nuclei in the core of such a star are pushed together, no energy is released. The iron does not burn, regardless of how hot the star's core becomes; the nuclear fires are extinguished; and the star has reached the end of life. It can never again shine by any slow nuclear-fusion process, and there is no energy left to support the core. The star has become bankrupt, completely spending all of its internal resources; there is nothing left to do but collapse.

Although a person may never recover from unexpected poverty, a supergiant star has a final moment of glory. The inert core collapses in less than 1 second, bounces, and then explodes as a supernova with the light of 1 billion suns (see Section 11.5). The shattered star and all of the elements made inside it then are dispersed into surrounding space. This material provides the seeds for future planets and stars, which explains where most of the heavy elements now found in the Earth and the Sun came from.

Origin of the Material World

What accounts for the origin of the chemical elements of which our everyday world is composed? No one could explain how these atoms were produced until the early twentieth century, when subatomic particles – the protons and neutrons – were found in the compact nucleus of the atom. In the early stages of these investigations, the English astronomer Arthur Eddington realized that the hot, dense interiors of stars might provide a location for element transformation. He wrote in 1920 that "I think that the suspicion has been generally entertained that the stars are the crucibles in which the lighter atoms which abound in the nebulae are compounded into more complex elements."[18]

Eddington was located at Cambridge University, where other scientists then were demonstrating that the nuclei of all atoms are composed of the nucleus of the hydrogen atom, the proton. Eddington soon convinced most astronomers that subatomic, or nuclear, energy must power the stars.

Within a decade, the great stellar abundance of hydrogen had been established, and Robert d'Escourt Atkinson showed how element synthesis in stars, starting with the proton-proton reaction, might account for both stellar energy and the origin of the elements. In his view, the observed relative abundance of the elements could be explained by the creation of the less abundant, heavy nuclei from the more abundant, lighter nuclei, particularly hydrogen and helium. The formation of heavy nuclei from the nuclear reactions of lighter nuclei is termed *nucleosynthesis.*

Baron Carl Friedrich von Weizsäcker proposed two mechanisms for nucleosynthesis, either within stars or in the primeval "fireball" explosion out of which the expanding universe arose. Today, we know that our material world indeed was synthesized in both places. Most of the heavy elements that now are found in the universe were created within former stars by various nuclear reactions at different times and under different physical conditions. However, all of the hydrogen and most of the helium now observed in the universe was synthesized in the big-bang "fireball" explosion. Our understanding of the details of these processes is intimately related to the observed element abundance in the Sun.

The Observed Abundance of the Elements

An important key to understanding how stars might synthesize the elements is obtained from their relative abundances, initially studied by chemists rather than astronomers. The American chemist William D. Harkins, for example, found an important clue when he noticed that elements of low atomic weight are more abundant than those of high atomic weight and that on average, the elements with even atomic numbers are about 10 times more abundant than those with odd atomic numbers and about the same number. These features led Harkins to conjecture that the relative abundances of the elements depend on nuclear rather than chemical properties and that heavy elements must have been synthesized from the lighter ones.

Decades later, astronomers showed that stars are the crucibles in which all but the lightest elements are formed by nuclear reactions that convert light elements into heavier ones, and that the systematic decline in the abundance of heavier elements can be attributed to the relative scarcity of stars that have evolved to the stage that creates them.

Two other American chemists, Hans E. Suess and Harold Clayton Urey, provided a detailed discussion of the elemental and isotopic abundances of the Sun and similar stars, calling attention to the many fluctuations that appear in the general trend of an exponential decline with increasing atomic weight. This discussion served as a major stimulus for modern ideas concerning stellar nucleosynthesis.

Synthesis of the Elements Inside Stars

The grand concept of nucleosynthesis in stars first was established definitively in two papers written by the English astronomer Fred Hoyle. He showed that both theoretical and experimental considerations were required, and he placed the concept of stellar nuclear reactions within the framework of stellar structure and evolution using the then-known nuclear data.

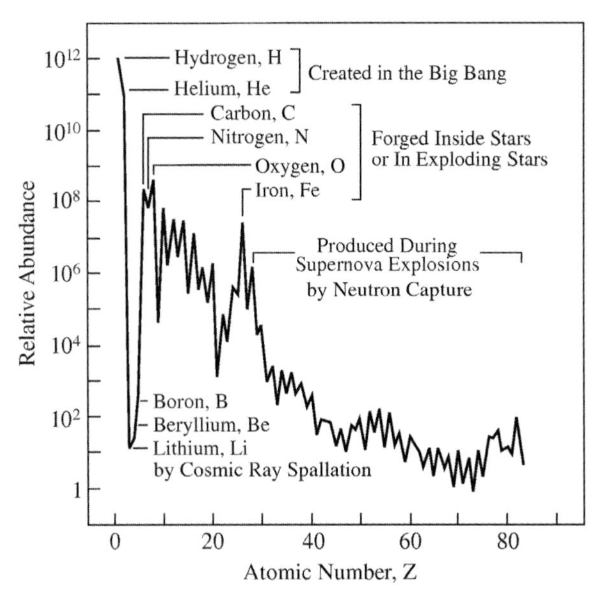

Figure 8.12. Abundance and origin of the elements in the Sun. The relative abundance of elements in the solar photosphere, plotted as a function of their atomic number Z, which is the number of protons in an atom's nucleus and roughly half the atomic weight. The abundance is plotted on a logarithmic scale and normalized to a value of 1 million million, or 1.0×10^{12}, for hydrogen. Hydrogen, the lightest and most abundant element in the Sun, was formed about 14 billion years ago in the immediate aftermath of the big bang that led to the expanding universe. Most of the helium now in the Sun also was created then. All of the elements heavier than helium were synthesized in the interiors of massive stars that no longer shine, and then were wafted or blasted into interstellar space where the Sun subsequently originated. Carbon, nitrogen, oxygen, and iron were created over long time intervals during successive nuclear burning stages in former massive stars as well as during their explosive death. Elements heavier than iron were produced by neutron-capture reactions during the supernova explosions of stars that lived and died before the Sun was born. The light elements boron, beryllium, and most of the lithium are believed to originate from heavier cosmic-ray particles that were stripped of some of their ingredients by collisions, in a process called *spallation*. The exponential decline of abundance with increasing atomic number and weight can be explained by the rarity of stars that evolved to later stages of life. (Data courtesy of Nicolas Grevesse.)

The detailed abundance data of Suess and Urey served as an inspiration for a comprehensive review of the major element-producing reactions in stars written by Geoffrey R. Burbidge, E. Margaret Burbidge, Fred Hoyle, and William Fowler. Their seminal publication, titled "Synthesis of the Elements in Stars" and published in the *Reviews of Modern Physics*, is known as the B^2FH paper, after the surnames of the four authors. It provided the fundamental framework on which most subsequent studies of stellar nucleosynthesis are based (Fig. 8.12).

As a star ages, it experiences successive core contraction and stable core burning stages, in which heavier elements are synthesized from lighter ones and the nuclear ash of one stage becomes the fuel of the next one. These core nuclear-burning reactions proceed at a progressively hotter, denser, and faster pace, such as successive hydrogen, helium, carbon, neon, oxygen, and silicon burning. Elements with an even number of protons in their nuclei, (e.g., carbon and oxygen) are more abundant because they are formed by nuclear reactions with helium, which contains two protons rather than one in its nucleus.

The fearsome B^2FH four created their origin scenario at Cambridge University, where Hoyle was in residence. The English husband-and-wife Burbidges had moved to the United States in 1955; Fowler, just recently graduated from the California Institute of Technology, was distinctly American, as evidenced by his cheerful disposition and pronounced love of steam engines and baseball.

Margaret Burbidge must have found it initially difficult in the male-dominated world of astronomers. At first, she could get viewing time at the Mount Wilson telescope only by pretending to be her husband's assistant. But Margaret triumphed in the end; I remember Margaret Thatcher, then serving as Education Minister, presenting an award to Margaret Burbidge at a meeting of the International Astronomical Union in Brighton, England. Burbidge's husband Geoffrey walked out of the ceremonial hall during the presentation; but we do not know why.

As realized by B^2FH, the temperature is not everywhere the same inside a star; therefore, its nuclear evolution is most advanced in the central regions and least or not at all advanced near its outer regions. Thus, the composition of a star could not be expected to be uniform throughout. At the end of its life, a massive star would be layered with successively thinner shells of helium, carbon, oxygen, and silicon surrounding an inert and nonburning iron core that exploded as a supernova, casting out the overlying layers containing heavy elements.

Although the general flow of the observed relative abundance of the more abundant elements can be explained by successive static, or nonexplosive, burning stages within stars, elements heavier than iron – as well as many of the detailed ups and downs of the abundance curve for lighter elements – were attributed to fast nuclear reactions during the explosive, supernova death of massive stars. The American hydrogen-bomb tests had indicated that heavy elements are created by rapid neutron bombardment during the explosions, and Fowler and Hoyle had shown that similar processes occur in supernovae.

B^2FH used the slow, s, neutron capture; rapid, r, neutron capture; and proton, p, capture to explain the details of the abundance curve. Some of the nuclear-burning reactions produced free neutrons, residing outside any atomic nucleus. These neutrons can slowly fuse with abundant heavy nuclei to produce the relatively rare ones, encountering no electrical obstacle because the neutron is electrically uncharged. Unhindered by electrical repulsion, the neutrons permit the extension of stellar nucleosynthesis from iron all the way to uranium. A free neutron also can decay into a proton, and the rapid capture of both neutrons and protons during supernova explosions helps to forge the heavier elements.

Big-Bang Nucleosynthesis

The synthesis of elements inside stars is an incomplete scenario for it does not explain the origin of any of the hydrogen and most of the helium in the observable universe. Moreover, because deuterium is destroyed rapidly inside stars, there also must to be another explanation for its cosmic existence. As proposed by the eclectic George Gamow and his colleagues, these elements must have been produced during the exceptionally hot and dense first moments of the big-bang explosion that gave rise to the expansion of the universe. In fact, it once was thought that all of the elements found today might have been created back then.

The relative abundance of elements produced during the detonation of atomic bombs was studied during World War II (1939–1945) to determine which nuclear reactions had occurred; Gamow therefore knew how elements were being produced during chain reactions in these explosions. He reasoned that similar nuclear reactions in the primeval "fireball," or big bang, built up the chemical elements we see today, and he applied the comparison to the cosmic abundances determined by astronomers, chemists, and other scientists. Working with his young colleague Ralph A. Alpher, he proposed that all of the elements were produced in a chain of nuclear reactions during the earliest stages of the expanding universe. They supposed that the original substance of the material universe, the cosmic "ylem," consisted solely of neutrons at high temperature. Some of these neutrons decayed into protons, and successive captures of neutrons by protons led to the formation of the elements.

This novel idea was published in a paper titled, "The Origin of the Chemical Elements," with Hans Bethe added as an author, even though he contributed nothing to the research or the writing. This was a pun on the first letters of the Greek alphabet – alpha, beta, and gamma, or α, β, and γ – for Alpher, Bethe, and Gamow.

Two years after the α-β-γ paper, the Japanese scientist Chushiro Hayashi showed that in the first moments of the expansion, the temperature was hot enough to create more exotic particles, such as neutrinos and positrons (e.g., the anti-matter particles of electrons). The mutual interaction of all of the subatomic particles present in the first moments of the big bang establishes the relative numbers of neutrons and protons, which in turn determine the amount of helium produced.

With this correction, modern computations by Robert Wagoner, Fowler and Hoyle, as well as David N. Schramm and others, conclusively demonstrated that all of the hydrogen and deuterium – and most of the helium – found in the cosmos today were synthesized in the immediate aftermath of the big bang. These processes are known as *big-bang nucleosynthesis*, and they indicate how much matter is now in the universe in both visible and invisible forms.

So, Gamow was partly right. The lightest and most abundant element, hydrogen – and therefore the majority of atoms that we see today – indeed were formed at the dawn of time even before the stars existed, in the immediate aftermath of the big bang that produced the expanding universe. All of the hydrogen that is now in the universe was created then, about 14 billion years ago, and so was most of the helium, the second most abundant element. When we buy a balloon inflated with helium, we are getting atoms created about 14 billion years ago; all of the hydrogen now found in stars and interstellar space also was created then.

Why weren't all of the heavier elements produced in the early stages of the big bang? By the time that the expanding universe became sufficiently cool to allow nucleosynthesis to occur, at about 1 billion K, it also became low enough in density that *three-body reactions* could occur only very infrequently. At higher temperatures, radiation destroys nuclei as fast as they are made; by the time the temperature has dropped to 1 billion degrees, the density was only sufficient to produce helium and small amounts of deuterium and lithium by the merger of two nuclei, known as *two-body reactions*. The rate of helium burning by the triple-alpha process is proportional to the square of the density at a given temperature; at 1 billion degrees, the density of the expanding universe is too low by many orders of magnitude for appreciable operation of the triple-alpha process.

In simpler terms, the expanding universe was not and is not in equilibrium, so it rapidly cooled down and thinned out to low density, making the simultaneous collision of three helium nuclei – the alpha particles – nearly impossible. The dense cores of giant stars, however, are in equilibrium for the long intervals of time needed to accrue noticeable amounts of carbon by the triple-alpha process. Only the very light elements, such as deuterium and helium, could be produced by two-body reactions during the big bang, whereas carbon had to be synthesized by three-body reactions in the interior of stars.

This completes our account of the origin of the elements inside stars and during the big bang, with one oversight: the under-abundant, light elements with atomic weights between hydrogen and helium. These light elements – boron, beryllium, and most of the lithium – probably were produced by *spallation reactions* in which energetic charged particles, known as cosmic rays, strip off the components of heavy nuclei to form light nuclei.

The First and Second Generation of Stars

Because big-bang nucleosynthesis produced no elements heavier than helium, the earliest stars had to be composed of the lightest abundant elements: hydrogen and helium. This first generation of stars, known as Population I, was untouched by heavier elements. Then, as a result of ongoing stellar alchemy, the most massive first-generation stars forged heavier elements in their cores, scattering them into space by winds or explosions as they died. These heavy elements, known as *metals* to astronomers, then were recycled and incorporated into second-generation stars. Because some of their material came from previous stars, these Population II stars are polluted somewhat by the heavier elements, the metals.

As Hans Bethe exclaimed in the last paragraph of his 1967 Nobel Lecture, "stars have a life cycle much like animals [including humans]. They get born, they grow, they go through a definite internal development, and finally they die, to give back the material of which they are made so that new stars may live."

Observations of stellar spectra confirm this scenario. Very old Population I stars, which formed a long time ago when the universe was young, have less than 0.1 percent of their mass in elements heavier than hydrogen or helium. We see these survivors of the earliest times in the oldest globular star clusters. Astronomers have not yet found the completely pure stars with absolutely no metals, but they are confident that the second-generation Population II stars contain a greater proportion of heavy elements, at 2 percent to 3 percent of their mass.

Cosmic Implications of the Origin of the Elements

During the billions of years before the Sun was born, massive stars reworked the chemical elements, fusing lighter elements into heavier ones within their nuclear furnaces. Carbon, oxygen, nitrogen, silicon, iron, and most of the other heavy elements were created this way. The enriched stellar material then was cast out into interstellar space by the short-lived massive stars, gently blowing out in their stellar winds or explosively ejected within supernova remnants.

The Sun and its retinue of planets condensed from this material about 4.6 billion years ago. They are composed partly of heavy elements that were synthesized long ago and far away in the nuclear crucibles of stars that lived and died before the Sun and planets were born. The Earth and everything on it spawned from this recycled material – the cosmic leftovers and waste products of stars that have disappeared from view.

Moreover, the first stars could not have had rocky planets like the Earth because there initially was nothing but hydrogen and helium. The only possible planets would have been icy balls of frozen gas. Without carbon, life as we know it could not evolve on these planets.

Perhaps the most fascinating aspect of stellar alchemy is its implications for life on the Earth. Most of the chemical elements in our bodies, from the calcium in our teeth to the iron that makes our blood red, were created billions of years ago in the hot interiors of long-vanished stars. Therefore, we are all made of "star stuff." If the universe were not very, very old, there would not have been time enough to forge the necessary elements of life in the ancient stellar cauldrons. The lightest element, hydrogen, needed for the water in our bodies, nevertheless was synthesized when the observable universe was very young, in the first instants of the big bang. Therefore, we are the offspring of both the stars and the big bang – true children of the cosmos.

9

The Material Between the Stars

9.1 Bright Stars Light Up Their Surroundings

Most of the universe seems dark and empty. Atoms, for example, are largely without substance, with most of their mass confined within a tiny nucleus. The space between the atoms also seems to be empty – a nothingness without anything – so the atoms can have somewhere to go, to move into. The room in which we are sitting is largely empty and so is the solar system, with the planets occupying only a miniscule fraction of its volume. The space separating two adjacent stars is more than 50 million times larger than the stars themselves. All of this emptiness provides a transparent structure to the world, giving form and shape to material things and defining their place and freeing them from their surroundings.

However, the unseen chambers of space are not empty. Our room contains the transparent air we breathe. In space, a 1-million-degree plasma always is streaming away from the Sun, enveloping the planets and filling interplanetary space with invisible electrons, protons, and magnetic fields.

Farther out, the space between the stars – which looks like an empty black void – is filled with cold atoms of hydrogen, wrapped in dark and deep shadows. Stars form out of this supposed emptiness and eventually return to the darkness from which they came. We see some of this interstellar material when the brightest stars illuminate nearby regions. The energetic starlight heats, ionizes, and lights up the rarefied surrounding material.

When telescopes were first trained on the night sky, astronomers found apparently diffuse, luminous places that are larger than stars (Figs. 9.1 and 9.2). They were called *nebulae*, the Latin word for "clouds." In the late eighteenth century, the French astronomer Charles Messier published a famous catalogue that listed the brightest nebulae as well as star clusters; they are now designated by M followed by the number in his list. Some of the nebulae remained diffuse, tenuous, and unresolved even with the biggest telescopes (Table 9.1).

Such objects were dubbed *emission nebulae* when the English astronomer William Huggins discovered that they display emission lines in their visible-light spectra. Lines of ionized hydrogen, ionized nitrogen, and a then-unidentified substance named *nebulium* were found.

Figure 9.1. Rosette Nebula. This large, circular emission nebula, or H II region, known as the Rosette Nebula, lies at a distance of about 5,200 light-years from the Earth and is about 130 light-years in diameter. Parts of this region include nebulae designated as NGC 2237, NGC 2238, and NGC 2239, as well as the open star cluster NGC 2244. The mass of the Rosette Nebula is estimated to be 10,000 solar masses. Hot O and B stars in the core of the Rosette Nebula exert pressure on the nearby interstellar material, triggering star formation, and heat the surrounding gas to a temperature of about 6 million K, causing it to emit x-rays observed from the Chandra X-Ray Observatory. (Courtesy of KPNO/CTIA.)

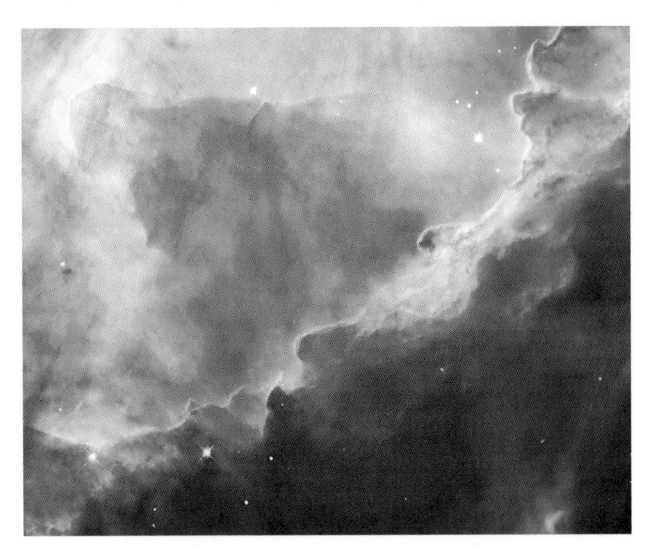

Figure 9.2. Turbulent interstellar gas. Glowing hydrogen gas in a small region of the Omega Nebula, which also is designated as M 17. The wave-like patterns of gas have been sculpted and illuminated by intense ultraviolet radiation from young, massive stars, which lie outside the picture to the upper left. The ultraviolet radiation heats the surface of otherwise cold, dark clouds of interstellar hydrogen. (A Hubble Space Telescope image, courtesy of NASA/ESA/J. Hester, Arizona State University.) (See color plate.)

Table 9.1. *Bright named emission nebulae*

Name	RA (2000) h	m	Dec. (2000) °	′	θ[a] ′	Distance (light-years)
Lagoon Nebula, M 8	18	03.8	−24	23	40 ×90	5,200
(Also known as NGC 6523, central region Hourglass Nebula, and star cluster NGC 6530)						
Omega Nebula, M 17	18	20.8	−16	11	37 ×47	5,000
(Also known as NGC 6618, Horseshoe Nebula, and Swan Nebula)						
Orion Nebula, M 42	05	35.4	−05	27	60 ×66	1,500
(Also known as NGC 1976, northwestern part is M 43, NGC 1982)						
Rosette Nebula, NGC 2237	06	33.8	+05	00	78 ×78	5,200
(Nebulous region includes NGC 2238, NGC 2239, NGC 2246, and cluster NGC 2264)						
Trifid Nebula, M 20	18	02.6	−23	02	27 ×29	5,200
(Also known as NGC 6514)						

[a] Angular diameter θ in minutes of arc, designated ′.

Astronomers turned to the hot stars embedded within the emission nebulae as the source of their luminosity and emission lines. The English astronomer Arthur Eddington, for example, showed that the ultraviolet radiation of bright blue stars, of spectral class O, will ionize and heat the surrounding gas to temperatures of about 10,000 K. The Dutch astronomer Herman Zanstra explained the hydrogen emission lines by supposing that hot, central stars ionize the surrounding hydrogen atoms and that visible light is emitted when the electrons freed by ionization recombine with protons to make hydrogen atoms.

Spectral lines of emission nebulae also arise from ionized atoms of oxygen and nitrogen undergoing "forbidden" transitions in the low-density nebulae; such transitions are improbable in the higher-density laboratory situation. They include lines designated [O II], [N II], [O III], and [N III], where O and N denote oxygen and nitrogen, respectively; II and III denote that the atom is missing one or two electrons, respectively; and the square brackets [ ] denote a forbidden line. Thus, in addition to hydrogen, the dominant spectral lines of emission nebulae are those of cosmically abundant oxygen and nitrogen as well as sulfur [S II] (Table 9.2).

As a luminous star's ultraviolet radiation moves out into surrounding space, it is absorbed and consumed; eventually, all of the available ultraviolet rays are used up in ionizing atoms

Table 9.2. *Intense spectral lines of emission nebulae*

Element	Wavelength (nm)	Element	Wavelength (nm)
[O II]	372.62	[N II]	654.81
[O II]	372.89	Hα	656.82
[O III]	436.32	[N II]	658.36
Hβ	486.1332	[S II]	671.6440
[O III]	495.891	[S II]	673.0816
[O III]	500.684		

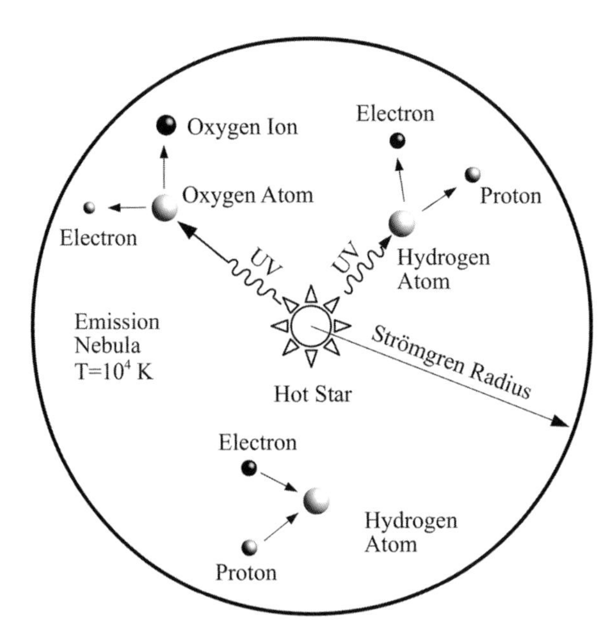

Figure 9.3. Spheres of ionization. Ultraviolet radiation, denoted by UV, from a hot star ionizes hydrogen and other atoms in its immediate vicinity, creating a nebulous region that radiates emission lines and contains abundant ionized hydrogen, denoted H II. They are known as emission nebulae or H II regions. In this figure, the size of the atoms, ions, electrons, and protons are exaggerated greatly to visualize them. The ionization by the UV creates numerous free electrons and protons that are not attached to atoms. They subsequently recombine to make atoms in a process of continued ionization and recombination. The free electrons can emit two types of radiation, illustrated in Figure 9.4. At large distances from the star, its ultraviolet rays are all absorbed and can travel no farther into surrounding space. This limits the radius of the emission nebula, or H II region, to the Strömgren radius at about 30 light-years.

close to the star. As a result, the rays cannot travel farther than the immediate vicinity of the star. This is why the night sky remains black outside the periphery of emission nebulae. As shown by the Danish astronomer Bengt Strömgren, the central star creates a sphere of ionized gas that envelops it. He found that although the interstellar hydrogen usually is electrically neutral, or un-ionized, the very hot O and B stars could generate enormous but sharply bounded spheres of ionization (Fig. 9.3). For some years, these regions were referred to as "Strömgren spheres," and their size still is designated as the Strömgren radius, which depends on the temperature of the star and the density of the surrounding material. For typical interstellar hydrogen densities of 10 million to 20 billion atoms per cubic meter, the Strömgren radius ranges from 1 to 300 light-years, for central stars with disk temperatures between 26,000 and 48,000 K.

Because hydrogen is by far the most abundant element in interstellar space, as it is in the stars, the emission nebulae contain large amounts of ionized hydrogen, designated H II; in the parlance of modern astronomers, they now are known as H II regions, pronounced "H two regions." They contain electrons that were freed from their atomic bonds during ionization (Table 9.3).

Consider a lone hydrogen atom, consisting of a single electron and a single proton, near a luminous star. The situation is becoming hot, so the electron finally gains enough energy

Table 9.3. *Physical properties of emission nebulae (H II regions)*

N = number density = N_e = electron density = 10^7 m^{-3} to 10^{10} m^{-3}
T_e = electron temperature = 10,000 K = 10^4 K
R_s = Strömgren radius = 0.65 light-years to 326 light-years = 0.2 pc to 100 pc ≈ (0.6 to 300) $\times 10^{16}$ m
T_{eff} = effective temperature of ionizing star, spectral class O9 to O5 = (3 to 5) $\times 10^4$ K

to break free of its former atomic bonds with the proton. The electron moves away at high speed due to its relatively low mass when compared to that of the proton.

If the free electron encounters another proton, the two particles attract one another by their opposite electrical charge, and one of two things can happen: The electron can continue moving into surrounding space, deflected from a straight path; or the electron can be captured into an enduring embrace, joining together with the proton to again form a hydrogen atom (Fig. 9.4).

When the electron continues on its way, it changes speed during the encounter with the proton, emitting electromagnetic radiation in the process. This radiation is known as *bremsstrahlung*, from the German word for "braking radiation"; it also is called *free-free radiation* because the electron begins and remains free. In H II regions, the free electrons emit *bremsstrahlung* radiation detected at radio wavelengths.

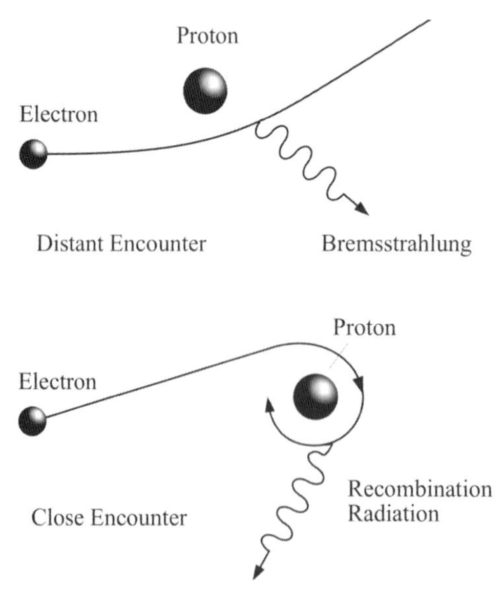

Figure 9.4. Radiative interactions between electrons and protons. When an electron moves rapidly and freely outside an atom, it inevitably passes near a proton in the ambient gas. There is an electrical attraction between the electron and proton because they have equal and opposite charge, and this pulls the electron toward the proton. If the interaction is distant, it bends the electron's trajectory and alters its speed. The electron then emits electromagnetic radiation known as *bremsstrahlung*, from the German for "braking radiation"; this also is called *free-free radiation* because the electron remains free and unattached to the proton. In a close encounter, the electron goes into orbit around the proton, forming a hydrogen atom and cascading down through allowed orbital energies. In this case, the electron emits recombination radiation, also known as *free-bound radiation*.

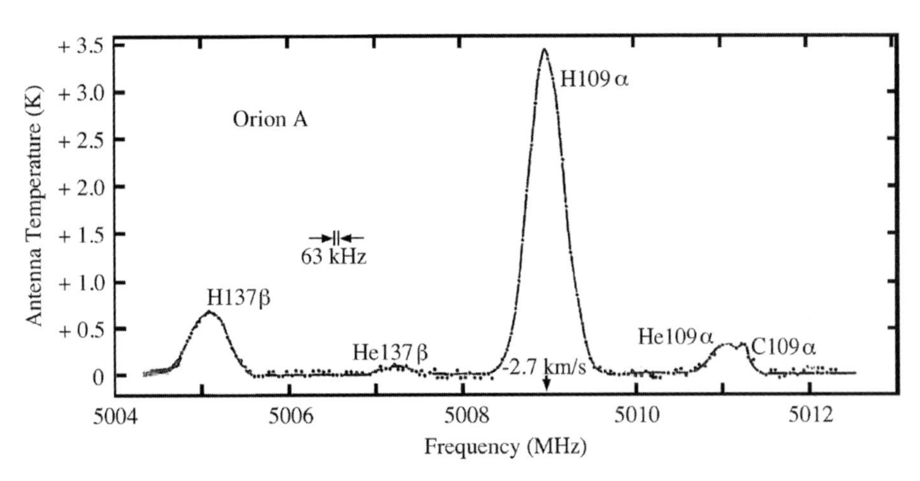

Figure 9.5. Recombination lines. Radio-frequency spectrum of the Orion Nebula, also designated as M 42, showing the hydrogen, H, and helium, He, recombination lines from the α (m − n = 1) and β (m − n = 2) transitions at high quantum numbers n = 109 and n = 137. The hydrogen lines are more intense than the helium or carbon lines because of the greater abundance of hydrogen. (Adapted from K. R. Lang, *Astrophysical Formulae*, Heidelberg: Springer Verlag, 1979.)

When a free electron is captured by a proton to make a hydrogen atom, the electron cascades down through the ladder of possible orbital energies, emitting radiation in the process. This radiation is known as *recombination radiation* because the two recombine to make an atom; it also is called *free-bound radiation* because the former free electron is bound once again. The free electrons in H II regions are always recombining with protons to make hydrogen atoms, detected at radio wavelengths (Fig. 9.5) as well as at optically visible wavelengths.

9.2 Dark Places Filled with Dust

The contrasting tones of light and dark accentuate our surroundings, helping us to see. They are what make sunrise, sunset, or a stormy day so beautiful. This interplay of light and shadow also can be seen in the celestial realm, where the dark places create shadowy silhouettes against nearby bright stars (Figs. 9.6 and 9.7).

It was the American astronomer Edward Emerson Barnard who contributed most to our early awareness of these dark regions, at a time when most astronomers thought that they were an unfathomed emptiness – literally, holes in the sky. His family was poor and, by the time he was nine years old, Barnard had begun to work as an assistant in a photographic studio, learning techniques that would be invaluable in his later career in astronomy.

Barnard became an accomplished portrait photographer and did not begin serious astronomical work until the age of 30, when he moved to the Lick Observatory on the isolated Mount Hamilton in California – where he systematically photographed the dark and bright regions of the Milky Way using wide-field portrait lenses. This work spanned 30 years: first at Lick and then at the Yerkes Observatory near Chicago, culminating in two stunning catalogues of the regions, which he noncommittally called dark markings. Their nebulous form and shape suggested to Barnard that the dark places were not empty; in some cases, they even seemed to interact with the bright regions that enfolded them.

Figure 9.6. Dark clouds in the Carina Nebula. Molecular clouds in the Carina Nebula contain so much gas and dust that they are opaque to optically visible light, forming dark and dense structures where new stars may be born. Energetic stellar winds and intense radiation from nearby massive stars are sculpting the outer edges of the dark clouds. This image is a composite of observations taken in light emitted by hydrogen and oxygen atoms using the Hubble Space Telescope. The Carina Nebula, also designated NGC 3372, is about 7,500 light-years away from the Earth and spans more than 300 light-years. (Courtesy of NASA/ESA/Hubble Heritage Project/STScI/AURA.) (See color plate.)

Figure 9.7. Cosmic eggs. A cloud of cool interstellar dust and molecular hydrogen gas may have embryonic, unseen stars embedded inside. Ultraviolet light from nearby hot stars uncovers dark globules of gas at the top of the cloud, each about the size of our solar system. They have been called "eggs," an acronym for evaporating gaseous globules. This cloud is located in the Eagle Nebula, also designated as M 16. (A Hubble Space Telescope image, courtesy of NASA/ESA/STScI/J. Hester and P. Schowen, Arizona State University.)

Whereas Barnard was interested chiefly in the peculiar shapes of the dark regions, the German astronomer Maximilian Wolf was concerned with measuring their distances and absorbing powers. By counting stars in an obscured and an adjacent unobscured region, Wolf was able to demonstrate that the dark areas absorb the light of distant stars.

Because Wolf could not detect any substantial difference in the colors of stars that lie outside of and behind the dark nebulae, he concluded that the dark regions must be composed of solid dust particles. Their scattering properties depend weakly on wavelength, unlike gas atoms that scatter light much more effectively at shorter wavelengths. If the scattering were due to smaller gas molecules, it would depend strongly on wavelength.

The interstellar absorption of starlight was demonstrated by Robert J. Trumpler's 10-year study of open star clusters at the Lick Observatory. The Swiss-born astronomer used the colors of the brightest stars in each open cluster to infer their luminosity, or *absolute magnitude*, which he combined with their observed brightness, or *apparent magnitude*, to determine their distances. However, when these distance estimates were combined with the measured angular diameters of the open clusters to infer their size, he found that the linear diameters increase with distance – and the difference was not trivial. In whatever direction he looked, more remote clusters seemed to be about twice as large as the closer ones. Moreover, the effect was systematic: The farther away a cluster was, the larger it appeared to be.

Concluding that this pervasive, systematic change in physical size was impossible, Trumpler instead assumed that all open clusters actually have the same linear diameters or physical extent. This meant that the initial distance estimates were overestimated due to the absorption of starlight by an amount that increases with distance. The greater the distance, the more the absorption, making the remote clusters look systematically fainter and even farther away than their actual distance.

Cosmic dust absorbs and scatters starlight, which means that interstellar space is not transparent. The removal of short-wavelength, blue-colored light that occurs when starlight passes through dust is known as *reddening* because it makes the observed star redder than it would be without the intervening dust. When radiation passes through a longer path, containing a greater density of interstellar dust, the amount of reddening increases, making distant stars appear redder. Reddening by the Earth's atmosphere is responsible for the red color of a sunset as well as the full red Moon seen during a total lunar eclipse.

The unfolding story of interstellar dust took an unexpected turn when two American astronomers, William A. Hiltner and John Scoville Hall, independently observed highly polarized light from reddened stars. The polarization was oriented along a common plane due to an alignment of the elongated dust particles by the interstellar magnetic field (Fig. 9.8).

What is this interstellar dust? The particles must be smaller than 1/10,000th of a meter across or they would completely block starlight and not scatter it; they also must be larger than gas molecules, the scattering of which depends strongly on wavelength. To weakly absorb and scatter starlight, interstellar dust particles must be comparable in size to the wavelength of visible light, or roughly 1 millionth of a meter across, which is about the same size as the particles of cigarette smoke or the dust motes seen when we look back toward a movie projector.

The combined effects of scattering and absorption of starlight by interstellar dust is called *extinction*. The shape of the curve of starlight extinction as a function of wavelength provides

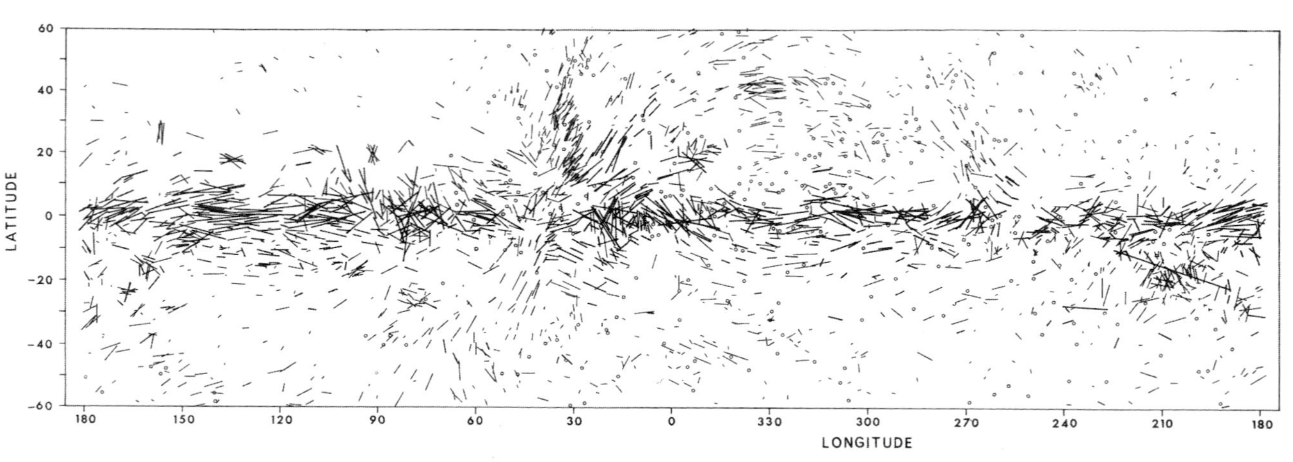

Figure 9.8. Polarized starlight. The light from nearly 7,000 stars is polarized, with the strength and direction of polarization designated by the short lines. The observations are plotted in galactic coordinates where the galactic plane, or the Milky Way, runs horizontally across the middle of the figure. The starlight polarization has been attributed to dust grains elongated along the interstellar magnetic field. (Courtesy of D. S. Mathewson.)

information on the size and composition of the interstellar dust grains; when observations were extended to ultraviolet and infrared wavelengths, more mysteries arose. The amount of extinction increases in a "bump" at ultraviolet wavelengths; it may be caused by graphite or a slightly less well-ordered form of carbon.

Ice, carbon, and silicates, or some combination of these, are most likely the principal ingredients of interstellar dust. Most of the solid silicate dust found in interstellar space probably came from the cool expanding outer envelopes of evolved giant stars that are rich in oxygen or carbon; the outer atmospheres of these stars exhibit the absorption bands of silicates.

Dust also produces strong infrared radiation that is studied using instruments aboard the Spitzer Space Telescope. Interstellar dust is measured from the emission at long infrared wavelengths, whereas dust being formed in old, evolved stars is observed using medium infrared wavelengths.

9.3 Interstellar Radio Signals

The American radio engineer Karl Jansky inadvertently discovered radio noise of cosmic origin in 1931, when radio waves were being used extensively for global communications. At that time, the Bell Telephone Laboratories assigned Jansky the task of tracking down and identifying natural sources of radio noise that were interfering with ship-to-shore radio communications. He constructed a rotating antenna that pointed sideways at the horizon and enabled identification of the interference, including the radio static produced by lightning discharges from distant thunderstorms.

Fortunately, the antenna's wide field of view also pointed partway up into the sky, thereby detecting a persistent extraterrestrial hiss of unknown origin that was comparable in intensity to terrestrial lightning. By observing the variation of its intensity as a function of direction and time of arrival, Jansky established that the radio source must lie outside the solar system. This serendipitous discovery of cosmic radio noise caused quite a stir in the general public, which was able to listen to his "star noise" and the "hiss of the universe" broadcast on local radio programs.

The public interest was appropriate, for the cosmic radio waves were tuning into worlds that cannot be seen in visible light. It is like standing by the shore of a lake on a moonless night: Only the sound of the moving waves indicates that something is out there. The radio noise was telling us about the invisible space between the stars.

However, the astronomical community almost completely ignored Jansky's results, most likely because he did not publish them in an astronomical journal and his radio techniques were outside the conventional methods of traditional astronomy. It was not until a decade later that amateur astronomer and radio engineer Grote Reber confirmed and extended Jansky's investigations; he published these new results in 1944 as an article on "Cosmic Static" in the *Astrophysical Journal*.

Because the radio emission of even the nearby Sun could not be detected with Jansky's antenna, normal stars were ruled out as a possible source of the cosmic radio emission. Their thermal radio emission would be exceptionally faint, when compared to that at visible wavelengths, and further diluted by the greater distance. Some nonthermal source of

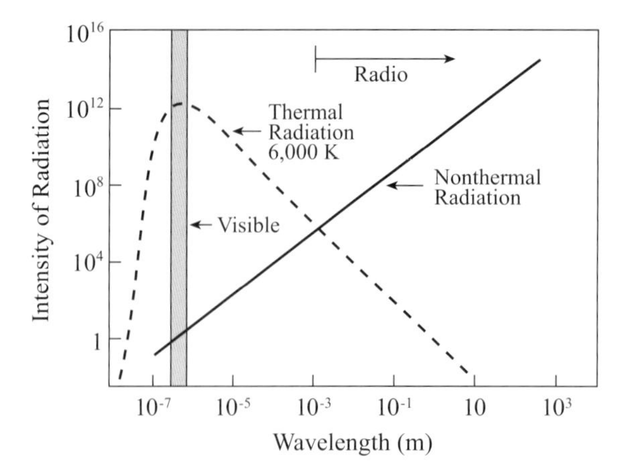

Figure 9.9. Spectrum of thermal and nonthermal radiation. When the intensity of thermal radiation is spread out as a function of wavelength, the resultant spectrum is most intense in a band of wavelengths that depends on the temperature. This peak occurs at visible wavelengths when the temperature is about 6,000 K. Such a hot gas emits radiation at longer wavelengths but at a lower intensity. In contrast, nonthermal radiation is more intense at longer radio wavelengths. High-speed electrons emit nonthermal synchrotron radiation in the presence of a magnetic field (see Fig. 9.10).

radiation was required that emitted most intensely at the longer radio wavelengths (Fig. 9.9).

As it turned out, the extraterrestrial radio signals were not coming from stars but instead from interstellar space. The cosmic emission is attributed to energetic electrons traveling at a speed close to that of light. They are similar to the cosmic-ray protons and electrons impinging on the Earth's atmosphere; however, only the electrons emit radio waves, and they are moving in the space between the stars. The electrons are called *relativistic* electrons because the equations of relativity apply at such high speeds. The speed of electrons is so great that their kinetic energy of motion cannot be attributed to the thermal energy of a hot gas at any plausible temperature.

High-speed electrons have been propelled to such large energies that it is impossible for them to bond with anything. However, it is another story when they encounter a magnetic field; they move away from or around the magnetism. To avoid the unpleasant situation, the high-speed electrons change direction and must swerve away from a magnetic field. A moving charged particle cannot move straight across a magnetic field but instead gyrates around it.

High-speed electrons spiral about the interstellar magnetic field and emit nonthermal synchrotron radiation at radio wavelengths (Fig. 9.10). The name is derived from the man-made synchrotron particle accelerator where it was first seen. Such nonthermal radiation is more intense at longer wavelengths. More energetic electrons have a shorter lifetime, expending their energy by synchrotron radiation at a greater rate than less energetic electrons. High-energy electrons also emit synchrotron radiation at a higher frequency or shorter wavelength than low-energy electrons. This accounts for the nonthermal radiation spectrum

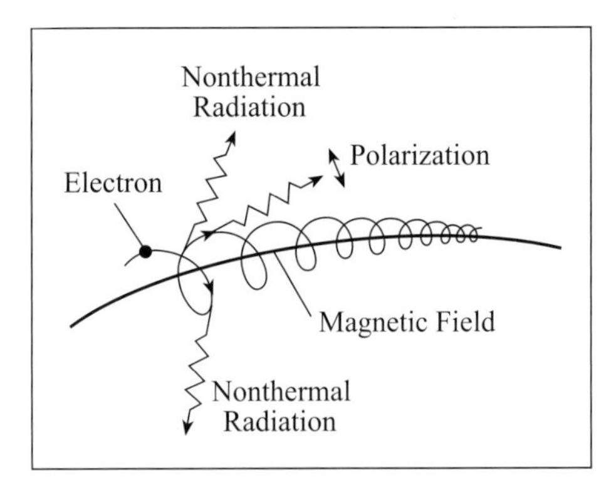

Figure 9.10. Synchrotron radiation. Electrons moving at velocities near that of light emit a narrow beam of synchrotron radiation as they spiral around a magnetic field. This emission sometimes is called nonthermal radiation because the electron speeds are much greater than those of thermal motion at any plausible temperature. The name *synchrotron* refers to the manmade ring-shaped synchrotron-particle accelerator where this type of radiation was first observed; a synchronous mechanism keeps the particles in step with the acceleration as they circulate in the ring.

in which the source of synchrotron radiation is most intense at longer rather than shorter wavelengths, where a collection of high-speed electrons will be depleted of those with highest energy first.

9.4 Cold, Rarefied Hydrogen Atoms Between the Stars

Only rare, hot, luminous stars can ionize nearby hydrogen, and we wonder what is in the vast dark places outside them, other than the dust already discussed. Stars still are forming out of something and because the stars are mostly hydrogen, there should be hydrogen atoms within interstellar space. However, it is so cold out there that we cannot detect any atoms. None of the atoms is ionized and the electrons are all in their lowest energy ground state, unable to change their atomic orbits. These electrons seem discontented with the situation, turning first in one direction and then in another but never changing orbits or leaving the atom.

Signals from the atomic electrons eventually were found by radio astronomers at a wavelength of 21 cm. The prediction and discovery of this radiation follows a paper trail of international scope, remarkable coincidences, and a professional courtesy of bygone times. It also is a tale of two graduate students and their ever-so-smart advisors.

Reading of the unexpected discovery of interstellar radio noise by Jansky and Reber, the Dutch astronomer Jan Oort asked his graduate student Hendrik C. "Henk" Van de Hulst to see if there were any spectral lines in the radio spectrum. He investigated the matter and predicted a radio wavelength spectral line that might be detected from interstellar regions of electrically neutral, or un-ionized, hydrogen atoms.

Van de Hulst realized that these regions, now designated H I and pronounced "H one," would be cold and that most of the atoms would be in their lowest energy state. The electron of the hydrogen atom in this state has two possibilities in the direction of its spin, or rotation;

a rare collision between two atoms could result in a change of the spin direction of an electron in one of the atoms. The atom then is in an unstable configuration, so its electron will soon return to the original spin direction. This releases a small amount of energy and produces radiation at a wavelength of 21 cm. As Van de Hulst pointed out, these spin transitions occur rarely in the tenuous interstellar gas; however, an observer might well detect them when looking through the vast extent of interstellar space.

In 1945, the prediction was published by Van de Hulst in an obscure Dutch journal, *Nederlands tijdschrift voor natuurkunde*, with an obtuse article titled "Radio Waves from Space: Origin of Radio Waves." The Soviet theorist Iosif S. Shklovskii confirmed the prediction, with greater detail, but in the Russian language. At about this time, Harold I. "Doc" Ewen, a graduate student at Harvard University, became interested in radio astronomy. His advisor, Edward M. Purcell, asked his wife, Beth, who was good with languages, to translate Shklovskii's paper about the 21-cm line. After reading it, Purcell encouraged Ewen to build a radio receiver to search for the transition. They also had a copy of Van de Hulst's original work, translated from the Dutch.

Ewen constructed a radiometer to detect the hypothetical radiation using electronic components scavenged from other places or bought with a $500 grant and $300 from Purcell. Much of the equipment was borrowed every Friday and returned each Monday from Harvard's Cyclotron Laboratory, using a wheelbarrow. Accurate measurements of the expected hydrogen-line wavelength, using terrestrial atomic hydrogen in a laboratory at Columbia University, enabled Ewen to tune his receiver to the precise wavelength of 21.106 cm. At Purcell's suggestion, the receiver was switched between the wavelength of the expected signal and an adjacent one, with a difference that might contain the expected 21-cm line. Such a wavelength-switched, or frequency-switched, receiver has since been widely adapted by radio astronomers to remove unwanted noise from an observed signal containing a spectral line.

The receiver was connected to a simple horn antenna constructed of plywood, lined with copper sheeting, and mounted on a ledge outside a window in the university's physics building. Shortly past midnight, when the Earth's rotation brought the plane of the Milky Way through the beam of the horn antenna, Ewen succeeded in detecting the 21-cm transition with the novel receiver.

As it turned out, Van de Hulst was in Cambridge, Massachusetts, as a visiting professor at Harvard College Observatory and the Australian radio astronomer Frank Kerr was visiting Harvard on a Fulbright grant. So Ewen and Purcell invited them over to describe their discovery and urge them to have their people confirm the result.

At the meeting, the Harvard team learned for the first time that the Dutch group at Leiden had been actively trying to detect the radio transition for several years. A description of the wavelength-switched receiver was provided to Van de Hulst, leading to the conversion of the Dutch system and, in a gracious move – which would be unheard of in today's competitive scientific world – Purcell insisted that publication of their discovery be delayed until the Dutch group confirmed it. Their discovery was published when it was confirmed by the Dutch astronomers Karl Müller and Oort as well as in a cable from Australian radio astronomers. A coordinated report from all three centers then was published in the journal *Nature* in 1951.

The detection of the 21-cm line revolutionized studies of the interstellar medium. The H I regions do not emit visible light; hence, they are invisible at optical wavelengths, but they

Table 9.4. *Physical properties of H I regions of interstellar atomic hydrogen*

N_H = density of hydrogen atoms = 10^6 to 10^8 m^{-3}
T_k = kinetic temperature = 10 to 100 K
R = radius = 3 to 33 light-years = 1 to 10 pc $\approx$ (3 to 30) $\times 10^{16}$ m
ν_H = frequency of 21-cm spin transition = 1.420205751768 $\times 10^9$ Hz $\approx$ 1,420 MHz
A_{mn} = probability of 21-cm spin transition = 2.8 $\times 10^{-15}$ s^{-1}

do emit radio waves that are 21 cm long. Typical parameters of these regions of interstellar atomic hydrogen are given in Table 9.4.

9.5 Molecular Cocktails in a Smoky Room

Soon after the discovery of interstellar atomic hydrogen at radio wavelengths, astronomers began to speculate about the possibility of detecting molecules with radio waves, which are sensitive to the coldest clouds of interstellar matter where molecules might survive. Observations of any molecule first required accurate measurements of the wavelength of its radio spectral features in the terrestrial laboratory. Furthermore, radio telescopes with surfaces accurate to a few centimeters or better had to be constructed for receiving the short-wavelength emission of molecules, and new methods of spectral analysis needed to be developed.

Charles H. Townes and his colleagues at Columbia University made the first precision laboratory measurements of the radio-frequency transitions of the hydroxyl (OH) molecule, composed of an atom of oxygen, O, and an atom of hydrogen, H. At an international symposium of radio astronomy, Townes presented laboratory measurements of the rotational transitions of other molecules that might be detected, including carbon monoxide and water.

Alan Barrett and his colleagues at the Massachusetts Institute of Technology then built the sophisticated equipment needed to obtain the first observations of interstellar OH, at the wavelength specified by Townes. The discovery was followed closely by intensive searches for OH and eventually resulted in the realization that some sources are as small as stars and act like cosmic masers ("maser" is an acronym for the "microwave amplification by stimulated emission of radiation.")

Townes moved to the University of California at Berkeley and, within a year, he and his graduate students discovered ammonia and water in interstellar space. This was followed soon by the detection of the embalming fluid formaldehyde, as well as carbon monoxide and hydrogen cyanide by other groups.

These discoveries had not been anticipated even a decade earlier because astronomers had overlooked the importance of interstellar dust grains in shielding molecules from destructive ultraviolet starlight and acting as a catalyst in forming complex molecules.

The most abundant cosmic molecule is molecular hydrogen, denoted H_2, which consists of two atoms of hydrogen, designated H. This molecule cannot be observed directly at radio wavelengths; however, the second most abundant interstellar molecule, carbon monoxide, maps out its distribution. As might be suspected, the interstellar molecules are composed of the most abundant atoms in the universe, beginning with hydrogen, but also carbon (C), oxygen (O), and nitrogen (N). They combine to form molecules such as ammonia, NH_3; carbon monoxide, CO; hydrogen cyanide, HCN; water, H_2O; and formaldehyde, H_2CO.

The early findings triggered an avalanche of molecular searches in which groups of young radio astronomers engaged in an exciting pursuit of previously unseen interstellar molecules. The net result was the discovery of an array of hundreds of interstellar molecules, including complex organic molecules such as ethyl alcohol, or ethanol – the substance that gives beer, wine, and liquor their intoxicating power. Although interstellar matter is generally sparse and tenuous, the molecules are concentrated within dark dust clouds that can be 1 million times as dense as a typical region of interstellar hydrogen atoms.

10

New Stars Arise from the Darkness

10.1 How the Solar System Came into Being

The Nebular Theory

Where did the Sun and its attendant planets come from? How and when did they form? The most likely explanation is provided by the *nebular theory*, which states that the Sun and planets formed together as a result of the gravitational collapse of an interstellar cloud of gas and dust – also known as the solar *nebula*, Latin for "cloud" (Fig. 10.1). This theory accounts for the orderly, aligned motions of the major planets. They all move in a narrow band across the sky, implying that their orbits all lie in nearly the same plane, which nearly coincides with the Sun's equatorial plane. All of the planets move in the same direction within their Sun-centered orbits, and both the Sun and most of the major planets rotate in this direction – Venus and Uranus are the exceptions.

The orbits of most of the planetary moons, or satellites, imitate those of the planets in being confined to the planet's equatorial plane and revolving about the planet in the same direction that the planet rotates.

This regular orbital arrangement of the solar system is not accidental. Even if 1 million million million, or 10^{18}, solar systems were made haphazardly and the planets and moons were thrown into randomly oriented orbits, only one of these solar systems would be expected to look like our own. So, it is exceedingly unlikely that the planets became aligned by chance.

Although Newton's laws and Kepler's laws describe the present motions of the solar system, they cannot explain the remarkable arrangement of its planets and satellites. Additional constraints are required, which describe the scenario before the planets were formed and set in motion. These initial conditions are provided by the *nebular hypothesis*, the basic idea of which was introduced in the mid-eighteenth century by the German philosopher Immanuel Kant in his book, *Allgemeine Naturgeschichte und Theorie des Himmels* (*Universal Natural History and the Theory of the Heavens*). Kant pictured an early universe filled with thin gas that collected into dense, rotating gaseous clumps. One of these primordial concentrations was the solar nebula. The Sun formed at the center of the spinning solar nebula and the planets formed from swirling condensations in a flattened disk revolving around it.

Figure 10.1. Formation of the solar system. An artist's impression of the nebular hypothesis, in which the Sun and the planets were formed at the same time during the collapse of a rotating interstellar cloud of gas and dust that is called the *solar nebula*. The center collapsed to ignite nuclear reactions in the nascent Sun, and the surrounding material was whirled into a spinning disk where the planets coalesced. (Courtesy of Helmut K. Wimmer, Hayden Planetarium, American Museum of Natural History.)

There is now so much evidence for the nebular hypothesis that it has acquired the status of a theory, whose basic tenets are still valid. The spinning solar nebula, attracted by its own gravity, fell in on itself 4.6 billion years ago, becoming increasingly denser until the middle became so packed, tight, and hot that the Sun began to shine. The planets formed at the same time, within a flattened rotating disk centered on the contracting proto-Sun.

This is the essence of the original nebular theory, which explains qualitatively the fact that the major planets and their large moons all revolve in the same direction within the plane that coincides with the equator of the rotating Sun. This regular, aligned pattern of motion is a natural consequence of the rotation and collapse of a solar nebula composed of gas and dust from which the Sun and planets were produced.

Planets of Rock, Liquid, and Ice

If the nebular theory is correct, we might expect that all of the planets would have the same composition as the Sun because they all formed from the same interstellar nebula. After all, they should have the same ingredients as the material from which they formed – and they do, but with a varying mix.

The abundance of elements in the giant planet Jupiter does indeed mimic that of the Sun, with a predominance of the lightest element, hydrogen. Unlike the Sun, the Earth is composed mainly of heavier elements, and this difference must be explained. It is related to the fact that there are two main types of major planets – the *terrestrial planets* and *giant planets* – which differ in size, composition, and distance from the Sun.

The four planets close to the Sun – Mercury, Venus, Earth, and Mars – are known as *terrestrial planets* because they are similar to the Earth. These inner planets are rocky and relatively compact and dense. In contrast, the four *giant planets* – Jupiter, Saturn, Uranus, and Neptune – which reside in the outer parts of the planetary system, are big, gaseous, and have relatively low mean mass densities. Unlike the inner terrestrial planets, rings and numerous satellites also encircle each of the outer giant planets.

A clue to these differences comes from the locations of the two types of planets when they originated. The terrestrial planets formed in the warm regions of the flattened solar nebula, close to the bright, young Sun, whereas the giant planets formed farther from the Sun in the colder outer regions of the solar nebula.

In the inner regions of the solar nebula, the higher temperatures vaporized icy material that could not condense, leaving only rocky substances of relatively high mass density to coalesce and merge to form the terrestrial planets. Also, the low total mass and high initial temperature of these planets, as well as their proximity to the Sun, did not allow them to capture and retain the abundant lighter gases – hydrogen and helium – directly from the solar nebula.

The rocky terrestrial planets were so hot in their formative stages, beginning about 4.6 billion years ago, that their interior rock and metal melted and gravity separated them by density. In a process known as *differentiation*, the denser material (e.g., iron) sank toward the center, whereas the less dense rocks (i.e., the silicates) remained closer to the surface. The planets then cooled from the outside in as time elapsed, so we now can walk across the Earth's solid surface; however, our planet still has a molten core due to heat generated by radioactive elements inside it.

At larger distances from the Sun, where the solar nebula was colder, icy substances condensed and combined with heavier substances to form the massive cores of giant planets. These cores became sufficiently massive to gravitationally capture some of the surrounding hydrogen and helium, which was pulled into the giant planets. The low temperatures at remote distances from the Sun thus enabled the giant planets to retain the abundant light gases and grow even bigger, with large masses and low mass densities. Jupiter's low mass density, for example, indicates that it is composed largely of hydrogen and helium, just as the Sun is.

When the masses of the Sun and planets are determined, we find that the Sun does not only lie at the heart of our solar system; it also dominates it, which means that most of the nebular mass outside the Sun never became part of any planet. Some 99.866 percent of all of the matter between the Sun and halfway to the nearest star is contained in the Sun. All of the objects that orbit the Sun – the planets and their satellites, the comets and the asteroids – add up to only 0.134 percent of the mass of our solar system. Relative to the Sun, the planets are insignificant specks, left over from its formation and held captive by its massive gravity.

Almost all of the hydrogen and helium gas that enveloped the newly formed Sun must have disappeared, vanishing into thin air (so to speak). The powerful winds of the young Sun apparently cleaned out the solar system, blasting away all of the remaining gases that had not condensed to make the planets. Some of the leftover rocky material, not found in the terrestrial planets, now is located in the asteroid belt, and some of the remaining ice is located in the distant precincts of the solar system where the comets reside. Most of the hydrogen and helium that was not near the giant planets or in the Sun also was blown away in the formative stages of the solar system.

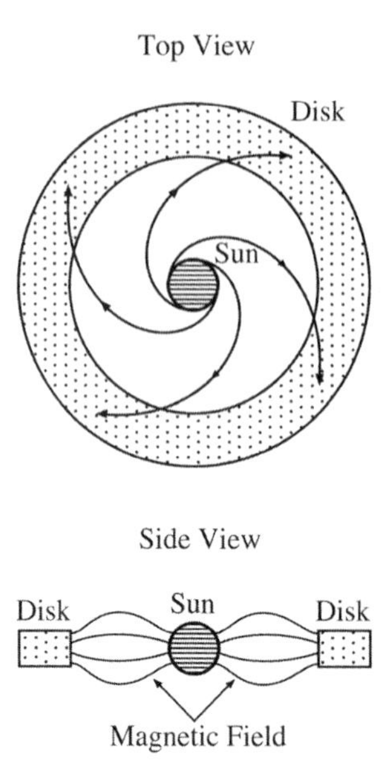

Figure 10.2. Magnetic brakes. The central, rapidly rotating Sun is connected to an ionized, slowly rotating disk by magnetic fields (*side view*). The magnetic field is twisted into a spiral shape (*top view*) and acts as a brake on the Sun's rotation, transferring angular momentum from the Sun to the proto-planetary disk.

A powerful solar wind and intense magnetic field during the Sun's youth may have conspired to produce the Sun's current slow rotation. A strong magnetic field, generated by the fast solar rotation in its early epochs, connected the Sun to the distant, slowly rotating material in the surrounding disk, acting as a magnetic brake to the solar rotation (Fig. 10.2). The rapidly rotating Sun swept the magnetic field by the slow-moving charged particles in the outer solar nebula, producing a drag that slowed down the Sun.

If the nebular theory is correct, it should apply to the formation of other stars, not only the Sun. The interstellar clouds, in fact, are even now in the process of creating young stars that are currently embedded in the interstellar gas and dust that spawned them. However, not every interstellar cloud is in the process of star formation. For the most part, the interstellar gas and dust is too hot and too tenuous to spontaneously collapse into stars, which is why there is still sufficient interstellar material around to create new stars after billions of years of continued star formation.

10.2 Star Birth

Giant Molecular Clouds

Interstellar clouds of gas and dust provide the raw material for new stars, but they do not form everywhere within interstellar space. Stars now are forming within particularly cold, dense, and massive regions known as *giant molecular clouds*. These contemporary incubators of newborn stars have temperatures as low as 10 K, span tens of light-years, and have a mass of up to 1 million solar masses, mainly in the form of hydrogen molecules. These giant molecular clouds are now the dominant star-forming component of the interstellar medium.

Table 10.1. *Physical properties of giant molecular clouds*

N_{H2} = density of hydrogen molecules = 10^{10} m^{-3} to 10^{12} m^{-3}
T_k = kinetic temperature = 10 K to 30 K
R = radius = 0.6 light-years to 32 light-years = 0.2 pc to 10 pc ≈ (0.6 to 30) × 10^{16} m
M = mass = 10^4 M$_\odot$ to 10^6 M$_\odot$ ≈ 2 × 10^{34} kg to 2 × 10^{36} kg

As many as 1 million million, or 10^{12}, hydrogen molecules can be packed into every cubic meter of such a giant molecular cloud (Table 10.1). In contrast, there are no more than 100 million, or 10^8, hydrogen atoms in 1 cubic meter of the interstellar material outside the giant molecular clouds; these atoms are about 10 times hotter, at about 100 K, than giant molecular clouds. The dust in the dark molecular clouds blocks the harsh ultraviolet radiation in space and enables chemical reactions to form complex, delicate molecules from the atomic constituents of the interstellar gas.

If a giant molecular cloud becomes sufficiently massive and dense, the mutual gravitation of its parts will overcome the outward gas pressure from inside, and the cloud starts falling in on itself. Once this gravitational collapse is underway, the giant cloud fragments into smaller components; and the pieces collapse until their cores become hot enough to ignite nuclear fusion, burning hydrogen to become stars like the Sun. Some clouds are even now in the process of creating stars (Fig. 10.3). Thus, stars are continually reformed, as new stars arise in the dark spaces between the old ones.

Figure 10.3. Mountains of creation. The infrared heat radiation of hundreds of embryonic stars (*white/yellow*) and windblown, star-forming clouds (*red*), detected from the Spitzer Space Telescope. The intense radiation and winds of a nearby massive star, located just above the image frame, probably triggered the star formation and sculpted the cool gas and dust into towering pillars. (Courtesy of NASA/JPL-Caltech/Harvard-Smithsonian CfA/ESA/STScI.) (See color plate.)

Gravitational Collapse

The English physicist and mathematician Sir James Jeans considered the stability conditions of a gas subject to perturbations in mass density, showing that a fluctuation greater than a critical size – now called the *Jeans length* – or a mass greater than a critical mass – known as the *Jeans mass* – will become unstable to gravitational collapse. This collapse occurs when there is insufficient gas pressure to support a large, massive interstellar cloud against the combined gravitational attraction of its component parts.

Giant molecular clouds can have a mass greater than the Jeans mass, but it is more difficult for clouds with a smaller mass to form stars. They require external compression to begin the collapse.

Triggering Gravitational Collapse

Why hasn't all of the interstellar gas and dust drawn together to make stars? In most regions of interstellar space, the temperatures are high enough and the densities low enough for a long-lived stable equilibrium between outward gas pressure and inward gravitational pull. Magnetic fields also can help an interstellar cloud resist gravity. The interstellar magnetic field can generate a magnetic pressure that is comparable to the gas pressure in its vicinity and can support a gas cloud against gravity,

For the most part, interstellar clouds merely swirl through space – too hot, agitated, and magnetic to collapse into stars. Compression by an external agent nevertheless can force an isolated cloud into gravitational collapse. Occasionally, gas clouds collide with one another, for example, generating shock waves that resemble chaotic surf near an ocean beach. The interstellar waves can compress the colliding clouds, initiating their gravitational collapse.

A spectacular type of external compression is provided by a nearby exploding star, or *supernova*. When a massive star exhausts its thermonuclear fuel, it can explode in one last burst of activity, ejecting a spherical shock wave that expands at a speed of 10,000 km s^{-1}. The wave produced by the detonation of a nuclear bomb is analogous to the shock wave of a supernova.

As proposed by the Estonian astronomer Ernst Öpik, who spent the second half of his career in Ireland, the shocks and expanding remnants of the explosion can trigger the collapse of a normally stable interstellar cloud. Like a giant snowplow, the shock wave pushes nearby interstellar gas and dust together, compressing clumps of matter to sufficiently high density for gravitational collapse to ensue.

The solar nebula once may have been so spread out that its weak gravity was not sufficient for it to collapse to form the Sun and the planets. Instead, the explosion of a nearby star may have triggered the collapse. Some elements found in meteorites recovered on the Earth are apparently the decay products of radioactive elements that must have been produced in such a stellar explosion no more than a few tens of millions of years before our solar system formed. If these elements were created before this time, they would not be around now, having decayed away.

Emission from hot, massive, young stars also can compress nearby gas and dust into gravitational collapse and the formation of new stars. Associations of bright O and B stars that were formed about 1 million years ago, for example, now are expanding and dispersing

Figure 10.4. Star-forming region. This multiple-wavelength portrayal combines infrared *(red)*, visible light *(green)*, and x-ray *(blue)* images of the bright, star-forming region designated NGC 346. It is located in the Small Magellanic Cloud that orbits our Milky Way Galaxy at a distance of about 210,000 light-years. Both wind-triggered and radiation-induced star formations are revealed, primarily by the infrared emission of the cold dust *(red)*, detected from the Spitzer Space Telescope. Young stars enshrouded by dust appear as red spots with white centers. The pressure of intense radiation from massive stars in the central regions of NGC 346 pushed against nearby gas, causing it to expand, and created shock waves that compressed nearby dust and gas into small new stars. Red-orange filaments surrounding the center of the image show where this process occurred. The supernova explosion of a very massive star apparently triggered the formation of even younger stars, seen as a pinkish concentration at the top of the image. Strong winds from this exploding star pushed dust and gas together about 50,000 years ago, compressing it into new stars. The x-rays *(blue)*, observed from ESA's XMM-Newton orbiting telescope, reveal very warm gas. The visible light *(green)* radiation was detected using the European Southern Observatory's New Technology Telescope. (Courtesy of NASA/JPL-Caltech/ESA/MPIA.) (See color plate.)

into space. The intense radiation and powerful winds associated with a previous and nearby generation of O and B stars could have triggered the collapse of neighboring material, giving rise to the expanding stellar associations. These newer, younger stars may trigger the formation of other stars in the future, in an ongoing process of sequential star formation.

Thus, stars do not form just anywhere. They are born either in cold, dense, molecular clouds or in proximity to exceptionally massive and short-lived stars by various mechanisms, including the pressure of stellar radiation, winds, and explosions (Figs. 10.4 and 10.5).

Protostars

A star in the process of formation commonly is called a *protostar*. Such an embryonic star shines by the release of gravitational energy during the collapse of interstellar material,

Figure 10.5. The North America Nebula in infrared. Clusters of relatively young stars, about 1 million years old, are found throughout this infrared image of the North America Nebula, also designated as NGC 7000. The infrared detectors aboard the Spitzer Space Telescope penetrated the dark clouds seen in optically visible light, viewing young stars in many stages of formation, including gas and dust cocoons, disks, and jets. The North America Nebula is about 1,500 light-years from the Earth and spans about 50 light-years. (Courtesy of NASA/JPL-Caltech.) (See color plate.)

but it has not yet begun to shine by nuclear fusion in its core. Protostars are exceptionally bright at infrared wavelengths, which can be used to detect them within dark clouds.

Once an interstellar cloud becomes sufficiently dense, by either external compression or within a giant molecular cloud, the mutual gravitational attraction of its parts will overcome the gas pressure and cause this cloud to start collapsing. As this protostar falls inward, the gas gains gravitational potential energy, much in the way a waterfall gains energy when its water moves toward the ground. Some of the energy of the protostar is converted into heat as the gas particles fall inward and collide with one another (Fig. 10.6).

When observations of young stellar clusters are combined with the theoretical studies of the Japanese astrophysicist Chiushiro Hayashi, the pre-main-sequence evolution of protostars of different masses can be deciphered. As illustrated in Figure 10.7, their tracks in the Hertzsprung–Russell (H–R) diagram initially move straight down and subsequently turn to the left and continue that way until the protostar arrives on the main sequence. The outward pressure of the hot gas, which is now heated by the nuclear-fusion reactions, prevents the star from collapsing further. It has settled down for a long rather uneventful life as a main-sequence star, the longest stop in its life history.

How long does it take for a collapsing protostar to become a star? One estimate for the time-scale on which clouds collapse is the freefall time. This is the time it would take a cloud to undergo gravitational collapse from its original shape to a single point, neglecting gas

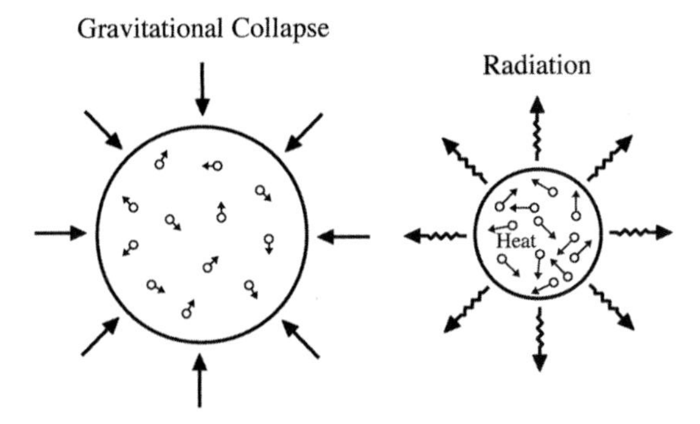

Figure 10.6. Gravitational collapse produces heat and radiation. The collapse of an interstellar cloud of gas and dust (*left*) compresses the cloud and heats it (*right*). When the cloud shrinks, gravitational potential energy is converted into heat as the gas particles fall inward and collide with one another. This also produces radiation that can carry off some of the energy. The velocities of the gas atoms are denoted by arrows that point in the direction of atomic motion and have lengths that increase with the speed of motion. Higher speeds occur in the compressed cloud, where the gas atoms move faster and in all directions.

pressure that counteracts this force. For the Sun, the unopposed freefall collapse time is about 40,000 years, but there are forces that oppose the collapse as it takes place, making it take about 1 million years for the Sun to undergo gravitational collapse to the main sequence. This is much shorter than the thermonuclear lifetime of about 10 billion years once the Sun

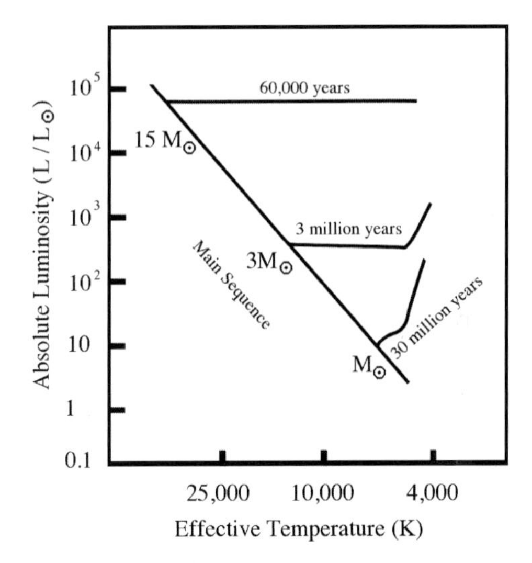

Figure 10.7. Protostars on the Hertzsprung–Russell diagram. Evolutionary tracks of protostars of various masses in the H–R diagram, ending with their arrival on the main sequence when stars have begun burning hydrogen in their core. The absolute luminosity, L, is given in units of the Sun's absolute luminosity, denoted $L_{\odot}$. The star mass is given in units of the Sun's mass, designated $M_{\odot}$. The mass values are specified along the main sequence, from upper left to lower right. High-mass stars, which have greater luminosity than low-mass stars, are found at higher points on the main sequence and take a shorter time to arrive there. The protostar lifetimes are given above the relevant track. During star formation, transport of a protostar's internal energy is dominated by either radiation (*horizontal lines*) or convection (*vertical lines*). Stars with lower mass ultimately have larger interior convective zones.

begins to shine by hydrogen burning in its core. When compared to the Sun, more massive stars have shorter freefall times to the main sequence and shorter thermonuclear life times on the main sequence as well.

Losing Mass and Spin

The early stages of star life can be the most active; for example, young stars can have strong stellar winds that drive away protostellar material that may still envelop them. Even now, billions of years after its birth, the Sun generates a solar wind; we suspect that its winds were more violent in its youth, clearing out the solar nebula when the planets were formed.

Hot, exceptionally luminous stars can generate immensely powerful winds with greater mass-loss than the current solar wind. Mass-loss rates of up to 1/10,000th, or 10^{-4}, solar masses per year have been observed; at this rate, a star of say 10 times the mass of the Sun would blow itself away in just 100,000 years.

Stellar winds are not confined to youth alone, for stars of advancing age also can produce strong winds. When a star of relatively low mass evolves into a red giant, the gravitational attraction at its inflated outer layers becomes much smaller than during the star's former life on the main sequence. This reduces the star's ability to hold onto its outer atmosphere and increases the likelihood that some of it will escape. As a result, red giant stars also have strong winds, sometimes losing a significant fraction of their mass during this stage of stellar evolution.

Younger stars also rotate faster than older stars. The rotation speed can be measured from the Doppler broadening of a star's spectral lines. Rapidly rotating stars have broad lines; slowly rotating stars have sharp and narrow lines. Observations of this line-broadening indicate that the rotation speed of main-sequence stars decreases from left to right on the H–R diagram from luminous, young stars to fainter, older ones. The older, late-type stars may rotate more slowly than early-type stars of relatively young age because of the magnetic braking proposed to account for the Sun's unexpectedly slow rotation.

10.3 Planet-Forming Disks and Planets around Nearby Stars

The Plurality of Worlds

The ancient Greeks imagined that all matter consists of tiny moving particles, both indivisible and invisible, which they called *atoms*, and that all material objects can be created by the coming together of a sufficient number of atoms. In the second century BC, the Greek philosopher Epicurus of Samos proposed that the chance conglomerations of innumerable atoms, in an infinite universe, should result in the formation of a multitude of unseen, Earth-like worlds.

The Roman poet Lucretius wrote about the plurality of worlds around 55 BC, declaring that innumerable particle seeds are rushing on countless courses through an unfathomable universe, making it highly unlikely that our Earth is the only planet to have been created and that all of those other particles are not accomplishing anything.

The belief in other worlds – some possibly inhabited – orbiting stars other than the Sun dates back at least as far as the late sixteenth century, when the Italian philosopher and priest Giordano Bruno reasoned that other planets would remain invisible to us because they are small and dim and would be hidden in the glare of their host star.

Bruno spent the last eight years of his life in the prisons of the Inquisition. He eventually was tried by the Catholic Church, bound to a stake, and burned alive in Rome, perhaps more for his heretical religious views – such as his doubts about the Immaculate Conception, the Holy Trinity, and Christ's divinity – than for his belief in an infinite universe filled with countless habitable planets circling other stars.

During the nineteenth and twentieth centuries, astronomers used telescopes to explore the distant reaches of the Milky Way, showing that it contains about 100 billion stars. More recently, hundreds of planetary worlds, once only imagined, have been observed orbiting nearby stars. It is reminiscent of Percy Bysshe Shelley's lines:

> Worlds on worlds are rolling over
> From creations to decay,
> Like the bubbles on a river
> Sparkling, bursting, borne away.[19]

Proto-Planetary Disks

Planetary systems probably formed around many stars as a result of the gravitational collapse of an interstellar cloud of gas and dust that created the stars, all in accordance with the nebular theory of the origin of the solar system. The collapsing cloud would have rotated faster and faster, giving spin to the in-falling material, which was flattened into a planet-forming disk with a star at the center. Because rotation imparts motion to the colliding material in the direction of spin, the random gas motions of the original cloud are changed into a rotating, flattened disk. The rotation prevents gas and dust from raining directly onto the central star, instead making it settle into a rotating disk from which planets can form. The direction in which the disk is spinning coincides with the direction of the new star's rotation as well as the direction of the orbits of any planets that may be formed in the disk.

Astronomers have discovered flattened, rotating disks of gas and dust around nearby stars. The first evidence for these planet-forming disks was obtained in the early 1980s with instruments aboard the InfraRed Astronomical Satellite (IRAS), using technology pioneered by the military to detect the infrared heat of the enemy.

The IRAS instruments detected excess infrared radiation from four nearby stars, beyond what would be expected from the star alone (Table 10.2). This implied the presence of a circumstellar disk of cool dust in orbit around the star, which would radiate at infrared wavelengths and produce the excess. The hotter stars would shine brightly in optically visible light and emit relatively little infrared.

The Spitzer Space Telescope recently used its powerful infrared vision to detect hundreds of stars with excess infrared radiation, suggesting that they harbor planet-forming disks. In fact, the youngest nearby stars usually are found embedded in the dense clouds of the interstellar gas and dust that spawned them.

The closest disk system to our own, surrounding the star Epsilon Eridani, contains two infrared-emitting belts: the first at approximately the same position as the asteroid belt in

Table 10.2. *Stars with an excess of infrared radiation detected from the IRAS satellite*[a]

Star	Luminosity ($L_\odot$)	Spectral Type	Mass ($M_\odot$)	Distance (light-years)
Vega	37	A0 V	2.1	25.3
Fomalhaut	18	A3 V	2.1	25.13
Epsilon Eridani	0.34	K2 V	0.82	10.49
Beta Pictoris	8.7	A6 V	1.75	63.4

[a] The luminosity is in solar units of $L_\odot = 3.828 \times 10^{26}$ J s^{-1}; the mass is in units of the Sun's mass $M_\odot = 1.989 \times 10^{30}$ kg.

our solar system; and the second, denser belt between the first one and a more remote ring of icy bodies similar to our own Kuiper belt.

The Hubble Space Telescope (HST) discovered flattened disks of dust swirling around at least half of the young stars in the Orion nebula, shining in reflected visible light. The high-resolution and sensitivity of the HST also have been used to obtain detailed images of dusty, planet-forming disks surrounding Sun-like stars, providing insights to the beginnings of our solar system (Fig. 10.8). The flattened, rotating disks suggest that the nebular hypothesis applies to them, and material in the disks is expected to coalesce into full-blown planets if it has not done so already.

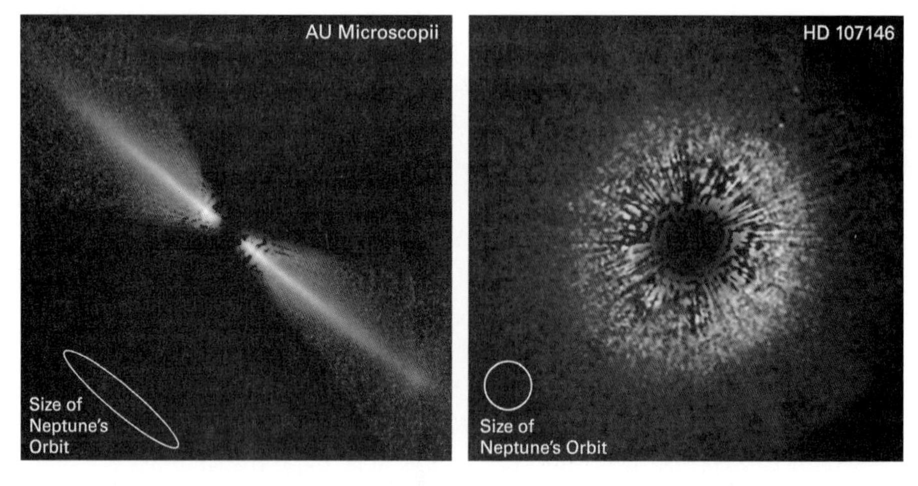

Figure 10.8. Dusty disks around Sun-like stars. Instruments aboard the Hubble Space Telescope have obtained these images of the visible starlight reflected from thick disks of dust around two young stars that still may be in the process of forming planets. Viewed nearly face-on, the debris disk surrounding the Sun-like star known as HD 107146 (*right*) has an empty center large enough to contain the orbits of the planets in our solar system. Seen edge-on, the dust disk around the reddish dwarf star known as AU Microscopii (*left*) has a similar cleared-out space in the middle. HD 107146 is 88 light-years away and is thought to be between 50 million and 250 million years old, whereas AU Microscopii is located 32 light-years away and is estimated to be just 12 million years old. (Courtesy of NASA/ESA/STScI/JPL/David Ardila – JHU [*right*], and John Krist – STScI/JPL [*left*].) (See color plate.)

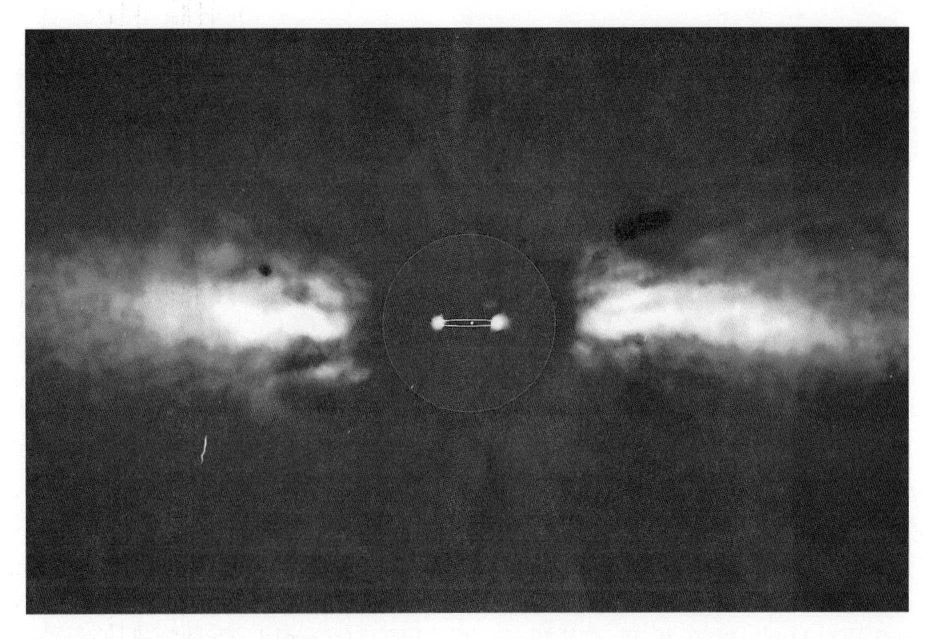

Figure 10.9. Exoplanet on the move. An exoplanet's orbital motion, denoted by the central white elliptical line, was imaged from an adaptive optics instrument attached to the VLT in Chile. The small white spot at the center shows the location of the host star, Beta Pictoris. Observations in 2003 are located at the left side of the planet's orbital ellipse and those in 2009 are on the right side. The larger dust disk surrounding the host star also is shown by the large flattened blue image at the left and the right. (Courtesy of ESO/A. M. Lagrange.) (See color plate.)

In the meantime, the circumstellar disk around one of the IRAS stars, Beta Pictoris, became the first to be imaged at visible wavelengths by using an occulting disk to block the star's bright light. Detailed observations of the disk were obtained more than a decade later with the HST; eventually, the ground-based Very Large Telescope (VLT), located in Chile, was used with adaptive optics to show that a Jupiter-sized world is moving around the star (Fig. 10.9). This giant planet, designated Beta Pictoris b, is located from its host star at a distance between 9 and 15 times the Earth–Sun distance of 1 AU, or at about the same distance as Saturn from the Sun at 9.539 AU.

Circumstellar dust around another IRAS star, Fomalhaut, also has been imaged with the HST. The sharp inner edge of the dust ring suggested that a planet was clearing out the material inside the ring. The HST detected the light of a Jupiter-size world orbiting Fomalhaut in the expected place, which is an enormous 115 AU from the star. The fantastic images of Beta Pictoris and Fomalhaut confirmed that infrared-emitting circumstellar disks are indeed signposts of planet formation, but they were obtained more than a decade after the even more astounding detection of the first planets orbiting a Sun-like star. These were also Jupiter-sized worlds but they were orbiting unexpectedly close to their host star.

The First Discoveries of Exoplanets

Individual planets shine by reflecting light that is much fainter than the light of the star that illuminates them. The visible light reflected by Jupiter, for example, is about 1 billion, or 10^9, times dimmer than the light emitted by the Sun, and that which is reflected by the Earth is 10 billion times fainter. As a result, planets are almost always too small and too faint to be seen directly in the luminous radiation of their nearby star. Their presence only recently has been inferred from their miniscule gravitational effects on the motions of the star around which they revolve or when they chance to pass in front of a star, momentarily blocking the star's light when viewed from the Earth. Such extrasolar planets that orbit around stars other than the Sun are called *exoplanets*.

The presence of an unseen planet orbiting a normal star like the Sun was first deduced by recording the way its gravity pulls at the star it orbits. Like two linked, whirling dancers, the planet and star tumble through space, pulling one another in circles. They both orbit a common center of mass where their gravitational forces are equal, somewhat like the equilibrium point of a seesaw, where the forces of two people balance. This fulcrum is closest to the heavier person, or to the massive star in the stellar case. So, the star moves in a much smaller circle, a miniature version of the planet's larger path. The more massive the planet and the closer it is to the star, the stronger the planet's gravitational pull on the star and the more the planet perturbs it.

To detect this tumbling motion, astronomers had to look for the subtle compressing and stretching of starlight as an unseen planet tugged on a star, pulling it first toward and then away from the Earth, causing a periodic shift of the stellar radiation to shorter and then longer wavelengths (Fig. 10.10). To measure the effect, astronomers must observe the wavelength of a well-known spectral feature, called a *line*, and measure the Doppler shift of its wavelength. A periodic wavelength change would measure the repetitive velocity change.

However, an orbiting planet produces an exceedingly small variation in the wavelength of spectral lines emitted from its star. Massive Jupiter, for example, makes the Sun wobble at a speed of only about 12 m s^{-1}. To detect the Doppler effect of a star moving periodically with this speed, astronomers would have to measure the wavelengths with an unheard accuracy of at least 1 part in 30 million and use a computer to search for periodic, back-and-forth wavelength changes.

Therefore, the effect could not be detected until sensitive spectrographs were constructed to precisely spread out the light rays; the enhanced light-collecting powers of electronic CCDs then were used to record the dispersed starlight. Because no single line shift is significant enough to be seen, computer software had to be written to add up all of the star's spectral lines, which shift together, and to combine them repeatedly at all possible regularities, or orbital periods, and with continued comparison to nonmoving laboratory spectral lines.

It took decades for astronomers to develop these complex and precise instruments. Then, in the 1990s, two Swiss astronomers from the Geneva Observatory in Switzerland, Michel Mayor and Didier Queloz, accomplished the seemingly impossible, discovering the first planet that orbits an ordinary star: the faintly visible, Sun-like star 51 Pegasi, only 48 light-years away from the Earth in the constellation Pegasus, the Winged Horse.

They had detected the back-and-forth Doppler shift of the star's light with a regular 4.23-day period, measured by a periodic change of the star's radial velocity of up to 50 m s^{-1} (Fig. 10.11). To produce such a quick and relatively pronounced wobble, the newfound planet must be large, with a mass comparable to that of Jupiter – which is 318 times heftier than the Earth – and it was moving in a tight close orbit around 51 Pegasi, at a distance of only 0.05 AU.

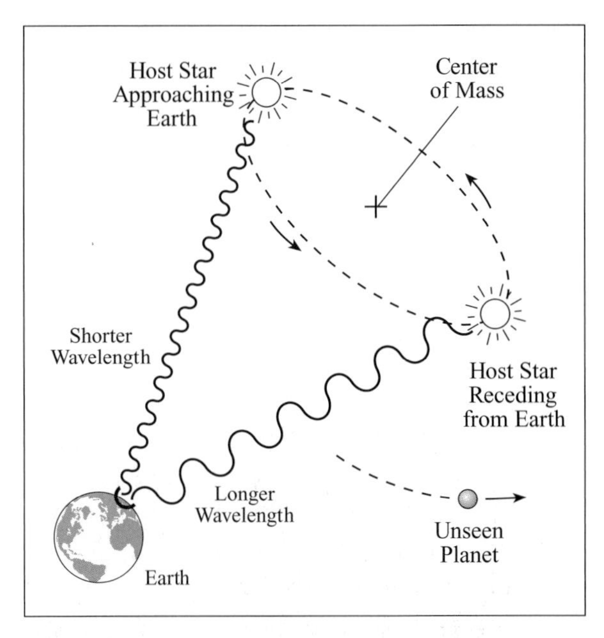

Figure 10.10. Starlight shift reveals invisible planet. An unseen planet exerts a gravitational force on its visible host star. This force tugs the star in a circular or oval path, which mirrors in miniature the planet's orbit. As the star moves along this path, it approaches and recedes from the Earth, changing the wavelength of the starlight seen from the Earth through the Doppler effect. When the planet pulls the star toward us, its light waves pile up slightly in front of it, shortening or "blueshifting" the wavelength that we detect. When the planet pulls the star away from us, we detect light waves that are stretched, or "redshifted." During successive planet orbits, the star's spectral lines are periodically shortened and lengthened, revealing the presence of the planet orbiting the star, even though we cannot see the planet directly.

Planets that are closer to a star move around it with greater speed and take less time to complete an orbit, in accordance with Kepler's third law. Thus, the Earth takes a year, or 365 days, to travel once around the Sun at a mean distance of 1 AU, whereas Mercury, the closest planet to the Sun, orbits our star in a period of 88 days at 0.387 AU. A short orbital period of only 4.23 days meant that the newfound planet is located at a distance of only 0.05 AU from its parent star, or about one eighth the distance between Mercury and the Sun. Thus, a completely unanticipated planet had been found, rivaling Jupiter in size and revolving around 51 Pegasi in an orbit smaller than Mercury's.

No one anticipated that a giant planet would orbit so close to its star. The intense radiation and powerful winds of the newly formed star were expected to keep any hydrogen from gathering together into a planet, explaining why Jupiter and the other giant planets were formed far from the Sun in the cold, outer precincts of our solar system. However, this was good for planet hunters, for the large mass of a giant world would produce a more pronounced velocity change than the smaller mass of an Earth-sized world, and the close orbit meant a short orbital period that might be detected in weeks instead of years.

Fewer than two weeks after the announcement of a giant planet circling 51 Pegasi, two American astronomers, Geoffrey W. Marcy and R. Paul Butler, used their own past observations to confirm the result. Now that they knew that giant planets could revolve unexpectedly

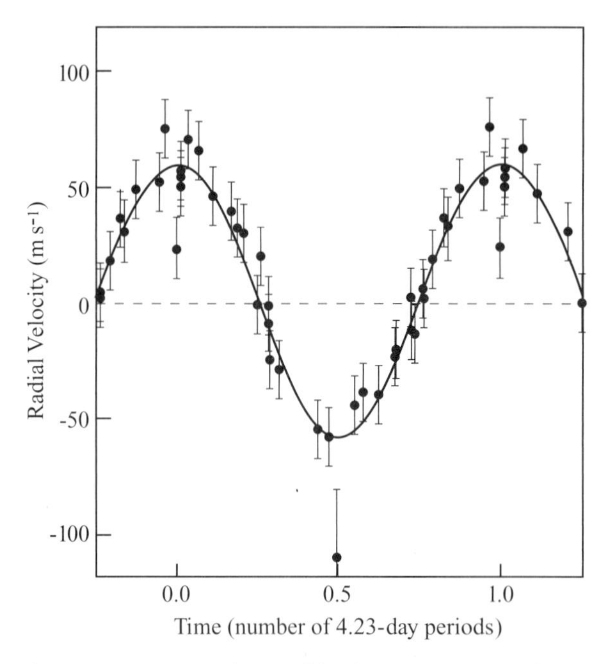

Figure 10.11. Unseen planet orbits the star 51 Pegasi. Discovery data for the first planet found orbiting a normal star other than the Sun. The giant, unseen planet is revolving around the solar-type star 51 Pegasi, located 50 light-years away from Earth. The radial velocity of the star, in units of meters per second, designated m s^{-1}, was measured from the Doppler shift of the star's spectral lines. The velocity exhibits a sinusoidal variation with a 4.23-day period, caused by the invisible planetary companion that orbits 51 Pegasi in this period. The observational data (*solid dots*) are fit with the solid line, whose amplitude implies that the mass of the companion is roughly 0.46 times the mass of Jupiter. The 4.23-day period indicates that the unseen planet is orbiting 51 Pegasi at a distance of 0.05 AU, where 1 AU is the mean distance between the Earth and the Sun. (Adapted from Michel Mayor and Didier Queloz, "A Jupiter-Mass Companion to a Solar-Type Star," *Nature* 378, 355–359 [1995].)

near a star, with short orbital periods, they used powerful computers to reexamine their observations of other nearby stars accumulated during previous years. They subsequently announced the discovery of two more Jupiter-sized companions of Sun-like stars.

These were astounding discoveries. In just a few months, astronomers had detected the first planets circling ordinary stars like our Sun. Other worlds were no longer limited to philosophical musings, scientific speculations, or artists' imaginations. After two millennia, a long-held dream has come true. We now can look up at the night sky and know that there definitely are unseen planets out there, orbiting perfectly ordinary stars that are now shining brightly.

Hundreds of New Worlds Circling Nearby Stars

After scientists realized that a large planet could be so near to its star, they knew where and how to look. By monitoring thousands of nearby Sun-like stars for years, American and European teams found hundreds of planets revolving about other nearby stars, most of them massive Jupiter-sized planets.

Some of the newfound worlds travel in nearly circular orbits, like those in the solar system, but they are much closer to their host star than Mercury is to the Sun. Dubbed "hot Jupiters" because of their size and proximity to the intense stellar heat, they are much too hot for human life to survive or water to exist. Their temperatures can soar to more than 1,000 K, far hotter than the surface of any planet in our solar system. Other newfound planets follow eccentric, oval-shaped orbits that deviate from a circular path, so they venture both near and far from their star. Many multiplanet systems also have been found as a result of longer and improved observations. The accelerating pace of discovery is documented at the extrasolar planets encyclopedia at http://exoplanet.eu/ and at http://planetquest.jpl.nasa.gov/.

Most of these worlds were discovered by the wobble they create in the motion of their host star, but some were discovered when they passed in front of the star, causing it to dim, or blink. If a planet happens to have a near edge-on orbit, as seen from the Earth, it periodically will cross directly in front of, or *transit*, its host star. Such a transit can be seen only if the orbit of the distant planet crosses the line of sight from the Earth, blocking a tiny fraction of the star's observed light and causing it to periodically dim repeatedly during the planet's endless journey around its star.

The size of a planet can be derived from the size of the dip. The fractional change in brightness, or *transit depth*, is equal to the ratio of the area of the planet to the area of the star. For the Earth and the Sun, as an example, the transit depth is 0.000084. The planet's temperature can be estimated from the characteristics of the star that it orbits and the planet's orbital period.

Searching for Habitable Planets

From a human perspective, the most interesting planets will be those as small as the Earth, in circular orbits at the precise distance from the heat of a Sun-like star to provide a haven for life. Scientists call this location a *habitable zone*, meaning that it could be inhabited – but not necessarily that it is. Such a planet might be detected by the transit method.

The orbital size can be calculated from the period of the repeated transit and the mass of the star. From the orbital size and the luminosity of the star, the planet's temperature can be calculated. This information would indicate whether the planet resides within the warm habitable zone – that is, the range of distances from a star where liquid water can exist on the planet's surface and life might exist. At closer distances, the water would boil away; at more remote distances, it would freeze solid.

The Kepler mission is specifically designed to detect hundreds of planets comparable in size to the Earth or smaller and located at or near the habitable zone. By measuring the brightness of 100,000 stars, it detects the periodic dimming of starlight produced when the planets pass in front of the stars. A transit by an Earth-sized planet produces a small change in the star's brightness of about 1/10,000, lasting for 2 to 16 hours.

The Kepler mission discovered several hundred new-planet candidates orbiting nearby stars. A few of the potential planets are nearly Earth-sized and orbit in the habitable zone of smaller, cooler stars than our Sun. Because these stars are less luminous than the Sun, the habitable zone is closer and planets within it have orbital periods that are shorter than our year, and they can be recognized in an observation time of a few years. The Kepler planet candidates require follow-up observations with the world's best ground-based telescopes to verify that they are actual planets.

In the meantime, the world's best telescopes are being employed to find new exoplanets using the velocity method. The European Southern Observatory's 3.6-meter telescope in La Silla, Chile, and the 10-meter Keck I telescope atop Mauna Kea in Hawaii have been used to discover many more exoplanets.

The atmospheres of transiting exoplanets also are being investigated using the Hubble Space Telescope, the Spitzer Space Telescope, and ground-based infrared telescopes. As a planet passes in front of and behind its star, astronomers can subtract the light of the star alone – when the planet is blocked – from the light of the star and planet together prior to eclipse. This isolates the emission of the planet and enables the detection of the infrared spectral signatures of gases in the planet's atmosphere. Water vapor and methane, for example, have been found in the atmosphere of at least one exoplanet.

11

Stellar End States

11.1 A Range of Destinies

No material object can exist forever, and stars are no exception. Although their lives may be measured in millions or billions of years, stars do age and die, and their existence is always transitory and impermanent. They all will cease to shine and return to the darkness from which they came.

Stars reach the end of their bright, shining lives when all of the available sources of subatomic energy have been exhausted. Their central thermonuclear reactions, which keep the star hot inside, are then turned off. It is akin to a hot fire that has burned away, and all that is left are warm embers that slowly cool and fade into ash.

There is no heat and outward pressure being generated inside such a star, so the internal support has been removed and it begins its ultimate contraction. Some stars finish their life quietly, cooling into dark remnants of their former splendor. Others explode and briefly shine with the light of 1 billion stars.

Nothing is completely destroyed, and a dying star always is being transformed into something else. The demise of a dying star results in the simultaneous creation of a new star from its collapsing core, with a final resting state that depends on the star's mass.

As in the beginning of their life, the central regions of all dying stars are subject to the unsupported, inward pull of gravity from all sides. There is no escape from it in any direction, and the entire stellar mass is compressed into an increasingly smaller radius. It only stops when some outward pressure grows sufficiently large to halt the contraction, which means that there is an enormous range in stellar size and mass density (Table 11.1). The mean mass density of the Sun, for example, is comparable to that of water, whereas the density of a neutron star is similar to that of the nucleus of an atom.

In its earliest stages, a protostar's gravitational contraction is stopped when the star begins to fuse hydrogen nuclei into helium nuclei, generating the internal heat and pressure that halts the formative collapse. A long time later, when the hydrogen runs out, gravity takes over again and compresses the core, heating it up until helium can be consumed in synthesizing carbon. Enough heat is then generated to balance the relentless force of gravity, and the star's outer atmosphere expands to giant or supergiant size.

Table 11.1. *Representative mass, radius, and mean mass density of the stars*[a]

Star	Mass ($M_{\odot}$)	Radius ($R_{\odot}$)	Mean Mass Density[b] (kg m^{-3})
Red giant star	1.2	100	0.0014
Sun	1.0	1.0	1,400
White dwarf star	0.6	0.01	0.84×10^{9}
Neutron star	1.5	0.00001	2.1×10^{18}
Black hole [c]	10.0	0.000004	0.2×10^{18}

[a] The mass is in units of the Sun's mass $M_{\odot} = 1.989 \times 10^{30}$ kg; the radius is in units of the Sun's radius $R_{\odot} = 6.955 \times 10^{8}$ m.

[b] The mean mass density $\rho = 3M/(4\pi R^{3})$ for a star of mass M and radius R.

[c] A representative radius for the black hole is taken as the Schwarzschild (or gravitational) radius, at which the escape velocity from the enormous gravity is equal to the speed of light.

What happens next depends on the mass of the star. Stars that have a mass comparable to the Sun's mass begin their ultimate collapse when the helium is gone, ending up as burned-out, Earth-sized white dwarf stars. The ultimate destinies of the rare, more massive, and luminous supergiants are explosive. They can leave a city-sized neutron star behind or be crushed into oblivion as a stellar black hole. Thus, there is a range of destinies: Giants turn into tiny white dwarfs; supergiants turn into even smaller neutron stars; and the bigger, heavier supergiants turn into black holes – all compressed into their end states by the never-ending force of gravity. The more massive the star, the smaller it eventually becomes.

11.2 The Winds of Death

Any star with a moderate mass, comparable to that of the Sun, eventually balloons into a red giant star. As the core nuclear reactions cease, the giant sheds its outer layers, which are blown away. All that remains of the outer part of the star is gas and dust – the dust of the dead.

From dust to dust, we might say, but stars also have an immortal aspect. The central regions of a red giant star collapse into another smaller white dwarf star – the final stage of its long life – and the material it has returned to the interstellar medium through the winds of death may be incorporated within future stars. They will be formed of material that has been enriched with heavy elements (e.g., carbon and oxygen), which were synthesized within the giant stars.

After discovering Uranus, the English astronomer William Herschel discovered a small glowing object that he designated a *planetary nebula* because of its round shape, which resembled the disks of planets as seen through a small telescope. However, planetary nebulae are not made of planets, and planets are not visible in them. The designation *nebula* is from the Latin word for "cloud," and it was used to distinguish the diffuse planetary nebulae, which have resolved disks, from unresolved, point-like stars.

Hershel and other astronomers soon discovered more of these objects, and four of them already had been included in Charles Messier's list of nebulae. The astronomers also realized

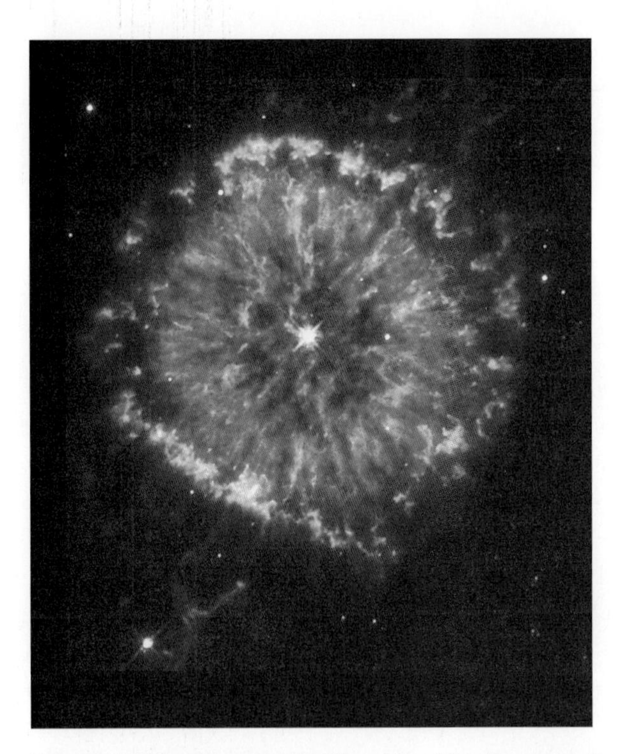

Figure 11.1. Planetary nebula. When a Sun-like star uses up its nuclear fuel, the star's center collapses into an Earth-sized white dwarf star and its outer gas layers are ejected into space. Such a planetary nebula is named after its round shape, which resembles a planet as seen visually in small telescopes, and is unrelated to planets. The shells of gas in the planetary nebula NGC 6751, shown here, were ejected several thousand years ago. The hot stellar core, exposed by the expulsion of the material surrounding it, has a disk temperature of about 140,000 K. Its intense ultraviolet radiation causes the ejected gas to fluoresce as a planetary nebula. (A Hubble Space Telescope image courtesy of the NASA/STScI/AURA/Hubble Heritage Team.) (See color plate.)

that a planetary nebula has a star at its center – the hot, exposed core of a dying red giant that illuminates the nebula (Fig. 11.1).

Another English astronomer, William Huggins, used his spectroscope to find a trio of emission lines in the Cat's Eye Nebula. When heated, a low-density gas radiates these emission lines; therefore, their presence indicated that the planetary nebulae contain hot, rarefied gas. However, at that time, no one knew how hot the gas was, or much about its elemental constitution.

The wavelength of one of the emission lines detected by Huggins coincided with hydrogen, but the chemical identification of the other two emission lines remained a mystery for more than a half-century. Although these green nebular lines were even stronger than hydrogen, they defeated attempts to identify them with elements known on the Earth. These spectral features initially were attributed to a previously unknown element called "nebulium," but the emission lines eventually were attributed to known elements that had become ionized by the ultraviolet light of the bright central stars.

When it was realized that the central stars of planetary nebulae are very hot, the English astronomer Arthur Eddington proposed that their ultraviolet starlight ionizes surrounding

Table 11.2. *Physical properties of planetary nebulae*

N = number density = N_e = electron density = (0.05 to 2.0) × 10^{10} m^{-3}
T_e = electron temperature ≈ (0.6 to 1.8) × 10^4 K
R = radius = 0.2 light-years to 0.8 light-years = 0.07 pc to 0.25 pc ≈ (2 to 7) × 10^{15} m
M = mass = $4\pi R^3 N m_P/3 \approx 10^{21}$ kg = 5 × 10^{-10} $M_\odot$
V_{exp} = expansion velocity = (1 to 9) × 10^4 m s^{-1}
τ_{exp} = expansion age ≈ 16,000 years

material and heats the nearby gas to about 10,000 K. The Dutch astronomer Herman Zanstra then showed that the hydrogen emission lines of planetary nebulae are produced when free electrons, which were liberated by the ionization by starlight, recombine with the protons to make hydrogen atoms, cascading through the atoms' various allowed electron orbits or energy levels and radiating the hydrogen emission line.

In a brilliant piece of detective work, the American astronomer Ira S. Bowen interpreted the two strong green emission lines as forbidden transitions of doubly ionized oxygen. His solution depended on the rarity of atomic collisions in the extremely tenuous planetary nebulae, which allows the occurrence of "forbidden" transitions. They are not actually forbidden but rather so improbable that they seldom take place in a higher-density laboratory situation, where an atom almost always is jostled by collisions into a different state before the forbidden radiation can be emitted. The observed emission lines indicate gas temperatures of about 10,000 K and electron or ion number densities of about 10 billion particles per cubic meter, denoted as 10^{10} m^{-3} (Table 11.2). This seems significant, but such densities are lower than the best vacuum used in a terrestrial laboratory.

The mass density and temperature of planetary nebulae resemble those of the emission nebulae; however, the planetary nebulae are about 10 times smaller and, unlike the emission nebulae, they also are expanding. A bright central star illuminates both types of nebulae, and they both emit similar spectral lines: those of ionized hydrogen and the forbidden emission lines of oxygen and nitrogen ions. Planetary and emission nebulae resemble the Earth's atmosphere in containing both oxygen and nitrogen, but the air we breathe is not so hot and it is not ionized.

As Zanstra realized, the intensity of the hydrogen emission line can be related to the temperature of the exciting star through the theory of the hydrogen atom and the Planck spectrum of thermal radiation. He found that these stars are enormously hot, and modern investigations show that they are the hottest stars known. The luminous central star radiates thousands of times more energy than the Sun and has a temperature of 100,000 K or more, much higher than any main-sequence star. This places it right off the scales of the Hertzsprung–Russell diagram, on the far left side (Fig. 11.2).

Powerful winds have blown the outer stellar layers away to reveal the star's hot interior. When a star becomes this hot, most of its radiation is at ultraviolet wavelengths, which brighten the surrounding nebula, but the star is relatively dim at the longer visible wavelengths and even may become invisible. However, as a young planetary nebula is blown outward by powerful winds, it slowly grows in size, thins out, and becomes transparent,

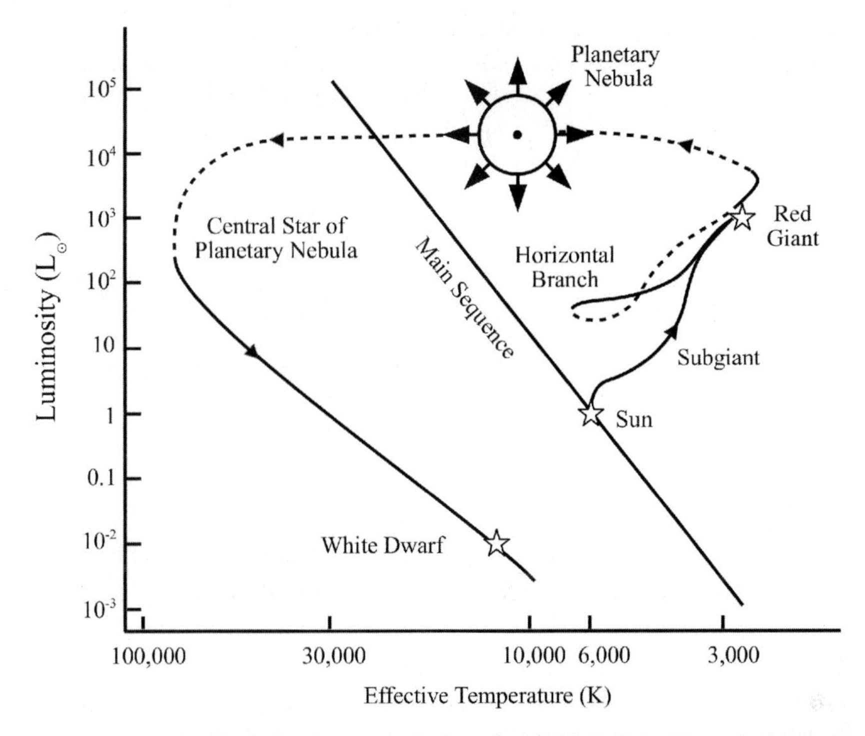

Figure 11.2. Formation of a planetary nebula and white dwarf star. The evolutionary track of a dying Sun-like star in the Hertzsprung–Russell diagram. When the star has exhausted its nuclear hydrogen fuel, which makes the star shine, it expands into a red giant star; after a relatively short time, the giant star ejects its outer layers to form a planetary nebula. The ejected gas exposes a hot stellar core, which collapses to form an Earth-sized white dwarf star that gradually cools into dark invisibility. The luminosity is in units of the Sun's luminosity, denoted $L_\odot$, and the effective temperature of the stellar disk is in units of degrees kelvin, denoted K.

revealing its source – the exposed core of a dying red giant. When more of the star's outer atmosphere is expelled and moves away from the central star, deeper and hotter layers of its interior are revealed.

When modern telescopes are used to zoom in and resolve the expanding gas and dust, they show that it has not been expelled in a single puff of stellar wind. Multiple blasts and gusts can occur, with different strengths and at various locations. Interactions between material moving away from the central star at different speeds and directions can produce many of the various shapes and forms of planetary nebulae (Figs. 11.3 and 11.4).

The observed expansion speeds of about 20 km s^{-1} and nebular dimensions of about 1 light-year across indicate that the expanding shells of gas were ejected about 16 thousand years before the expansion and size were measured. Their luminescent gas will continue to expand and disperse into interstellar space, cooling into invisibility and becoming indistinguishable from their surroundings in about 20,000 years. That is a relatively brief existence, only about 1 millionth of their former stellar lifetime of many billions of years. As a result, the planetary nebulae are much less numerous than the stars.

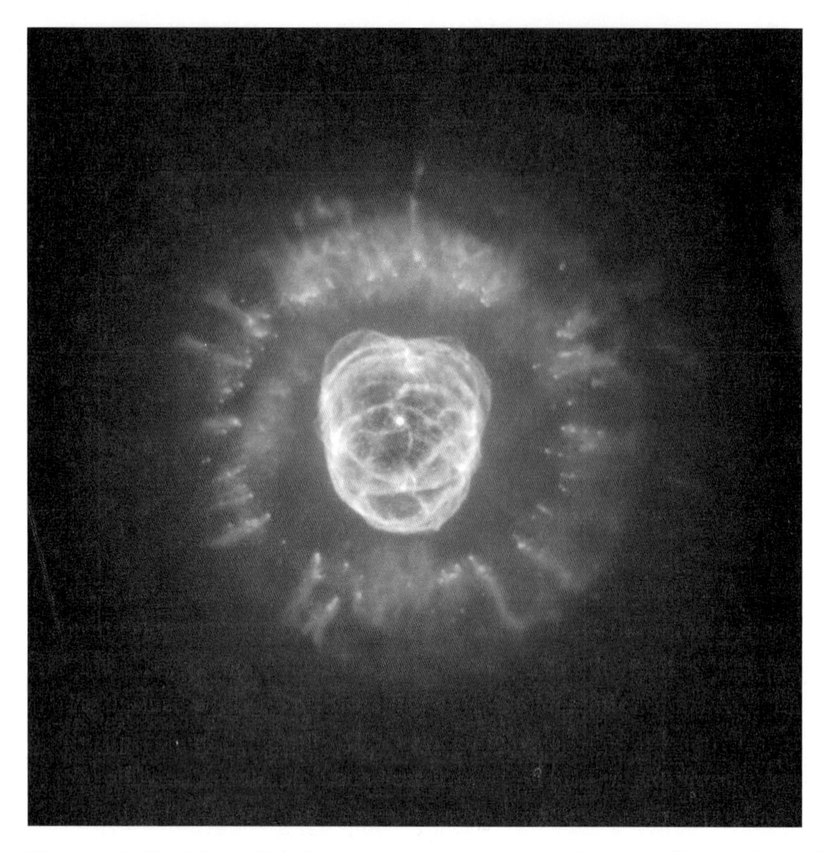

Figure 11.3. The Eskimo Nebula. About 10,000 years ago, a dying Sun-like star began flinging material into nearby space, producing this planetary nebula that is formally designated as NGC 2392. When first observed more than two centuries ago, it was dubbed the "Eskimo" Nebula because it resembled a face surrounded by a fur parka like those worn by Eskimos. It is located about 5,000 light-years from the Earth. This detailed image, obtained by the Hubble Space Telescope, reveals several episodes of ejection from the central star, including an outer ring of objects that are shaped like teardrops pointing outward and elongated, filamentary bubbles, each about 1 light-year in diameter. Dense material enveloping the star's equator has blocked ejected material, and intense winds moving at about 420 km s^{-1} have swept material above and below the equatorial regions. The bright central region contains another wind-blown bubble. (Courtesy of the NASA/Andrew Frucher/ERO Team [Slyvia Baggett/STScI/Richard Hook, ST-ECF, and Zolan Levay/STScI].)

Despite their infrequent appearance on cosmic time-scales, thousands of planetary nebulae are known. Some are listed in Table 11.3 with their popular names and their number in the New General Catalogue (NGC). Table 11.3 also provides the celestial position, distance, radius, expansion velocity, and apparent visual magnitude and temperature of the central star for the planetary nebulae. The distances often are uncertain, and the radii are between 0.1 and 3 light-years, but it is the popular names that are so fascinating. They describe the resemblance of planetary nebulae to everyday objects, such as the twin weights of a dumbbell, a fur parka enveloping an Eskimo's face, the head of an owl, a cat's eye, or simply a ring.

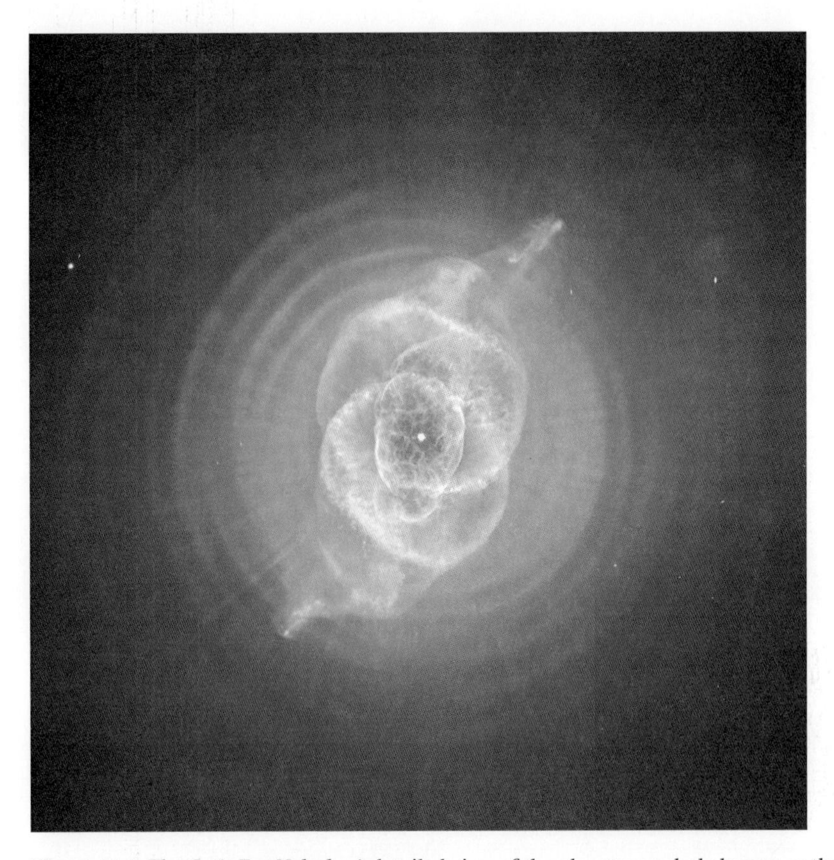

Figure 11.4. The Cat's Eye Nebula. A detailed view of the planetary nebula known as the Cat's Eye Nebula and formally designated NGC 6543. It reveals concentric rings, jets of high-speed gas, and shock-induced knots of gas. At least 11 large concentric shells mark the dense edges of spherical bubbles of gas and dust that have been ejected from a dying Sun-like star in regular, explosive pulses at 1,500-year intervals. The formation of more complex inner structures is not well understood. (A Hubble Space Telescope image courtesy of NASA/ESA/HEIC/the Hubble Heritage Team, STScI/AURA.) (See color plate.)

Stars that produce the planetary nebulae end up as small, dense white dwarf stars. In the very distant future, our Sun will become one, as will the majority of other stars. Their discovery was entirely unexpected.

11.3 Stars the Size of the Earth

The Discovery of White Dwarf Stars

The first white dwarf to be known is a companion of a much brighter star, 40 Eridani, also known as Omicron Eridani from its Greek letter designation. The fainter star is designated 40 Eridani B to distinguish it from the bright member, A, of the pair.

The American astronomer Henry Norris Russell recalled the discovery of the hot and faint nature of 40 Eridani B, which occurred during his visit to the Harvard Observatory. Russell thought it would be a good idea to obtain the spectrum of certain parallax stars, and

Table 11.3. *Bright named planetary nebulae*[a]

Catalog Designation	Popular Name[c] (Nebula)	RA (2000) h	m	Dec. (2000) °	′	D[b] (ly)	R (ly)	V_{exp} (km s^{-1})	m_V[d]	T[d] (K)
NGC 650–1	Little Dumbbell	01	42.4	+51	34.5	2,400	0.85	39	17.5	175,400
NGC 2392	Eskimo	07	29.2	+20	54.7	3,000	0.18	53	10.5	65,000
NGC 3242	Eye of Jupiter	10	24.8	−18	38.5	1,600	0.09	30	12.1	90,000
NGC 3587	Owl	11	14.8	+55	01.0	2,000	0.98	30	16.0	112,000
NGC 3918	Blue Planetary	11	50.3	−57	10.9	3,260	0.16	25	13.2	–
IC 3568	Lemon Slice	12	33.0	+82	34.0	4,500	0.4	–	–	–
MyCn18	Hourglass	13	39.6	−67	22.9	10,100	0.33	10	14.4[e]	–
Menzel 3	Ant	16	17.2	−51	59.2	4,140	0.26	–	17.6[e]	–
M 2–9	Butterfly	17	05.6	+10	08.6	5,542	0.20	31	15.7	–
Hen 3–1357	Stingray	17	16.4	−59	29.6	18,000	0.08	–	15.0	–
NGC 6369	Little Ghost	17	29.3	−23	45.6	2,000	0.15	41	15.9	58,000
NGC 6543	Cat's Eye	17	58.6	+66	38.0	3,000	0.20	20	11.0	50,000
NGC 6720	Ring	18	53.6	+33	01.8	2,300	0.47	30	15.7	150,000
NGC 6751	Dandelion	19	05.9	−05	59.6	≤8,000	≤0.43	40	13.9	76,000
NGC 6826	Blinking Eye	19	44.8	+50	31.5	3,600	0.25	11	10.7	47,000
NGC 6853	Dumbbell	19	59.5	+22	43.3	1,200	1.0	28	14.0	160,000
NGC 7009	Saturn	21	04.2	−11	21.8	2,000	0.38	20	13.0	90,000
NGC 7027		21	07.0	+42	14.2	2,900	0.13	18	16.3	185,000
NGC 7293	Helix	22	29.6	−20	50.2	715	2.0	25	13.4[e]	110,000
NGC 7662	Blue Snowball	23	25.9	+42	32.1	2,500	0.20	30	13.2	110,000

[a] The distance, D, and radius, R, are in units of light-years, abbreviated "ly". (Data courtesy of James B. Kaler, University of Illinois.)

[b] The distances of some planetary nebulae are not well known.

[c] Other names: Little Dumbell Nebula = M 76, Barbell Nebula, Cork Nebula; Eskimo Nebula = Clownface Nebula; Eye of Jupiter Nebula = Ghost of Jupiter Nebula, Eye Nebula; Owl Nebula = M 97; Blue Planetary Nebula = The Southerner Nebula; M 2–9 = Minkowski 2–9; Butterfly Nebula or Twin Jet Nebula; Ring Nebula = M 57; Dandelion Nebula = Dandelion Puff Ball Nebula; Dumbbell Nebula = M 27; Blue Snowball Nebula = Snowball Nebula.

[d] The estimated apparent visual magnitude, m_V, and temperature, T, of the central star.

[e] Apparent blue magnitudes.

Edward C. Pickering, director of the observatory, asked for the name of one of these stars. Russell replied that the faint companion of Omicron Eridani was an example. Pickering remarked: "Well, we make rather a specialty of being able to answer questions like that." So he telephoned the office of Williamina Fleming and asked her to look up the star's spectral classification. Mrs. Fleming once had been Pickering's maid. After famously stating that his maid could do a better job than his male assistants, Pickering had hired her to do the clerical work of the observatory.

In about a half hour, she reported that the spectrum of 40 Eridani B exhibited strong hydrogen absorption lines, implying that it was a hot, white star of spectral type A. Russell was flabbergasted and baffled about what it meant. Unlike other stars of this spectral class, 40 Eridani B was much fainter than its bright companion and therefore of much lower

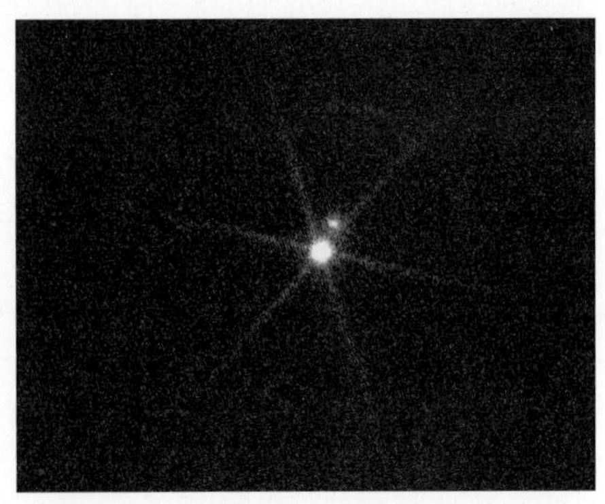

Figure 11.5. Sirius A and B. The brightest star in the night sky, called the Dog Star and also designated Sirius A, has a faint companion, Sirius B. This binary-star system is located only 8.6 light-years from the Earth. Sirius A is a main-sequence star with twice the mass of the Sun and a disk temperature of 9,940 K. Its white-dwarf companion, Sirius B, has 98 percent of the mass of the Sun and a disk temperature of 25,200 K. Due to its small size, comparable to that of the Earth, Sirius B is 1,000 times less luminous than Sirius A. The faint white dwarf can be seen in the lower left of the Hubble Space Telescope image taken in optically visible light (*left*): the cross-shaped diffraction spikes and concentric rings around the much brighter Sirius A are artifacts produced by the telescope imaging system. Sirius B is somewhat brighter in x-rays due to its higher temperature, as indicated in the Chandra X-Ray Observatory image (*right*). (Courtesy of NASA/H. E. Bond, and E. Nelan, STScI/M. Barstow, and M. Burleigh, U. Leicester, and J. B. Holberg, U. Az [*left*] and NASA/SAO/CXO [*right*].)

luminosity than other A-type stars. Pickering thought for a moment and with a kindly smile said, "I wouldn't worry. It's just these things which we can't explain that lead to advances in our knowledge."[20]

The American astronomer Walter S. Adams drew attention to the A0 spectral type of 40 Eridani B, which suggested a disk temperature of about 10,000 K, and noticed that it was surprising that such a hot star should exhibit such a very low luminosity.

Unfortunately, 40 Eridani B is so far away from its bright companion, with an orbital period of at least 7,300 years, that its mass could not be inferred from its orbital motion. However, this was not the case for Sirius B, the second white dwarf to be discovered. Adams showed that it also is a low-luminosity star of spectral class A0. It is a member of a binary-star system, with a luminous companion designated Sirius A, the brightest star in the night sky (Fig. 11.5).

The irregular motion of Sirius A first suggested the presence of its dim companion. The bright star swerves from side to side of a straight-line trajectory, and this swerving motion was attributed to the gravitational attraction of a nearby, unseen companion. The masses of Sirius A and B have been estimated from the orbital motion of the pair; the mass of Sirius A is 2.02 solar masses and the mass of Sirius B is 0.978 times that of the Sun.

What Adams did not point out explicitly was that the high disk temperature in combination with the low total luminosity meant that both 40 Eridani B and Sirius B must be very small – only about the size of the Earth. Furthermore, the rather ordinary mass of Sirius B meant that the average mass density of the star must be enormous – about 1 million times the average mass density of the Sun.

These stars now are known to be the inner, collapsed leftovers of dying red giant stars, exposed by the planetary nebulae that carried off the outer stellar atmospheres. Like a butterfly, a white dwarf star begins life by casting off a cocoon that enclosed its former self.

Unveiling White Dwarf Stars

Having depleted hydrogen in their cores, the central regions of solar-mass stars contract to become hot enough to fuse helium into carbon and oxygen. Simultaneously, the stars brighten and their outer parts expand to giant proportions; however, they soon reach the end of their life. There is not enough mass in most stars to generate a temperature hot enough to fuse carbon into neon, at about 1 billion K, and an inert carbon-oxygen core is surrounded by an inner helium-burning shell and an outer hydrogen-burning shell.

After shedding most of its outer material to form a planetary nebula, these giant stars leave behind a hot core of carbon and oxygen. Because it cannot generate additional heat by nuclear reactions, the core collapses to form a white dwarf with about the Sun's mass compressed to 1/100th of its former size and about the same radius as the Earth. Such a collapsed star is called a "white dwarf" because initially it is white in color and it is relatively small for a star.

The concept of a white dwarf as the exposed carbon-oxygen core of a former giant star initially was difficult to reconcile with the fact that the emitted light of the first white dwarfs contains strong spectral lines of hydrogen and it does not include the oxygen lines found in the surrounding planetary nebula. The French astronomer Evry Shatzman explained this paradox when he noted that the hydrogen resides in a thin outer atmosphere and that the heavy elements such as carbon and oxygen had to sink out of sight into the dense stellar interior. These elements are drawn by gravity into the unseen interior of the white dwarf and remain hidden by a hydrogen-rich layer that is only about 1/10,000th of the star's mass.

An inert, nonburning star composed mainly of carbon and oxygen represents the final destiny, the end state, of main-sequence stars with a mass of between 0.5 and 8 solar masses, and it accounts for most of the observed white dwarf stars. Main-sequence stars of lower mass, below 0.5 solar masses, bypass the giant stage – never becoming hot enough inside to fuse helium – and they collapse directly into white dwarfs composed of helium. However, these stars have such a low luminosity and temperature that they burn hydrogen slowly and have not yet exhausted their internal supply, even during the 14-billion-year age of the universe.

When first exposed, the white dwarf can be very hot because it was previously in the hot interior of a giant star (Fig. 11.6). It may have an initial temperature of 200,000 K due to its hot origin and the collapse that created it, but because there is no thermonuclear fuel, there is nothing left to heat a white dwarf star. It resembles a jewel that sparkles but generates no warmth.

The dense, burned-out cinder is destined to become the frozen remnant of a once-powerful star, a mere shadow of its former self. It slowly radiates away its remnant heat, cooling and

Figure 11.6. Hot white dwarf revealed. The expanding material in planetary nebula NGC 2440 has been cast off by a dying Sun-like star in episodic outflows and in different directions, revealing a central white dwarf star. It is one of the hottest stars known, with a disk temperature of about 200,000 K, which is more than 30 times hotter than our own Sun's photosphere. Ultraviolet radiation from the hot star has excited oxygen and nitrogen ions in the nebular gas, making them glow and fluoresce. Clouds of dust form long dark streaks pointing away from the central star. A much larger, cooler cloud of gas and dust remains dark and unseen in this visible-light image; it can be detected by its infrared radiation. (Courtesy of the NASA/Hubble Heritage Team/AURA/STScI.)

fading as it ages. This then is the way the Sun will end. It ultimately will fade away, like a dying ember in a fireplace.

A white dwarf's temperature will drop from an initial high of up to 200,000 K to an observed low that is colder than the Sun, steadily becoming fainter and dimmer. This cooling and dimming is so slow that the oldest white dwarf has not yet chilled to the point of invisibility. That is, observations indicate that the oldest white dwarf stars, which are remnants of the earliest stars, have not yet had time to cool to the lowest possible luminosity that might be observed. The amount of time that the oldest white dwarf has cooled is estimated to be about 9 billion years, which when combined with its former lifetime as a main-sequence and giant star gives a rough estimate to the age of the universe of about 14 billion years.

The High-Mass Density of White Dwarf Stars

The rather ordinary stellar mass of a white dwarf has been compressed to a star the size of the Earth and a remarkably high mean mass density up to 1 million times of the Sun's. At the time of their discovery, some astronomers thought that such a high density was impossible. As Arthur Eddington stated:

> We learn about the stars by receiving and interpreting the messages which their light brings to us. The message of the Companion of Sirius when it was decoded ran: "I am composed of material 3,000 times denser than anything you have ever

Table 11.4. *Physical properties of white dwarf stars*

M_{wd} = mass of white dwarf star $\approx 0.6\ M_\odot \approx 1.2 \times 10^{30}$ kg
M_{cwd} = critical upper mass limit for white dwarf star = $1.4\ M_\odot \approx 2.3 \times 10^{30}$ kg
R_{wd} = mean radius of white dwarf stars = $0.01\ R_\odot \approx 6 \times 10^6$ m $\approx R_E$ = radius of Earth
ρ_{wd} = mass density of white dwarf star $\approx 10^9$ kg m^{-3}
V_{esc} = escape velocity of white dwarf star = $(2GM_{wd}/R_{wd})^{1/2}$ = 9,000 km s^{-1} = 0.03 c
L_{wd} = absolute luminosity of white dwarf star $\approx 10^{-3}\ L_\odot \approx 3.8 \times 10^{23}$ J s^{-1}
T_{wd} = effective temperature of white dwarf star's visible disk = 4×10^3 K to 7×10^4 K
B_{wd} = surface magnetic field strength of white dwarf star = 10^2 tesla to 10^4 tesla
z_g = gravitational redshift of white dwarf star = $GM_{wd}/(R_{wd}c^2) \approx 0.02 = V_r/c$
(or radial velocity $V_r = z_g c \approx 60$ km s^{-1}, where c is the speed of light)

come across; a ton of my material would be a little nugget that you could put in a matchbox." What reply can one make to such a message? The reply which most of us made in 1914 was – "Shut up. Don't talk nonsense."[21]

However, Eddington already had realized that there is nothing inherently absurd about the high mass density of white dwarf stars. Because all of the electrons are stripped away from their atomic nuclei in the hot stellar interiors, the free electrons can be packed closely with the bare nuclei, within the former space of the empty atoms.

As Eddington previously pointed out, the high mass densities implied a strong gravitational redshift, and he predicted that it might be observed for Sirius B once allowance was made for the orbital motion of the double-star system. The gravitational redshift is the Doppler shift of a spectral line caused by the loss of energy in overcoming the gravity of the emitting object. Adams apparently confirmed the effect, providing an independent confirmation of the small size of white dwarf stars that had been inferred from their high temperature and low luminosity. This result was substantiated with greater clarity for Sirius B, 40 Eridani B, and other white dwarf stars using ground-based telescopes as well as the Hubble Space Telescope.

The small size and high mass density of the white dwarf stars also have been substantiated by measurements of their magnetic-field strength. During gravitational collapse, magnetic flux is conserved, and the magnetic-field strength increases with the inverse square of the decreasing radius. The dipolar magnetic field of the solar disk is about 0.01 tesla; for a white dwarf 1/100th the radius of the Sun, it will be amplified during collapse to 100 tesla. These intense magnetic-field strengths were confirmed for white dwarf stars by observations of their spectral lines or their wavelength-dependent, circularly polarized light. These and other physical properties of white dwarf stars are given in Table 11.4.

11.4 Crushed States of Matter

Nuclei Pull a White Dwarf Together as Electrons Support It

The matter deep inside a white dwarf star is completely ionized and composed of equal numbers of atomic nuclei and electrons. Because most white dwarfs are the crushed remnants of red giant stars, which previously fused helium into carbon, their collapsed cores consist mainly of carbon nuclei and electrons. Stars that are somewhat more massive than the Sun leave behind white dwarfs containing oxygen nuclei. Therefore, white dwarf stars contain various amounts of carbon nuclei or oxygen nuclei, depending on the star, and it is these nuclei that supply the mass and gravity of a white dwarf star.

Why does the core collapse stop at the white-dwarf stage? In other words, what is holding up the star? There are no internal nuclear reactions to provide energy, generate heat, and create pressure to oppose gravity. As the white dwarf radiates away the heat left over from its former life in a red giant star, it eventually might cool down to a temperature of absolute zero, and there would be no motion or thermal energy left to support the white dwarf star.

When the material cooled enough, it could be expected that the electrons would return to their former orbits around the nuclei, making larger atoms, which would force the nuclei apart and make the white dwarf expand in size. However, the stars have no internal energy to push against gravitation and accomplish this feat; therefore, it seemed that a white dwarf star could not become that cold.

This paradox was not resolved until the development of quantum mechanics and the realization that it is the electronic properties of the crushed matter that hold up a white dwarf. The high-speed motions of the densely packed electrons, rather than the nuclei, produce an outward pressure that holds the gravitational forces at bay. These motions are not due to the star's internal temperature or heat; in fact, the internal pressure of a white dwarf star is unaffected by temperature.

The quantum-mechanical description of a very dense, crushed state of matter is statistical, and it is related to the quantum numbers that specify the state of subatomic particles. That is, the properties of the particles (e.g., energy or location) can take on only specific quantized values and no others. This situation is related to two principles that govern the quantum state of the very small: (1) The *uncertainty principle* essentially states that at any given time, we cannot know exactly both where a particle is and where it is going; and (2) the *exclusion principle* forbids the existence of two or more particles in exactly the same quantum state.

The Austrian physicist Wolfgang Pauli proposed the exclusion principle, which states that two identical subatomic particles cannot occupy the same quantum state at the same time. It applies to electrons, protons, and neutrons, and it dictates how electrons behave in an atom, occupying their various energy levels. Each electron in an atom has its own space, which prevents the electrons from either joining together in the same location or falling into their atomic nucleus. The exclusion principle ensures the very existence of atoms.

It also means that the free electrons in a white dwarf star, which are not attached to atoms, cannot be in precisely the same place at the same time. They instead resist being squeezed into one another's territory, darting away at high speeds just to keep their own space, somewhat like active dancers in a very crowded nightclub. This provides the pressure of the crushed state of matter, which is caused by the electrons' resistance to crowding.

The Italian physicist Enrico Fermi first worked out the statistical description of a large number of identical subatomic particles based on the exclusion principle. He specified the conditions in which all of the particles have the least possible energy without violating Pauli's exclusion principle, which specifies that every particle cannot occupy the very lowest energy, called the *ground state*, at the same time. In Fermi's solution, all of the states up to a certain limiting energy can be occupied, and all of those above that energy are not occupied.

Under such conditions, the collection of subatomic particles, or gas, is said to be *degenerate*, which has nothing to do with a lowly perverted character, as in a degenerate person with impaired virtue and moral principles. In mathematics, a degenerate case is a limiting one in which a class of objects changes its nature so as to belong to another, usually simpler, class; for example, a point is a degenerate circle. As shown by Ralph H. Fowler, who was Eddington's colleague at Cambridge University, it is the degenerate pressure of the electron gas that supports a white dwarf star.

Fowler applied Fermi's statistical description to such a dense star, showing that its electrons are completely degenerate. They produce an outward push, known as *degeneracy pressure*, to keep their own space and support the star. Not only is the electron degeneracy pressure strong enough to withstand the crushing gravity of a white dwarf, but it also is independent of the temperature of the electrons and involves no nuclear reactions. Because the pressure does not depend on temperature, it will persist even if the star cools to absolute zero, without the electrons ever rejoining the nuclei. As Fowler showed, individual electrons in the crushed matter even at a temperature of absolute zero still would have a kinetic energy comparable to the thermal energy of particles in an expanded gas the temperature of which is as large as 10 million K.

Because the equation of state of the degenerate electron gas is unaffected by temperature, any heating by hypothetical nuclear reactions will increase the temperature and rate of those reactions. The temperature would continue increasing until the star exploded, so we conclude that white dwarf stars do not shine by nuclear reactions. Their light must come from the slow leakage of the heat contained in the nondegenerate nuclei. Eventually, the white dwarf star will fade into a gigantic black molecule, a frozen star.

Radius and Mass of a White Dwarf

A higher-mass white dwarf will be squeezed into a smaller space by its gravity, so the star's radius decreases with increasing mass. For a large enough mass, we might imagine that the star's radius would become very small, perhaps even shrinking to almost zero; however, this is preposterous and there must be a limit to the mass.

As the white dwarf shrinks in size, its mass density becomes higher as does the degeneracy pressure of its electron gas. Under extreme compression, however, the average speed of the electrons increases and eventually approaches the speed of light. This means that there is an upper limit to the mass that can be supported by their pressure. After all, we cannot make an electron or anything else move faster than the speed of light.

The limiting mass for a white dwarf star is determined under high-speed, relativistic conditions, when the electrons approach the speed of light. As the German-Estonian astrophysicist Wilhelm Anderson showed, the increase of electron mass with velocity must be taken into account by using the equations of special relativity.

Both Anderson and the English physicist Edmund C. Stoner found that a star could contract only until the gravitational potential energy becomes insufficient to balance the increase in the kinetic energy of the electrons. Accordingly, for stellar masses larger than about 1 solar mass, there can be no equilibrium white dwarf configurations. This is common sense because the electrons will move at greater speeds with increasing stellar mass and density, but the electrons cannot move faster than the speed of light.

During his voyage from India to Cambridge University for his graduate studies, the Indian astrophysicist Subrahmanyan Chandrasekhar subsequently derived the equation of state of a degenerate electron gas in the extreme relativistic limit. His results indicated that a white dwarf star cannot be more massive than a critical mass of about 1.46 solar masses in modern determinations. More massive stars collapse under their own weight to form a neutron star or a black hole at the endpoints of stellar evolution. Although both Anderson and Stoner previously called attention to the existence of this upper mass limit, it is known now as the *Chandraskhar limit* because he was the first to derive the detailed equilibrium conditions in which degenerate electron gases support a dense star's gravity.

In 1983, the Nobel Prize in Physics was awarded equally to Chandrasekhar, for his theoretical studies of the physical processes of importance to the structure and evolution of the stars, and to William Fowler, for his theoretical and experimental studies of the nuclear reactions of importance in the formation of the chemical elements in the universe.

11.5 Stars That Blow Up

Guest Stars, the Novae

For at least 2,000 years, astronomers, hunters, mariners, and others familiar with the brightest stars must have been amazed by a *nova*, or "new star," that would appear suddenly at a place in the sky where no star previously had been seen, and then slowly fade away until it disappeared back into invisibility, without a trace. The Chinese called them "guest stars" or "visiting stars" because they were not permanent members of the celestial sphere, instead appearing suddenly and then departing abruptly, like uninvited guests.

Every 20 years or so, a nova is bright enough and close enough to be conspicuous without the aid of a telescope, attracting the attention of both astronomers and the superstitious. Like good wine, they are specified by the year of their occurrence, and their location is specified by the constellation in which they appear. Nova Aquilae 1918 was the brightest of the twentieth century, at apparent visual magnitude −1.1; Nova Herculis 1934 has historical importance; and Nova Cygni 1992 is the brightest nova in recent history.

New stars do not appear suddenly out of nothing, like a bird flying out of a magician's hand or seemingly plucked out of thin air. Something of substance must have been at a nova's location before it appeared, to supply the energy of its outburst. By the mid-twentieth century, it was realized that this "mysterious something" was an inconspicuous star that previously had been recorded during systematic surveys of the dark night sky, using large telescopes and photographic exposures to record the dim light of faint stars.

So, the bright, short-lived novae are neither new nor temporary but instead existing stars that suddenly increase in brightness by as much as 100,000 times, returning to their original states after several months or a few years.

Telescopic observations have also enabled the detection of faint dwarf novae. They brighten repeatedly, on a time-scale from days to decades, although by a smaller amount and

Table 11.5. *Physical properties of some novae*[a]

Star Name	Year	m_{max}	L_{max} $(L_{\odot})$	D (ly)	P_{orb} (hours)	M_1 $(M_{\odot})$	M_2 $(M_{\odot})$	V_{exp} (km s^{-1})
Classical Novae								
GK Persei	1901	+0.2	$10^{5.3}$	1,500	47.92	0.9	0.25	1,200
V603 Aquilae	1918	−1.4	$10^{5.6}$	800	3.31	0.66	0.2	265
DQ Herculis	1934	+1.4	$10^{3.9}$	316	4.65	0.62	0.44	315
V1974 Cygni	1992	+4.4	$10^{4.9}$	19.53	0.83	–	–	
			10,430					
Dwarf Nova								
SS Cygni	*b*	8.3	≈10	541	6.603	0.60	0.40	–
Recurrent Nova								
RS Ophiuchi	*c*	4.5	0.1	≥2,000	455.7 d	1.4	–	–
					1.4	–	–	

[a] Maximum visual magnitude, m_{max}, and maximum luminosity, L_{max}, in units of the Sun's luminosity $L_{\odot} = 3.828 \times 10^{26}$ J s^{-1}; distance in light-years, abbreviated "ly"; orbital period P_{orb}; white dwarf mass, M_1, and companion mass, M_2, in units of the Sun's mass $M_{\odot} = 1.989 \times 10^{30}$ kg; and expansion velocity V_{exp}.

[b] The dwarf nova SS Cygni undergoes frequent and regular outbursts every 7 to 8 weeks, with an apparent visual magnitude of $m_{min} = 12.2$ at minimum and $m_{max} = 8.3$ at maximum. More than 800 outbursts have been observed since its discovery in 1896.

[c] The recurrent nova RS Ophiuchi erupted in 1898, 1907, 1933, 1945, 1958, 1967, 1985, and 2006. It is a binary-star system with a red giant star in a 455.7-day orbit around a white dwarf star of mass near the Chandrasekhar limit.

with lower luminosity than the classical novae whose explosive outbursts are visible without the aid of a telescope. More than 900 outbursts of the dwarf nova SS Cygni, for example, have been observed since its discovery in 1896. It varies from apparent visual magnitude of 12.2 at minimum to 8.3 at maximum, every 7 to 8 weeks.

The properties of some of the classical novae, visible by the unaided eye, are listed in Table 11.5 with the dwarf nova SS Cygni.

What Makes a Nova Happen?

A major new understanding of novae occurred when a few American astronomers began to examine the total light and spectra of ex-novae, long after the intense light of the nova outburst had faded to a relatively weak level. It then was discovered that a nova is not one star but rather two stars very close together. Twenty years after the 1934 eruption of Nova Herculis, for example, Merle F. Walker found that this nova is an eclipsing binary system with a remarkably brief orbital period of only 4.6 hours. The shortness of the period indicated that the two stars are very close together, practically touching one another. Nearly a decade later, Walker was able to show that Nova T Aurigae 1891 also is an eclipsing binary system with a short period of 4.8 hours.

Alfred H. Joy examined the absorption and emission lines of the dwarf nova SS Cygni, identifying it as a binary-star system with a short orbital period of 6.6 hours. The emission lines originated in a blue-white dwarf star, whereas the absorption lines came from a red main-sequence star the size of which was estimated to be roughly half the distance between

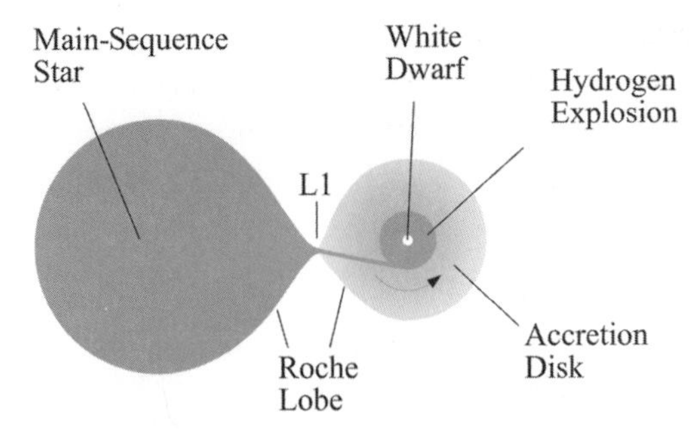

Figure 11.7. Nova. A classical nova is a thermonuclear explosion that occurs on the surface of a white dwarf star that is in a close orbit with a main-sequence star. The strong gravitational attraction of the white dwarf pulls its nearby companion into an elongated shape, the outer edge of which is designated the *Roche lobe*. Some of the hydrogen in the outer atmosphere of the main-sequence star spills over at the inner Lagrangian point, denoted L1, where the gravitational pull of the two stars is equal. This hydrogen spirals into a rotating accretion disk and down to the white dwarf, igniting an explosion, like a colossal hydrogen bomb.

the two stars. They were so close to one another that mass could spill from the red star into the blue star, and the nova process might be related to this mass flow.

During the ensuing decade, Robert P. Kraft demonstrated that membership in a short-period binary system is a necessary condition for a star to become a nova of either the classical or dwarf type. One of the stellar pair was usually a blue-white dwarf star; the other red component was usually a cool main-sequence star of spectral type G, K, or M or one of the same spectral types that is aging and expanding into a red giant. The short orbital period indicated that the two stars are so close that hydrogen flows from the red companion onto the white dwarf, reviving the "dead" star and giving it a brief new life in a cataclysmic nuclear explosion. The term *cataclysmic variable star* is used now to designate such close binary-star systems, in which one of the components – conventionally called the *primary star* – is a white dwarf that accretes matter from its secondary companion. The category includes classical, dwarf, and recurrent novae.

The immense gravity of the white dwarf star distorts the shape of its nearby companion, stretching it into an elongated configuration (Fig. 11.7). This is a tidal effect in which the side of the companion star that is nearest the white dwarf is pulled toward it, and the companion's center is pulled away from the side that is farthest from the white dwarf. Material that is pulled close enough to the white dwarf can pass outside the gravitational control of the red companion and go into orbit around the white dwarf, eventually spiraling into it.

The region in which the companion star retains gravitational control of its substance is known as its *Roche lobe*, named after the French astronomer Édouard Roche who described it more than a century ago. When the outer hydrogen atmosphere of the close companion star overflows its Roche lobe, it does not fall directly into the white dwarf star. In the absence of an intense magnetic field, the gas is accreted into an orbiting disk that resembles the proto-planetary disks that circle protostars. The gas streams onto the magnetic poles when there is a strong magnetic field directing the flow.

In either case, the hydrogen is pulled slowly into the white dwarf, which compresses and heats the gas to very high temperatures. The accumulating hydrogen is a potential bomb that remains harmless until detonated, like a stick of dynamite.

The little disturbances, the tiny perturbations, of the accreting matter build up until the white dwarf cannot tolerate it any longer. The star heats up, is eventually pushed beyond tolerance, and unleashes vast power in an explosive outburst. A similar thing can happen when you live with someone for a long time; the endless minor irritations eventually become overwhelming and you explode in heated anger.

A thin layer of hydrogen slowly builds up on the white dwarf as its companion keeps feeding matter into it. The pressure and temperature rise until hydrogen fusion suddenly is ignited, at about 10 million K. A runaway thermonuclear explosion then occurs, and a white dwarf that normally would cool and fade away if left alone suddenly shines as brightly as 100,000 Suns. The explosion subsides after a few weeks, but the overflow of the companion continues.

The envelope of the white dwarf star is thrown off during the nova explosion, at high speeds of up to several thousand kilometers per second; despite the violence, however, the amount of material ejected is only about 0.000005 of a solar mass. The white dwarf therefore can regain its composure, retain its stability, and potentially generate additional novae as its companion continues to feed matter into it. An example of such a recurrent nova is RS Ophiuchi, which has exploded into a bright nova state at least six times between 1898 and 2006.

Eventually, the hydrogen may build up until it pushes the white dwarf above its limiting mass; the entire star then explodes, not only the thin outer atmosphere. This is another story that involves a supernova that suddenly and unpredictably brightens with the light of 1 billion Suns.

A Rare and Violent End, the Supernovae

On rare occasions, an entire star is annihilated and suddenly becomes so bright that it can be seen easily in daylight rather than just at night like the novae. The Chinese emperor's astronomers in the Sung dynasty recorded one on July 4, 1054, near the constellation now known as Taurus, the Bull. The Chinese chronicles indicate that the new star initially was brighter than everything in the night sky except the full Moon; could be seen during the daytime for three weeks after its first appearance; and remained visible in the night sky for 22 months, without the aid of telescopes, which had not yet been invented. The "new" star of 1054 was definitely far brighter and longer lasting than any other guest star.

More than four centuries passed before the unheralded appearance of other exceptionally brilliant guest stars, and this time they shook the very foundations of European thought. As Aristotle taught, heavenly bodies were supposed to be eternal, pure, changeless, incorruptible, and perfect, unlike anything on the Earth. Yet, in a span of just 32 years, two new daytime stars could be seen by almost anyone in the Earth's Northern Hemisphere who happened to look up. Each star remained fixed in the heavens for about a year and then disappeared from view.

Both events also were discovered at a time before telescopes were invented. The Danish astronomer Tycho Brahe witnessed the first visitor in 1572, which initially was brighter than the planet Venus. Perhaps because of the excitement caused by his discovery, Brahe built an observatory in which detailed measurements were made of stars and the planets. Johannes

Table 11.6. *Historical supernovae visible to the unaided eye*[a]

Explosion Date	m_{max}	M_{max}	L_{max} ($L_{\odot}$)	Visible (Months)	Type	Remnant Name	D (ly)	θ (′)	R (ly)
SN 185	−8.0	−20.2	$10^{10.0}$	8	Ia	RCW 86	9,100	45	56
SN 386	+1.5	–	–	3	–	–	–	–	–
SN 393	−1.0	−11.0	$10^{6.3}$	8	II/Ib	*b*	3,000	70	30
SN 1006	−7.5	−19.2	$10^{9.6}$	21	Ia	PKS 1451−41	7,200	31	32
SN 1054	−6.0	−17.5	$10^{8.9}$	22	II	Crab Nebula	6,500	7	6.6
SN 1181	−1.0	–	–	6	–	*c*	>8,000	–	–
SN 1572[d]	−4.0	−16.4	$10^{8.5}$	16	Ia	Tycho	11,500	8.3	14
SN 1604[e]	−2.5	−16.4	$10^{8.5}$	12	Ia	Kepler	20,000	3.2	9

[a] Maximum apparent visual magnitude, m_{max}, maximum absolute magnitude, M_{max}, and maximum luminosity, L_{max}, in units of the Sun's luminosity $L_{\odot} = 3.828 \times 10^{26}$ J s^{-1}; length of visibility by the unaided eye; supernova type; supernova remnant name; distance D in light-years, abbreviated "ly"; angular diameter θ of supernova remnant in minutes of arc, denoted ′; and remnant radius R in light-years.

[b] Supernova remnant RX J1713.7–3946.

[c] Radio source 3C 38. The radio and x-ray pulsar J0205+6449 may not be associated with SN 1181.

[d] Supernova explosion also known as Tycho's star, supernova remnant 3C 10.

[e] Supernova explosion also known as Kepler's star, supernova remnant 3C 358.

Kepler, who used Brahe's observations of planets to determine the laws of their motion, spied another bright new star as it lit up the heavens in 1604.

The exceptionally brilliant guest stars of 1054, 1572, and 1604 were much brighter and longer lasting than the conventional novae known at the time or subsequently. They therefore have been dubbed *supernovae*, a term coined by the Swiss astronomer Fritz Zwicky.

Now that we know the distances, a supernova also refers to a stellar outburst the maximum luminosity of which exceeds by factors of several billions the luminosity of our Sun (Table 11.6). This is millions of times the peak luminosity of a classical nova. Moreover, unlike novae, the new breed of exploding stars has nothing conventional to return to after the explosion. They require so much energy that the mass of an entire star is annihilated.

In 1934, Walter Baade and Fritz Zwicky communicated to the United States National Academy of Sciences a remarkable pair of papers on supernovae. In one paper, they showed that the enormous energy emitted in the supernova process corresponds to the total conversion of an appreciable fraction of a star's mass into energy. In the second paper, they predicted that a supernova explosion will accelerate charged particles to very high energies and that supernovae that occur only once in a millennium can account for the energetic cosmic-ray particles that now rain down on the Earth's atmosphere from all directions in outer space. In this more speculative paper, the two Caltech astronomers also speculated that the collapsing core of the explosion might become a neutron star of very small radius and extremely high mass density.

It took a half-century for astronomers to realize that Baade and Zwicky were correct on all counts. Then, with characteristic "modesty," Zwicky claimed to have made the most concise triple prediction ever made successfully in science.

Table 11.7. *Characteristics of supernova types*[a]

Characteristic	Type Ia	Type Ib	Type II
Optical Spectrum	No hydrogen Si II at 615.0 nm	No hydrogen He I at 587.6 nm	Hydrogen present at 656.3 nm
Maximum Luminosity	$10^{9.8}$ $L_{\odot}$	$\approx 10^{9.1}$ $L_{\odot}$	$10^{9.1}$ $L_{\odot}$
Ejection Velocity	$\geq 10^4$ km s^{-1}	$\geq 10^4$ km s^{-1}	$\leq 10^4$ km s^{-1}
Ejected Mass	≈ 1 $M_{\odot}$	≈ 1 $M_{\odot}$	≈ 5 $M_{\odot}$
Progenitor Star	White Dwarf	Wolf-Rayet	Supergiant
Progenitor Star Mass	1 $M_{\odot}$	4 $M_{\odot}$ to 7 $M_{\odot}$	≥ 8 $M_{\odot}$

[a] Maximum luminosity in units of the Sun's luminosity $L_{\odot} = 3.828 \times 10^{26}$ J s^{-1}; mass values in units of the Sun's mass $M_{\odot} = 3.854 \times 10^{30}$ kg.

The initial evidence for supernovae was extraordinarily sparse. In the Milky Way, they are seen at intervals of roughly 100 years, which is about 3,000 times less common than dwarf novae; successive generations of astronomers may have to wait centuries to observe a supernova.

In the early decades of the twentieth century, astronomers discovered numerous faint novae in spiral nebulae, which suggested that the spirals were very distant if these "new" stars were like the classical novae seen in the Milky Way. At this distance, an exceptionally bright nova, observed in the nearest spiral Andromeda, would have the luminosity of a supernova.

When the enormous distances to the spiral nebulae were confirmed, it was realized that they are not nebulae at all but instead galaxies that each contain about 100 billion stars. Moreover, at maximum, a supernova briefly will outshine 1 billion stars in the same galaxy.

Because there are many of these extragalactic spirals, now called galaxies, Zwicky realized that a systematic photographic survey quickly would catch at least one star in the act of supernova explosion – and he was right. He detected the first one with a camera attached to a modest telescope placed on the roof of a building at Caltech.

By observing the spectra and fading light of supernovae in distant galaxies, astronomers subsequently found that there are two methods for stars to come to such a violent end. Modern identifying characteristics of supernovae of different types are given in Table 11.7, including the important Type Ia and Type II.

The peak light output from Type I supernovae typically is one or two magnitudes more luminous than that of the fainter Type II supernovae, which fade more slowly. We now know that the decay of radioactive elements produced during high-temperature Type II explosions heats the expanding gas and produces the optically visible light.

Walter Baade used Tycho's observations of the decaying light from the brilliant 1572 supernova to classify it as Type Ia; Kepler's meticulous observations of the 1604 event indicated a Type II light variation. Both of these supernovae were observed before telescopes were invented and spectra were obtained from cosmic objects.

Although the two types of supernovae release comparable amounts of total energy during their explosion, Rudolph Minkowski subsequently demonstrated that there is a radical difference in the mass and kinetic energy of their ejected material. The expanding shells of Type Ia contain roughly 1 solar mass of material, whereas those of Type II events are about five times more massive. The large mass difference suggests that the progenitor stars of Type Ia supernovae are less massive than those of Type II.

Both types of supernovae involve the explosive conversion of a star's entire mass into energy but by different physical mechanisms. A Type Ia stellar explosion is due to external causes. It involves a white dwarf star pushed into nuclear explosion by too much mass overflow from a nearby companion. The other Type II supernova is an internal event, which occurs during the gravitational collapse of the iron core of a massive star that has depleted all of its energy. The nuclear supernovae of Type Ia and gravity-powered ones of Type II are believed to occur with about equal likelihood in the Milky Way, at the rate of 1 every 50 to 100 years.

When a Nearby Star Detonates Its Companion

Like the novae, there is one type of supernova that gets assistance from the outside, being pushed over the edge into explosion. It is an awesome nuclear bomb that shatters an entire star. Such a supernova – now known as Type Ia and characterized by the absence of emission from hydrogen – occurs in a close binary-star system, with a white dwarf star – the shrunken dense remnant of a former low-mass star – circling a main-sequence star.

The English astrophysicist Fred Hoyle and his American colleague William A. "Willy" Fowler introduced the detailed mechanisms for this type of supernova. When the nearby companion stars expands as a result of its normal evolution, hydrogen from its outer atmosphere spills onto the white dwarf. The overflow, for example, might happen when the ordinary visible companion runs out of core hydrogen fuel and swells into a red giant star. As the hydrogen overflow continues, a steady increase in the mass of the white dwarf will compress and heat the star. As the increasing mass approaches the upper mass limit for a white dwarf star at 1.46 solar masses, the rise in internal temperature ignites a nuclear explosion.

Because the white dwarf is supported against gravity by temperature-independent, degenerate electron pressure, adding heat to the star's interior increases the temperature but not its pressure; therefore, the white dwarf does not expand and cool in response. Instead, the increased temperature initiates the fusion of carbon nuclei in a runaway nuclear explosion that obliterates the star in a few seconds.

In other words, the added mass detonates a carbon bomb, triggering explosive nuclear reactions that quickly spread throughout the star and completely shatter it. The entire star explodes into a Type Ia supernova that shines with the light of billions of Suns, and there is nothing left.

Because every one of these explosions is triggered at the same mass limit, under similar conditions, and also because the star is completely destroyed, Type Ia supernovae are expected to produce about the same maximum light output every time they occur, at about 5 billion times more luminous than the Sun with little variation. Astronomers use this bright, uniform luminosity as a "standard candle" to measure the distances to their host galaxies located far beyond the Milky Way in the remote parts of the observable universe, thereby determining the pace of its expansion.

Stars That Blow Themselves Up

There is more than one way to explode a star, and some of the supernovae are gravity-powered, catastrophic outbursts from very old massive stars. This method of shattering a star applies to an isolated star with the right mass – between about 8 and 20 times the Sun's mass – that blows itself apart, self-destructing in a cataclysmic, one-time act of stellar suicide. This Type II supernova, which exhibits hydrogen in its spectra, follows the creation of an iron core within an evolving, massive supergiant star.

The bright explosion is a symptom of old age, the last convulsions of an exceptionally massive star that has exhausted all of its subatomic energy. The material in the core of such a massive supergiant star is not degenerate in the mathematical sense like white dwarf stars. Instead, nuclear reactions proceed in the advanced burning stages at ever-increasing central temperatures until an iron core is produced and all of the available nuclear fuel has been exhausted. Deprived of these resources, the iron core collapses under its own weight into a neutron star or black hole in less than 1 second, and an explosion blows away the rest of the in-falling matter.

At the end of its bright life – with its nuclear fires spent and the heavy hand of gravity bearing down – a massive dying star emits one last spurt of energy. It reminds one of the Welsh poet Dylan Thomas, who wrote "Do not go gentle into that good night, old age should burn and rave at close of day; rage, rage against the dying of the light."[22]

When the iron nuclei in the core of such a star are pushed together, no energy is released. The iron does not burn, regardless of how hot the star's core becomes. So, there is no longer any energy being generated to sustain the star's structure. Then, a massive star, having burned brightly for perhaps 10 million years, can no longer support its own crushing weight and the iron core collapses.

The central iron core can be crushed into a ball no bigger than New York City in less than 1 second, accruing energy from its in-fall. Electrons are squeezed inside the iron nuclei, combining with their protons to make neutrons. The material is compacted to nuclear density, and the center collapses to form a neutron star. If the collapsing core is more massive than about 3 solar masses, however, there is no end in sight and the collapse proceeds to the formation of a black hole.

Having lost the supporting core, to the eventual formation of a neutron star or black hole, the surrounding material first plunges in toward the center, like a building with the foundation removed. When reaching mass densities approaching that of an atomic nucleus, the collapsing core bounces back and a powerful shock wave pushes out against the rest of the star. With the help of a dense shower of neutrinos produced in the collapsing core, the star's outer layers are torn apart and expelled into deep space at supersonic speeds. The doomed star suddenly increases in brightness 100 million fold, becoming a Type II supernova that briefly outshines up to 1 billion of its neighboring stars combined (Fig. 11.8).

This type of supernova hurls into surrounding space all of the elements synthesized inside the star and residing in shells surrounding the iron core before its collapse. During the high-temperature explosions, the supernova also produces vast amounts of other heavy elements, including radioactive elements such as uranium. Newly formed radioactive nickel, for example, eventually decays into iron, producing most of the iron now found in the

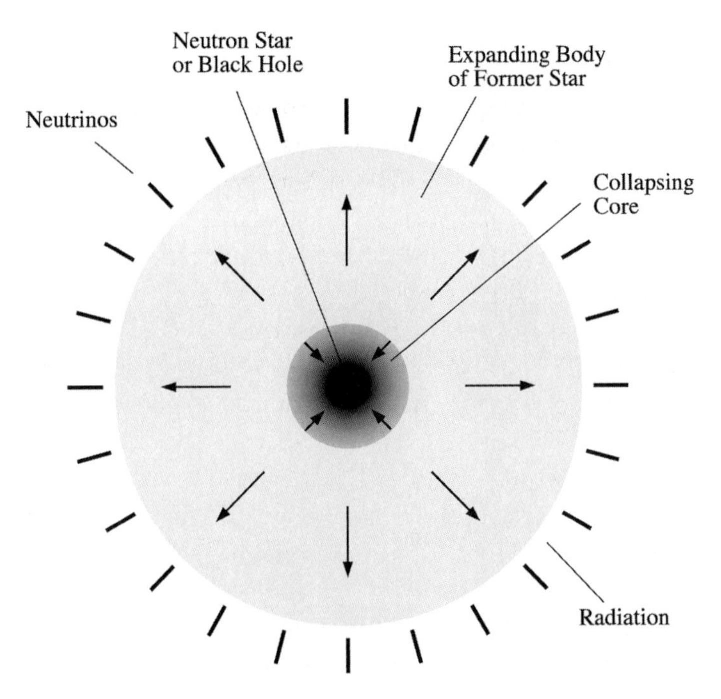

Figure 11.8. Type II supernova. In this type of supernova explosion, an isolated star blows up and its shattered remains are propelled into surrounding space. Radio and x-ray radiation from the expanding supernova remnant can be observed for thousands of years after the explosion. The core of the star is compressed by gravitational contraction into a neutron star or a black hole. Neutrinos emitted from the collapsing core remove most of the supernova energy and assist shock waves in pushing the stellar remains into an expanding remnant.

universe. Such an event occurred within a nearby satellite galaxy of the Milky Way in 1987, which led to new insights about how these explosions occur.

Light of a Billion Suns, SN 1987A

For more than three and a half centuries, nobody was fortunate enough to see a supernova by the unaided eye. Then, late in the evening of February 24, 1987, astronomers discovered one (Fig. 11.9), which was designated SN 1987A (SN for supernova; 1987A denotes the first one discovered that year). The star exploded 168,000 years before that night in the Large Magellanic Cloud, one of the two satellite galaxies of the Milky Way visible from the Earth's Southern Hemisphere. It generated an intense burst of light that peaked in visual brightness at the third magnitude.

Astronomers at the Las Campanas Observatory, a barren mountaintop near La Serena, Chile, were the first to notice the new star when Ian Shelton photographed it at 3 o'clock in the morning using a small 0.25-meter (10-inch) telescope placed in an unheated shed. The discovery was relayed to the International Astronomical Union's clearinghouse for such events, which relayed the news to astronomers throughout the world. By the next evening, nearly all major radio and optical telescopes south of the Equator were observing the supernova. One month later, it was featured on the cover of *Time* magazine with just one word:

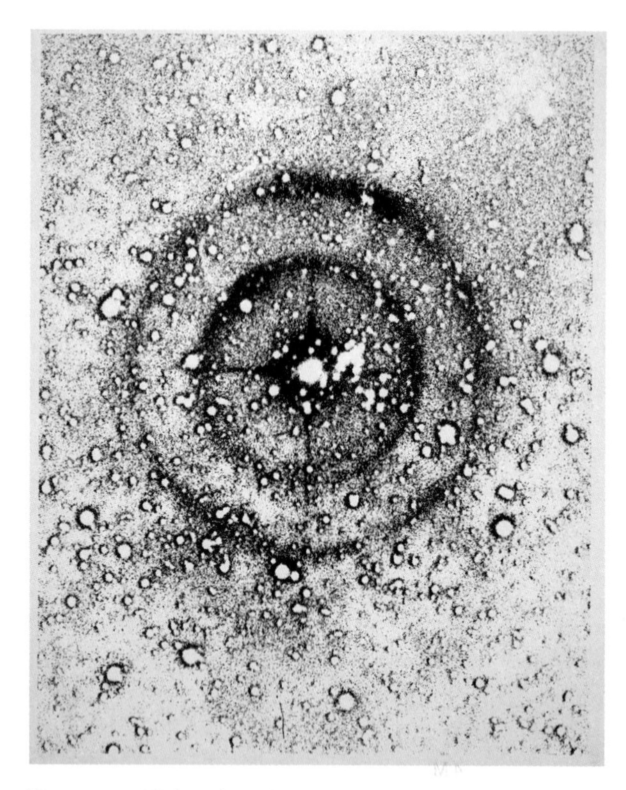

Figure 11.9. Light echoes from SN 1987A. Two complete rings of light surround the exploded star SN 1987A in this negative image taken with the 3.9-m (153.5-inch) Anglo-Australian Telescope on July 15, 1988. The initial flash of light from the supernova explosion has been reflected off clouds of interstellar dust and observed 14 months after the explosion was brightest, somewhat like an echo of sound. These light echoes arise in two thin sheets of microscopic dust grains located about 470 light-years (*inner ring*) and 1,300 light-years (*outer ring*) in front of the supernova. The rings have been made more prominent by photographically subtracting an image taken 3 years before the supernova exploded, canceling much that existed previously. Stars, however, are still visible as faint haloes. (Courtesy of David Malin and the Anglo-Australian Observatory.)

"BANG!" Astronomers were still watching its expanding debris years after the exploding star hurled it into space.

SN 1987A was a gravity-powered, iron-catastrophe Type II supernova, generated during the core collapse of a former blue supergiant star that began life with a mass of about 20 times that of the Sun. The physical properties of this progenitor star and its subsequent supernova explosion are listed in Table 11.8.

One of the more interesting observations of SN 1987A was made from beneath the Earth's surface, when massive subterranean instruments detected a few neutrinos emitted during the explosion. Although solar neutrinos had been observed coming from nuclear reactions that power the Sun, all other stars are so far away and the number of their neutrinos striking the Earth is so low that they had never been detected coming from any other cosmic object. For several seconds, however, the relatively nearby supernova explosion SN 1987A generated such enormous amounts of the elusive neutrinos that a small number was detected 3 hours before the first visible sighting of the supernova.

Table 11.8. *Supernova SN 1987A*[a]

Progenitor Star
Name: Sanduleak $-69^\circ 202$
Spectral Type: B3 Ia (blue supergiant)
Location: Large Magellanic Cloud
Distance: 168,000 light-years
Radius: 3×10^{10} m $\approx 50\ R_\odot$
Effective Temperature = 16,000 K
Luminosity = 4.6×10^{31} J s$^{-1} \approx 10^5\ L_\odot$
Mass $\approx 16\ M_\odot$
Neutrino Burst
Number of neutrinos detected: 19 anti-neutrinos
Energy of each neutrino: ≈ 20 MeV = 3.2×10^{-12} J
Neutrino flux at the Earth: 5×10^{14} m^{-2}
Number of neutrinos emitted: $\approx 10^{58}$ neutrinos
Energy released in neutrinos: $\approx 10^{48}$ J
Duration of neutrino burst: ≈ 10 s
Neutrino luminosity: $\approx 10^{48}$ J s$^{-1} \approx 10^{22}\ L_\odot$
Visible Explosion
Peak visible luminosity: 3.8×10^{33} J s$^{-1} \approx 10^7\ L_\odot$
Velocity of ejected material: $\approx 10^7$ m s^{-1}
Mass of ejected material: $\approx 8 \times 10^{30}$ kg $\approx 4\ M_\odot$
Kinetic energy of ejected material: $\approx 4 \times 10^{44}$ J

[a] The symbols $R_\odot$, $L_\odot$, and $M_\odot$, respectively, denote the radius, luminosity, and mass of the Sun.

Only 19 neutrinos were detected flashing through the underground darkness. Yet, even this small number signaled the presence of an awesome energy, vastly exceeding the amount contained in the light and expanding debris of the supernova (see Table 11.8). They indicated that 10 billion trillion trillion trillion, or 10^{58}, neutrinos were produced by the exploding star, emitting a total neutrino luminosity comparable to that of the optically visible luminous output of more than 2.5 billion billion stars like the Sun.

The neutrinos detected from SN 1987A solved one of the thornier problems in understanding such a supernova explosion. It was known that a stellar collapse generates tremendous amounts of energy, but there was difficulty explaining how that energy was transferred from the collapsing core into the outer layers of the star in sufficient amounts to produce an explosion. The core might rebound, sending shock waves propagating into the surrounding material, but computer simulations indicated that the shock waves could not blow away the rest of the star. They always became stalled when encountering the in-falling matter from the outer layers, like a vehicle trying to get up an icy slope during a winter storm.

Unlike the stalled shock waves, the flood of escaping neutrinos carries tremendous amounts of energy far away from the stellar core, a very small fraction of which gets caught in the in-falling outer layers of the collapsing star, heating up the gas to a temperature of more than 10 billion K. This produces a buoyant, convecting bubble of energy that reverses the in-fall and powers the explosion.

Three hours after the initial collapse and generation of neutrinos in SN 1987A, its heated bubble expanded, driving shock waves before it, and burst through the surrounding material, breaking the star apart and hurling its pieces into space, which produced the dazzling light of the supernova. This explains why the neutrinos were detected 3 hours before any light was seen.

Will the Sun Explode?

Is an explosion forecast for the Sun's future? It is going out alone, passing into its final resting state unaccompanied by a close companion. Although it will end up as a dense, high-gravity white dwarf star, even the nearest star still will remain far beyond the dead Sun's gravitational embrace, never orbiting it. Moreover, our Sun is nowhere near massive enough to ever explode by itself. Thus, no nova or supernova is expected when the Sun's light dies. It will go quietly into the oblivion of permanent night.

11.6 Expanding Remnants of Shattered Stars

In their explosive death, stars that go supernova blast their outer layers into surrounding space, expelling much or all of the stellar material at supersonic speeds of up to 1/10th of the speed of light. A strong shock wave forms ahead of the ejected material, colliding with the surrounding interstellar gas and heating it up to temperatures of tens of millions of K. The high-temperature material emits intense x-rays that have been observed with instruments aboard spacecraft located above the Earth's obscuring atmosphere, thereby recording the debris of cataclysmic stellar explosions that occurred even thousands of years ago, before recorded history.

Like the explosions that cast this material out, there are two types of supernova remnants. They can be the remains of a white dwarf star sent into explosion by a nearby companion in a Type Ia supernova or the explosive debris of a single massive star that has expired in a Type II supernova.

The Chandra X-Ray Observatory, for example, has imaged the Tycho supernova remnant (Fig. 11.10), named for Tycho Brahe who reported observing the original explosion in 1572. This remnant was formed by a Type Ia supernova, when a white dwarf was sent into annihilation by overflow from a nearby companion star. An arc of x-ray emission in the supernova remnant was attributed to material blown off the companion star, which otherwise survived the destruction of its neighbor and now is moving within the remnant more quickly than its neighbors as the result of the explosion. The properties of the arc and remaining star indicate that the former white dwarf star and its companion once orbited one another in a five-day period at a separation of less than 1/10th of the mean distance between the Earth and the Sun, or 0.1 AU.

Supernova remnants often emit intense radio radiation. The majority of these radio supernova remnants appear as bright rings, or shells, in projection against the sky. The intense radio radiation cannot be produced by a hot gas, like x-rays, but instead is emitted by electrons accelerated to high speeds by the supernova explosion and spiraling in a magnetic field.

Figure 11.10. Tycho supernova remnant. The expanding remains from a Type Ia supernova that occurred in 1572. It is named for the Danish astronomer Tycho Brahe (1546–1601), who recorded observations of its brightness in that year. The circular supernova remnant is located at a distance of about 13,000 light-years and is about 20 light-years across. It is bounded by an expanding shock wave and consists of ejected material moving away from the explosion and the interstellar material it sweeps up and heats along the way. The explosion left a hot cloud of expanding debris (*green* and *yellow*). The location of the blast's outer shock wave is seen as a blue sphere of very energetic electrons. Newly synthesized dust in the ejected material and heated preexisting dust from the area radiate at infrared wavelengths (*red*). Foreground and background stars in the image are white. This image is a composite of an x-ray image (*blue, green,* and *yellow*) taken from the Chandra X-Ray Observatory; an infrared image (*red*) taken from the Spitzer Space Telescope; and a visible light image (*white*) taken with the 3.5-m (138-inch) Calar Alto telescope located in southern Spain. (Courtesy of MPIA/NASA/Calar Alto Observatory.) (See color plate.)

A beautiful example of an expanding shell-like supernova remnant is the Cassiopeia A remnant, abbreviated Cas A. It is located at roughly 10,000 light-years away in the direction of the constellation Cassiopeia and is the brightest radio source in the sky (Fig. 11.11). The expanding shell of Cas A is also a strong source of x-rays, emitted by a 50-million-K gas (Fig. 11.12).

Despite its radio and x-ray brilliance, the remnant is faint at optically visible wavelengths, which nevertheless indicate that it is rich in oxygen and now expanding at a speed of about 5,000 km s^{-1}. The Type II supernova explosion would have been observed as a daytime star around 1680, when its light should have reached the Earth, but there are no historical records of the event. Thick clouds of interstellar dust, or material ejected from the massive star's outer layers, may have absorbed the light and rendered the explosion optically invisible.

The most spectacular example of a Type II supernova remnant is the Crab Nebula (Fig. 11.13), also one of the brightest radio sources in the sky (Table 11.9). As a New Orleans drummer might have said of a beautiful, spirited jazz singer, "she has everything." The Crab Nebula, also designated M 1, NGC 1952, or Taurus A, is the remnant of a supernova explosion

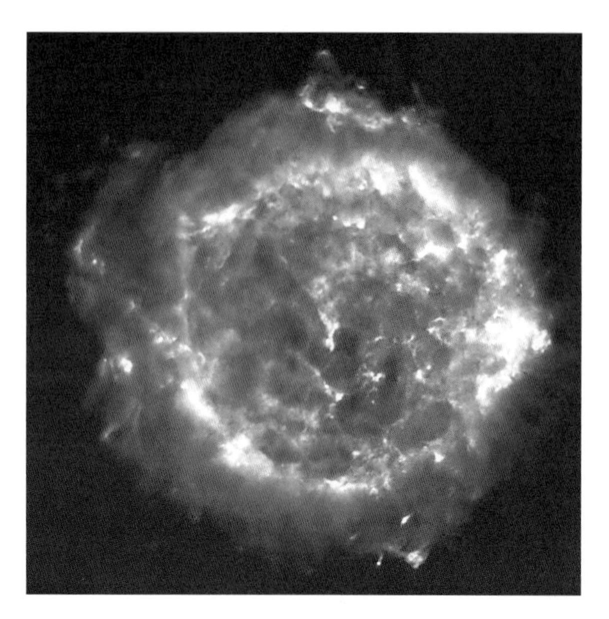

Figure 11.11. Radio image of Cassiopeia A supernova remnant. This Type II supernova remnant is the brightest radio source in the sky, other than the Sun, and is designated 3C 461 from its number in the third Cambridge catalogue of bright radio objects. It also is known as Cassiopeia A, the most intense radio source in the direction of the constellation of Cassiopeia. Also called Cas A, it is the remnant of a supernova explosion that occurred about 1680, or roughly 330 years ago as observed from the Earth, at a distance of about 10,000 light-years. It has a radius of about 8.6 light-years and is expanding at a velocity of roughly 5,000 km s^{-1}. (A VLA image courtesy of NRAO/AUI.)

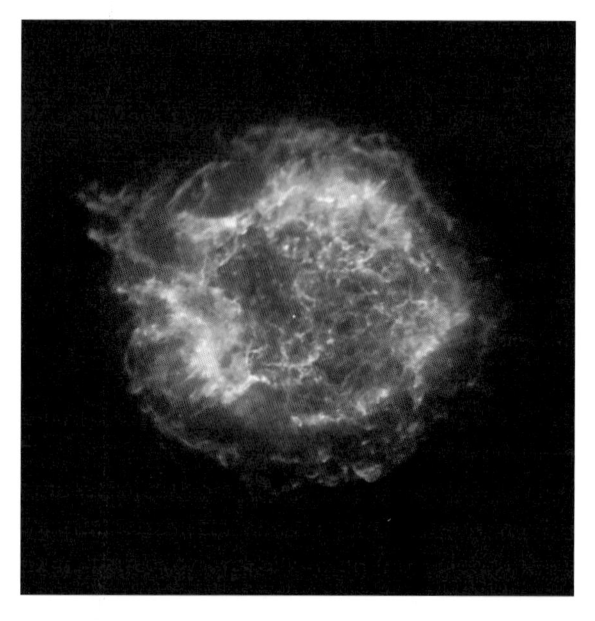

Figure 11.12. X-ray image of Cassiopeia A supernova remnant. The expanding supernova remnant Cassiopeia A has a temperature of about 30 million K and therefore is a luminous x-ray source, seen in this image from the Chandra X-ray Observatory. Still visible in x-rays, the tiny point-like source near the center of Cas A is a neutron star, the collapsed core of the star that exploded about 330 years ago, as observed from the Earth, and that gave rise to the expanding debris. (Courtesy of NASA/CXC/MIT/U. Mass. Amherst/M. S. Stage, et al.)

Figure 11.13. The Crab Nebula supernova remnant. The optically visible light of the Crab Nebula, designated as M 1 and NGC 1952, consists of two distinct parts: (1) a system of expanding filaments forms an outer envelope in which emission lines occur at well-defined wavelengths; and (2) an inner amorphous region that emits continuum radiation at all wavelengths. A Type II supernova explosion observed nearly 1,000 years ago, in 1054, ejected the filaments. The expanding filaments shine mostly in the light of hydrogen (*orange*) but also include the light of neutral oxygen (*blue*), singly ionized sulfur (*green*), and doubly ionized oxygen (*red*). The blue-white continuum glow concentrated in the inner parts of the nebula is the nonthermal radiation of high-speed electrons spiraling in magnetic fields. This continuum emission is powered by a spinning neutron star, the southwesternmost (*bottom right*) of the two central stars. The neutron star is the crushed, ultradense core of the exploded star. It also is a radio pulsar that acts like a lighthouse spinning 30 times a second. (Courtesy of NASA/ESA/J. Hester and A. Loll, Arizona State University.) (See color plate.)

that was observed in 1054. It is about 6,500 light-years away, has a diameter of 11 light-years, and expands at a speed of about 1,500 km s^{-1}.

The visible nebula consists of two distinct parts that emit radiation differently. A tangled, oval-shaped network of red and green filaments, seen in the light of bright emission lines from ionized atoms, encases the inner blue and milk-white continuum radiation. The filamentary remnants of the explosion contain about 4 solar masses of material, consisting mostly of ionized helium and hydrogen along with lesser amounts of carbon, oxygen, nitrogen, iron, neon, and sulfur. The progenitor star that was annihilated to make the explosion had an estimated mass of between 9 and 11 solar masses, some sent into invisibility within surrounding space and the rest remaining within a central neutron star.

In addition to an expanding shell, the radio nebula has a filled center that coincides with the inner visible light. These types of supernova remnants are named *plerions*, from the Greek word *pleres* for "full" or "filled."

The inner radiation contains practically all of the energy emitted by the Crab Nebula, and it is 1,000 times more intense at radio wavelengths than at optically visible wavelengths. It is impossible to reconcile the observed radio emission with the optical emission through the

Table 11.9. *The Crab Nebula supernova remnant*[a]

Expansion Center	R.A. (1950.0)>	$= 05^h\ 31^m\ 32.2^s \pm 0.1^s$
	Dec. (1950.0)	$= +21°\ 58'\ 50'' \pm 1''$
Explosion	Date	= 1054 A.D.
Progenitor Star Mass	M	$\approx 9\ M_\odot$ (main sequence)
Ejected Mass	M_{ejc}	$= 2$ to $3\ M_\odot$
Neutron Star Mass	M_N	$= 1.4\ M_\odot$
Pulsar Position	R.A. (1950.0)	$= 05^h\ 31^m\ 31.405^s$
	Dec. (1950.0)	$= +21°\ 58'\ 54.39''$
Pulsar Period	P	= 0.033 326 323 455 s
Pulsar-Period Derivative	dP/dt	$= 421.288 \times 10^{-15}$ s s^{-1}
Angular Extent	θ	$= 4.5' \times 7.0$’
Distance	D	$= 2.0 \pm 0.1$ kpc
		= 6,500 light-years
Linear Radius	R	$= 1.3$ pc $\times$ 1.4 pc
		4.1×6.1 light-years
Maximum Radial Velocity	V_{rmax}	$= 1{,}450 \pm 40$ km s^{-1}
Maximum Proper Motion	$\mu\,max$	$= 0.222 \pm 0.002\ ''$ yr^{-1}
Maximum Expansion Velocity	V_{exp}	= 1,500 km s^{-1}
Radio Flux Density	S	= 1,040 Jy at 1 GHz
Spectral Index	α	$= -0.30$ (integrated radio)
X-Ray Luminosity	L_X	$= 10^{30.38}$ J s^{-1}
Total Luminosity	L	$= 10^{31.14}$ J s^{-1}

[a] The mass values are in units of the Sun’s mass $M_\odot = 1.989 \times 10^{30}$ kg.

thermal radiation of a hot gas at any temperature. As pointed out by the Russian astronomer Iosif Shklovskii, both the radio and optical emission of the Crab Nebula come from synchrotron radiation emitted by high-energy electrons spiraling around magnetic fields at nearly the speed of light.

Electrons of extremely high energy emit optically visible light, whereas electrons of slightly lower energy radiate at radio wavelengths. Because the more energetic electrons lose their energy faster and also radiate at shorter wavelengths, the synchrotron-radiation mechanism provides a natural explanation for the nonthermal spectrum of the Crab’s radiation, which is more intense at longer wavelengths. At every wavelength of observation, from x-rays to radio waves, the bulk of radiation from the Crab Nebula is accounted for by the synchrotron-radiation mechanism.

Despite the successes of the synchrotron-radiation theory, explaining the origin of the energetic electrons that gave rise to the radiation remained a fundamental difficulty. The electrons radiating at optically visible wavelengths will dissipate their energy by synchrotron radiation in about 180 years, and the more energetic electrons that produce the short-wavelength x-rays should lose their energy and disappear in less than a year. Because the supernova radiation was emitted more than 900 years ago, the high-speed electrons producing the synchrotron radiation cannot be survivors of the original explosion. Instead, some unknown source must be continuously replenishing the energetic electrons.

The explanation was found in 1967, when the Italian astronomer Franco Pacini noticed that a rapidly spinning neutron star with a strong magnetic field could sustain the Crab Nebula's luminous glow; in the same year as Pacini's prescient idea, the first radio pulsar was discovered. The following year, pulsars were attributed to rapidly rotating neutron stars, and a pulsar was found at the center of the Crab Nebula, spinning 30 times a second. The nebula is powered by the pulsar wind that is composed of charged particles accelerated to nearly the speed of light by the rapidly rotating, intense magnetic field of the spinning pulsar. This brings the discussion to early speculations about the possible existence of neutron stars and the subsequent discovery of radio and x-ray pulsars.

11.7 Neutron Stars and Pulsars

Stars Made of Neutrons

Walter Baade and Fritz Zwicky proposed the possibility that neutron stars might exist just two years after the discovery of the neutron, by James Chadwick in 1932. They speculated that a supernova explosion is driven by the gravitational energy released when a massive star runs out of fuel and its core collapses, but the explosion may not completely destroy the stellar core. A dense cinder could remain at the center or, in their prescient words: "With all reserve we advance the view that a super-nova represents the transition of an ordinary star into a *neutron* star, consisting mainly of neutrons. Such a star may possess a very small radius and an extremely high density."[23]

In the meantime, World War II (1939–1945) was fought. Zwicky incorrectly accused Baade, a German "alien," of being a Nazi and threatened to kill him if he showed up on the Caltech campus. From then on, Baade refused to be left alone in a room with Zwicky, who entertained visitors by doing one-armed pushups. Most of Zwicky's colleagues experienced his eccentric and sometimes abrasive behavior; Zwicky called them "spherical bastards" because he thought they were bastards no matter how you looked at them.

Zwicky often had original, insightful ideas – even though that did not make him popular – and astronomers soon speculated about how a neutron star might be made. If a collapsing star is more massive than 1.4 solar masses, the inward force of its gravitation will overcome the outward degenerate electron pressure that halts the collapse at the white-dwarf stage. The electrons will be pushed into direct contact with the atomic nuclei to make neutrons, which become packed together in a star of only 12 kilometers in radius, about the size of New York City. The mass density of the neutron gas is enormous, comparable to that of the nucleus of an atom. A teaspoonful of this material would weigh about 5 billion tons. The properties of neutron stars are given in Table 11.10.

The American physicist J. Robert Oppenheimer and his Canadian graduate student George M. Volkoff found that neutron stars have a limiting mass of their own. The stars are supported by degenerate neutron pressure and, at such high mass densities, the effects of gravitation on space-time must be considered. Under these conditions, a stable neutron star can exist only in a finite range of masses, now believed to be between 1.4 and 3.0 solar masses. When a star's iron core weighs more than 3 solar masses, there is no end in sight and it collapses into a black hole.

Table 11.10. *Physical properties of neutron stars*

M_{ns} = mass of neutron star $\approx 2\ M_{\odot} = 3.978 \times 10^{30}$ kg
M_{cns} = critical upper mass limit for neutron star $\approx 3\ M_{\odot} = 5.967 \times 10^{30}$ kg
R_{ns} = radius of neutron star $\approx$ 12 km = 1.2×10^4 m
ρ_{ns} = mass density of neutron star $\approx 5 \times 10^{17}$ kg m^{-3}
V_{esc} = escape velocity of neutron star = $(2GM/R)^{1/2}$ = 210,000 km s^{-1} = 0.70 c
E_{ns} = binding energy released to form a neutron star = $GM_{ns}{}^2/R_{ns} \approx 8.8 \times 10^{46}$ J
P = rotation period = 0.001 to 10 s
B = magnetic-field strength = 10^8 tesla

Although important in hindsight, these early considerations about the possibility of neutron stars did not evoke much interest at the time. They would have remained a speculative curiosity if it were not for the serendipitous discovery of radio pulsars.

Radio Pulsars from Isolated Neutron Stars

Pulsars were discovered accidentally during a survey of the *scintillations*, or "twinkling," caused when radio radiation from cosmic sources passes through the Sun's winds. When the radio waves are viewed through the wind-driven material, they blink on and off, varying on time-scales of a few tenths of a second – in much the same way that stars twinkle when seen through the Earth's varying atmosphere.

To study these effects, Antony Hewish and his colleagues at Cambridge University built a large array of antennas connected to a radio receiver and chart recorder with a time constant of 0.1 second, the time-scale of the scintillations. When examining the charts in July 1967, graduate student Jocelyn Bell found a strong fluctuating signal in the middle of the night, when the array was pointed away from the Sun, and the effects of the solar wind should have been small.

What could be causing the mysterious fluctuations? Further investigations led to the astonishing detection of periodic radio pulses, with an exceedingly precise repetition period of 1.3372795 second; the first radio pulsar had been detected. No one had foreseen its existence, and no known astronomical object kept time so accurately.

We now know that the term *pulsar* is misleading for the compact stars do not pulsate – they rotate – but the name has stuck. It designates repeating pulses of radio emission rather than a pulsating star.

By the time the discovery was ready for publication, evidence of other radio pulsars was found in the existing chart recordings; within three weeks, a second paper announced the discovery of three additional radio pulsars. This triggered searches for other previously unknown pulsars with large radio telescopes using rapid time sampling rather than the long integration times formerly used. In less than a year, the list of pulsars was expanded to more than two dozen, and a pulsar was detected at the position of the very star thought to be the neutron- star remnant of the Crab Nebula supernova explosion.

The pulsars probably could have been discovered many years earlier, when other large radio antennae were constructed, but radio astronomers were used to adding up signals over long time intervals to detect faint cosmic radio signals. The long time resolutions precluded

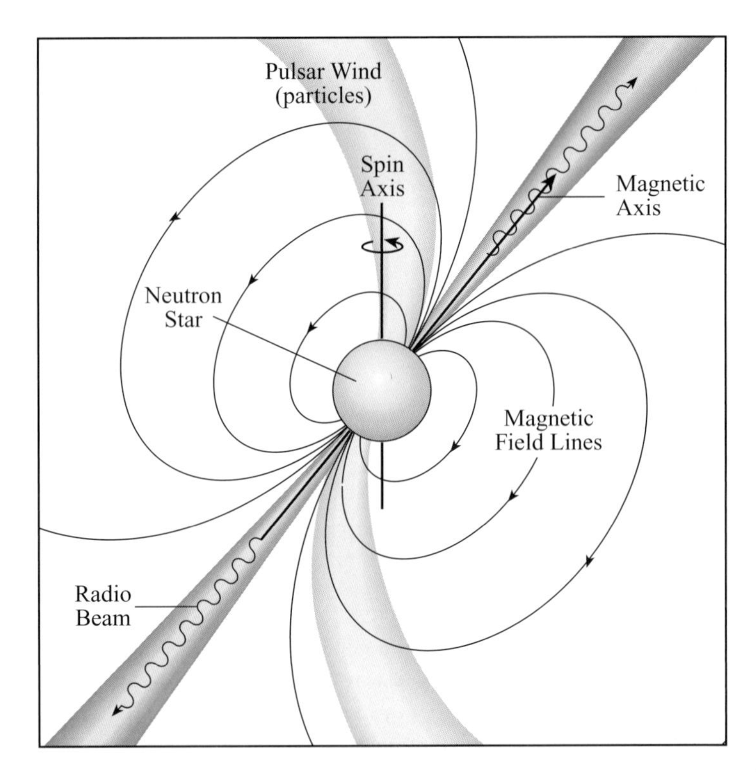

Figure 11.14. Radio pulsar. A spinning neutron star has a powerful magnetic field the axis of which intersects the north and south magnetic poles. The rotating fields generate strong electric currents and accelerate electrons, which emit an intense, narrow beam of radio radiation from each magnetic polar region. Because the magnetic-field axis can be inclined to the neutron star's rotation axis, these beams can wheel around the sky as the neutron star rotates. If one beam sweeps across the Earth, a bright pulse of radio emission, called a *pulsar*, is observed once every rotation of the neutron star.

detection of the pulsars. They are relatively faint radio sources when averaged over their period because there is no emission between the brief radio pulses. If time resolutions comparable to the pulsar-burst durations of milliseconds had been used, the intense radio bursts would have been detected easily. It is because Hewish specifically designed a new type of radio telescope for a study of the rapidly changing solar wind effects that the radio pulsars were discovered accidentally.

The extreme regularity of the periodic radio bursts suggested that they are controlled by the rotation of a massive body, and the short duration of the pulsar bursts suggested that their radiation originates in a body that cannot be much larger than the Earth. That is, the size should be smaller than the product of the speed of light and the burst duration; otherwise, it may be violating nature's upper speed limit, which is that of light.

The Austrian-born American astronomer Thomas Gold proposed that radio pulses are produced by a rapidly rotating neutron star with an intense magnetic field. He assumed that a pulsar would emit radio radiation in a beam, like a lighthouse, oriented along the magnetic axis (Fig. 11.14). An observer sees a pulse of radio radiation each time the rotating beam flicks across the Earth. Because the neutron star's beam could be oriented at any angle,

Figure 11.15. Pulsar in the Crab Nebula. A neutron star, or radio pulsar, at the center of the Crab Nebula is spinning at 30 times a second, accelerating particles up to the speed of light, and flinging high-speed electrons out into the Crab Nebula. This image combines optically visible light (*red*) from the Hubble Space Telescope and an x-ray image (*blue*) taken from the Chandra X-Ray Observatory. It shows tilted rings and waves of high-energy particles that appear to have been flung outward from the central pulsar, as well as high-energy jets of particles in a direction perpendicular to the rings. The ring-like structures are x-ray regions where high-energy particles slam into the nebular material. The innermost ring is about 1 light-year across. (Courtesy of NASA/HST/CXC/SAO/J. Hester, ASU.)

the beams of many pulsars would miss the Earth and several therefore would remain forever unseen.

Gold suggested definitive observational tests of his ideas. He noticed that a spinning neutron star gradually loses its rotational energy and slows down, successfully predicting that this would cause a slow lengthening of the radio-pulsar periods with time. He also predicted that radio pulsars with much shorter periods would be found, as they were.

The Crab Nebula subsequently was found to contain a radio pulsar spinning 30 times a second, making the supernova remnant glow in much the same way that Franco Pacini suggested before the discovery of radio pulsars. Astronomers have used the powerful Chandra X-ray Observatory to trace out the jets, rings, winds, and shimmering shock waves generated by the highly magnetized, rapidly spinning neutron star (Fig. 11.15). Moreover, the period of the Crab Nebula pulsar is increasing with time, just as Gold predicted. Its period of 0.033 second is growing longer by 0.0000000364 second each day. The loss of rotational energy inferred from the period increase of the pulsar is exactly what is needed to keep the Crab Nebula shining at the present rate for about 1,000 years, ever since the observation of the supernova explosion that was associated with the pulsar's birth.

Although most radio pulsars are alone in space without a nearby companion, some binary pulsars have been discovered, the most famous being PSR 1913+16 with a period of 0.05898 second; the PSR designates pulsar and 1913+16 specifies its position in the sky. It was found by Russell A. Hulse and Joseph H. Taylor, Jr., as a result of a deliberate high-sensitivity computerized search for new radio pulsars at the Arecibo Observatory in Puerto Rico. The discovery of a radio pulsar that is a member of a double-star system indirectly suggested the emission of gravitational waves, which had never been seen before (Focus 11.1) .

Individually stars are bound together so tightly that a supernova that leads to the formation of a neutron star in a binary-star system may not disrupt its companion; the two stars can remain together, as evidenced by the binary pulsars. This is important for understanding

Focus 11.1 Gravitational Waves from a Binary Radio Pulsar

Soon after the discovery of the binary radio pulsar PSR 1913+16, astronomers started to measure the expected small, steady lengthening in its 58.98-millisecond period, but it was unexpectedly increasing and decreasing every 7 hours and 45 minutes. This regular change meant that the pulsar was in orbital motion, with the pulses being compressed when the pulsar approached the Earth and pulled apart when moving away. The radio pulsar was in rapid orbit with another neutron star that did not emit detectable radio pulses, perhaps because its radiation beam is not aimed at the Earth. Precise timing of the radio pulses permitted measurements of the mass of the pulsar and its silent companion. They weighed in at 1.44 and 1.39 times the mass of the Sun, as would be expected for two neutron stars.

More important, after four years of measurements and the analysis of about 5 million pulses, Taylor and his colleagues found that the orbital period was slowly becoming smaller, implying a more rapid orbital motion and a slow shrinking of the average orbit size. The two stars are drawing closer, approaching one another by about 1 meter a year. This rate of orbital decay is the change expected if their orbital energy is being radiated away in the form of gravitational waves.

Albert Einstein predicted gravity waves as ripples in the curvature of space-time. They travel at the speed of light, as does electromagnetic radiation. However, whereas electromagnetic waves move through space, gravity waves alternately squeeze and stretch the very fabric of space-time. They are so weak and their interaction with matter so feeble, that Einstein himself questioned whether they ever would be detected. The gravitational radiation loss of the orbital energy of PSR 1913+16 nevertheless exactly matches the amount predicted by Einstein's theory, providing clear and strong evidence for the existence of gravitational radiation – for which Hulse and Taylor received the 1993 Nobel Prize in Physics.

However, neither Einstein nor anyone else ever predicted that two neutron stars would be found that emit gravitational waves detected by timing the pulsar emission of one. Taylor, Hulse, and their colleagues did not set out to find a binary neutron star, much less detect gravitational waves. It was another serendipitous discovery that makes astronomy so wonderfully unexpected and surprising.

A second binary, millisecond radio pulsar, designated PSR J1614–2230, has been found. It is attributed to a neutron star in orbit around a white dwarf star, and it has been used to test aspects of relativity theory other than gravitational waves.

pulsars that also have been detected at x-ray wavelengths. Unlike most radio pulsars, they are members of close binary-star systems rather than single, isolated neutron stars.

X-Ray Pulsars from Neutron Stars in Binary-Star Systems

Because x-rays are absorbed in our atmosphere, cosmic x-ray sources must be observed with instruments launched above the obscuring air, in rockets or satellites. By the mid-twentieth century, brief, 5-minute rocket flights had shown that the Sun radiates detectable x-rays, and it was thought that lunar material also might emit them when illuminated by solar x-rays.

Riccardo Giacconi's group at the American Science and Engineering Company (AS&E) concluded that x-rays emitted by conventional stellar objects other than the Sun would be too faint to be detected with existing instruments. They designed the sensitive equipment needed to detect the Moon's x-rays and to search for other unknown sources of x-ray radiation.

They unexpectedly found the first known discrete x-ray source outside of the solar system, which led to the discovery of a new class of cosmic objects and new physical processes. This pioneering rocket flight set the stage for a host of rocket and satellite observations of discrete x-ray sources, including x-ray stars that are 1,000 times brighter in x-rays than the Sun at all wavelengths that are 1,000 time more luminous in x-rays than in visible light. The x-rays signaled the presence of 1-million-degree gas spiraling from a close companion star into a neutron star or black hole.

Giacconi was awarded the 2002 Nobel Prize in Physics "for pioneering contributions to astrophysics, which led to the discovery of cosmic x-ray sources." He shared the prize with Raymond Davis, Jr., and Masatoshi Koshiba, who were recognized for their detection of cosmic neutrinos.

One of the brightest sources in the newly discovered x-ray sky, designated Centaurus X-3, pulses in x-rays every 4.84 seconds. The rapid pulsating variation was discovered shortly after the launch of the first dedicated x-ray satellite, on December 12, 1970 from the offshore San Marco platform, off the coast of Kenya. Because this date coincided with the seventh anniversary of the independence of Kenya, the satellite was given the name *Uhuru*, the Swahili word for "freedom."

After analyzing a year of observations of Centaurus X-3, the *Uhuru* scientists found a regular pattern of intensity changes of the x-ray pulses, which increased and decreased in strength with a much longer period of 2.087 days and systematic changes in the timing of pulses in the same period. These effects were attributed to a companion star, which was orbiting the x-ray source and regularly eclipsing its emission.

During the next year, accurate measurements showed that the average pulsation period of Centaurus X-3 was getting shorter, which meant that its rotation was speeding up, not slowing down like radio pulsars. This indicated that the rotational energy of the x-ray pulsar was increasing, rather than decreasing, with time.

The gain in rotational energy is interpreted in terms of matter drawn in from a nearby visible companion star (Fig. 11.16). Because matter is being pulled toward the surface of an x-ray–emitting neutron star, instead of being expelled from it, the neutron star is torqued up to a faster rotation. The material spirals in at the same direction as the neutron star's rotation. When it lands, it gives the neutron star a sideways kick, increasing its rotational

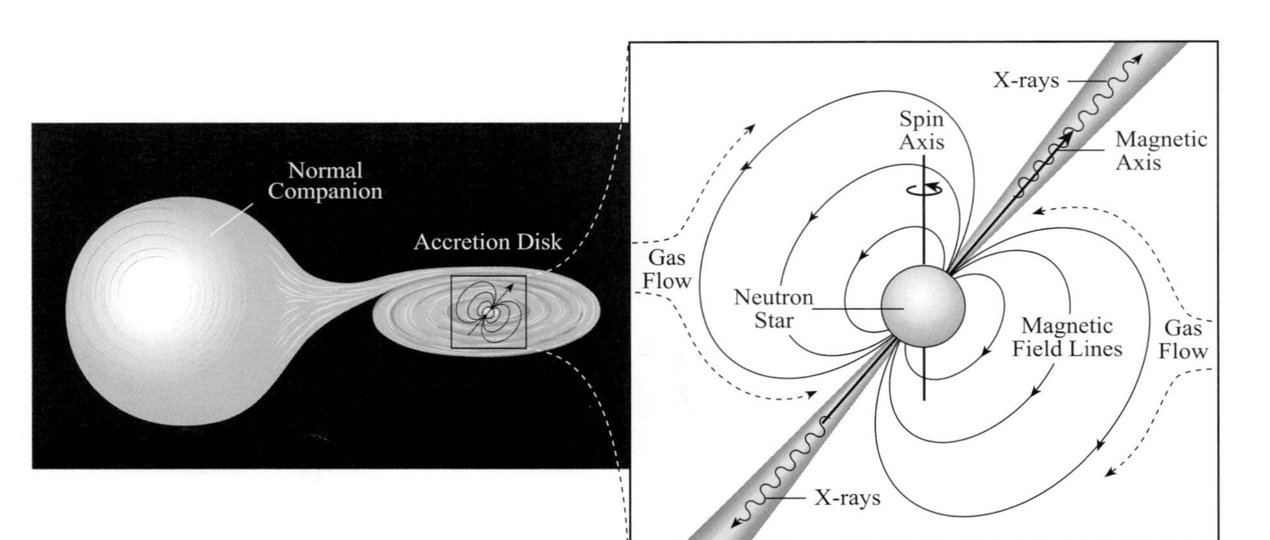

Figure 11.16. X-ray pulsar. The outer atmosphere of an ordinary star, detected in optically visible light, spills onto its companion, an invisible neutron star. The flow of gas is diverted by the powerful magnetic fields of the neutron star, which channel the in-falling material onto the magnetic polar regions. The impact of the gas on the star creates a pair of x-ray hot spots aligned along the magnetic axis at each magnetic cap. Because the magnetic-field axis can be inclined to the neutron star's rotation axis, the x-ray radiation from the hot spots can sweep across the sky once per rotation, which is observed as periodic x-rays if one of the hot spots intersects the observer's line of sight.

energy, speeding it up, and causing the rotation period to become shorter as time goes on. Because most radio pulsars are not members of binary-star systems, they expel material, lose rotational energy, slow down, and have periods that lengthen.

Because the two stars are orbiting rapidly around one another, the gas from the ordinary star does not fall directly onto the neutron star but instead shoots past the neutron star, swinging around it to form a whirling disk of hot gas, known as an *accretion disk*. It spirals around and down onto the central neutron star, like soapsuds swirling around a bathtub drain. The inner portions of the swirling accretion disk revolve more rapidly than the outer portions, just as the closer planets orbit the Sun at faster speeds than the more remote planets.

The rapidly spinning inner parts of the disk constantly rub against the slower-moving outer parts. This viscous friction heats up the accretion disk and causes the material in it to spiral inward. The closer the material moves toward the central neutron star, the hotter the in-falling gas becomes, eventually reaching temperatures of millions of K and emitting luminous x-rays. The intense magnetic field of a rotating neutron star acts as a funnel to guide in-falling matter onto a neutron star's magnetic north and south poles, creating an x-ray pulsar (see Fig. 11.16).

11.8 Stellar Black Holes

Imagining a Stellar Black Hole, the Point of No Return

John Michell, a British clergyman and natural philosopher, suggested more than two centuries ago that certain stars could remain forever invisible. He reasoned that a star might be so massive and its gravitational pull so powerful that light could not escape it. He wrote: "All light from such a body would be made to return to it by its own proper gravity."[24]

French astronomer and mathematician Pierre-Simon de Laplace popularized the idea in his *Exposition du systeme du monde* in the late eighteenth century. He subsequently showed that light could never move fast enough to escape the immense gravitational attraction of some compact stars. Their matter might be so concentrated, and the pull of gravity so great, that light could not emerge from them, making these stars forever dark and invisible.

When it was realized that light travels at a definite speed of roughly 300,000 km s^{-1}, a black hole could be defined as any object the escape velocity of which exceeds the speed of light. This is the velocity required for escape from an object's gravitational pull. All stars that we can observe have escape velocities smaller than the speed of light, which is why we can observe their light. The Sun, for example, has an escape velocity of 600 km s^{-1}. Compressing a star into a smaller size raises its escape velocity. When a dead stellar core of approximately the Sun's mass collapses into a white dwarf star, its escape velocity increases to about 9,000 km s^{-1}, or 3 percent of the speed of light. At a neutron star's radius, the escape velocity becomes about 70 percent of the speed of light.

If a massive star has consumed all of its available nuclear fuel and its core mass exceeds the upper limit to a neutron star's mass, at about 3 solar masses, the stellar core's gravity will overcome the degenerate neutron pressure of even nuclear matter – and there is no known force that can halt the collapse. The smaller the collapsing core becomes, the larger is its escape velocity, until it exceeds the speed of light and a stellar black hole is formed. It is black, or rather invisible, because no light can leave it, and it is a hole because nothing that falls into it can escape. It collapses without end.

In other words, if the core of a dead star is sufficiently massive, there is nothing left to hold back the inexorable force of gravity. The core continues to collapse forever, vanishing into deep, total darkness and eternal absence from the directly observable universe. The name "black hole" has entered everyday language as a common metaphor – as a place where something might become forever lost, like childhood, a former love, or a memory. A black hole also might remind us of our fears of being consumed, as in a job or marriage, or even destroyed, as in death.

Observing Stellar Black Holes

Since a black hole is invisible, and it does not absorb, emit, or reflect radiation, how do we know it is there? We detect a black hole by its gravitational effect on the motion of a visible star.

With remarkable foresight, the Reverend John Michell speculated in the late eighteenth century that the unseen star might betray its presence by its gravitational effects on a nearby, luminous star in orbit around it. In modern extensions of this idea, a black hole may be detected if it is in a tight, close orbit with a visible star whose outer atmosphere spills over into the dominant gravitational influence of the black hole. This material swirls around and down into the black hole – like new snow falling into the dark of night, orbiting faster and faster as it gets closer – as a result of the ever-increasing gravitational forces. The rapidly moving particles collide as they are compressed to fit into the hole, heating the material to temperatures of millions of K. At these temperatures, the gas emits almost all of its radiation at x-ray wavelengths. It is analogous to the accretion disks of binary x-ray pulsars, except that the invisible companion is a black hole rather than a neutron star.

So, the way to find a stellar black hole is to look for two stars that are in close orbit, one a normal visible star and the other unseen except for its x-rays. The mass, velocity, and orbital period of the visible star can be used to determine the mass of its orbital partner, which emits no visible light. If that mass is noticeably greater than the upper mass limit for a neutron star, set at about 3 solar masses, the unseen star is thought to be a stellar black hole. Any normal star with this mass would be very bright and easily seen through a telescope, but a black hole is dark, emitting no detectable visible light.

The archetype of a stellar black hole is Cygnus X-1, located in the constellation Cygnus and one of the first x-ray sources to be discovered. It is accompanied by a bright blue supergiant star of spectral class O, located at a distance of about 6,000 light-years from the Earth. Observations of the continuously and periodically shifting spectral lines of this bright visible star – by the English astronomers B. Louise Webster and Paul Murdin and independently confirmed by the Canadian astronomer Charles Thomas Bolton – indicated that it is revolving every 5.60 days about an invisible companion of more than eight times the mass of the Sun for the unseen companion. It emits no light and its mass is greater than that of the upper mass limit for a neutron star (i.e., approximately 3.0 solar masses); therefore, by elimination, it must be a stellar black hole.

The latest measurements of the Cygnus X-1 binary-star system (in 2011) indicate that the visible supergiant star has a mass of 19.2 solar masses and the unseen companion has a mass of 14.8 solar masses, which confirms that its mass is well above the upper mass limit for a neutron star. In addition, the two stars are separated by only 0.2 AU, or 20 percent of the distance from the Earth to the Sun and about half the separation of Mercury from the Sun. Like other supergiant stars of its spectral type, the visible star is thought to be shedding mass in a stellar wind at a rate of about 2.3 solar masses every 1 million years. Due to the proximity of the invisible companion, a significant portion of this wind is being drawn into the black hole to form its x-ray–emitting accretion disk. We now know of many stellar black holes identified in this way. The orbital properties of visible companions of cosmic x-ray sources indicate masses beyond the neutron-star limit.

Describing Black Holes

The outer edge of a black hole can be defined as the radius at which the escape velocity, required to escape from its gravitational pull, is equal to the speed of light. This radius is known as the *gravitational radius*, or the *Schwarzschild radius*, in recognition of the German astronomer Karl Schwarzschild, who first recognized its significance.

The Schwarzschild radius, marks the event horizon – literally, a horizon in the geometry of space-time beyond which no events can be seen – just as the Earth's horizon is the boundary for our vision. Nearby space then is said to curl into a black hole, carrying light and matter and any other form of energy with it. They are so intensely wrapped around a black hole that it becomes a cocoon disconnected from the outside and cut off forever from the rest of the universe.

What happens when something goes inside the event horizon? Any object passing through this boundary cannot stay still, for there is no known force that can overcome the powerful gravitational pull. It is a one-way street, a path of no return, where one can go in but cannot come out. It's a mirror image of your birth, when you came out into the world from a place you can't return to.

The event horizon also resembles human death. Once you have crossed that line, you also can't return. You are gone, lost, consigned to oblivion forever. For a star, only its gravitation remains; for humans all that is left is memory by other persons and perhaps a soul.

Theoretical considerations suggest that something might come back out of a black hole, but it would bear no resemblance to anything that entered a black hole. That reminds one of the song about Lynn, Massachusetts: "Lynn Lynn, the city of sin. You never come out the way you went in."

Black holes are mysterious objects. They cannot be observed directly because any radiation they might emit cannot escape. A black hole's presence can be inferred only from indirect, circumstantial evidence, using measurements in the accessible parts of the universe – the visible stars – to make inferences about the dark places that cannot be observed.

12

A Larger, Expanding Universe

12.1 Where Does the Milky Way End and How Does It Move?

A Fathomless Disk of Stars

On a clear, moonless night, we can look up and see a hazy, faint luminous band of light that stretches across the sky from one horizon to the other; it is known as the Milky Way (Fig. 12.1). According to ancient Greek myth, the goddess Hera, Queen of Heaven, spilled milk from her breasts into the sky. The Romans called the spilt milk the *Via Lactea*, or the "Milky Way." It also is designated as our Galaxy, derived from the Greek word *galakt-* for "milk," the celestial milk from Hera's breast.

The Milky Way has been long thought to be a path to the heavens, the road to paradise. In his *Metamorphoses*, the Roman poet Publius Ovidius Naso (Ovid) described it as a passageway to the homes of the gods. Sixteen centuries later, the English poet John Milton described the Milky Way as a broad and ample road, the dust of which is gold.

We are immersed within the Milky Way, viewing it edgewise from inside. When gazing directly into the band of starlight, we cannot see through to stars at the center or distant edges of the Milky Way, but if we look up and outside the thin disk of stars, we can look beyond them. It is similar to living in a city: We notice buildings all around us but none when we look up into the sky.

When Galileo Galilei turned one of the first telescopes toward the Milky Way, he found that it contains many otherwise unseen stars, which are too faint to be seen by the unaided human eye. Astronomers subsequently built increasingly larger telescopes, which collect more starlight and enable us to see the dim, golden beacons of fainter stars. They discovered more dim stars located between or beyond the brighter ones, which make the Milky Way look like a continuously distributed band of light when observed by the unaided eye.

The German-born English astronomer William Herschel spent much of his life trying to determine the shape and size of the Milky Way; he constructed the biggest telescopes at the time, with the largest mirrors and greatest light-gathering power. By counting the number of stars of different observed brightness in various directions in the night sky, he hoped to determine the places at which the stars disappeared, thereby plumbing the depths of the Milky Way.

Figure 12.1. The Milky Way. A panoramic telescopic view of the Milky Way, the luminous concentration of bright stars and dark intervening dust clouds that extends in a band across the celestial sphere. We live in this disk and look out through it. Our view is eventually blocked by the buildup of interstellar dust, and the light from more distant regions of the disk cannot get through. The center of the Milky Way is located at the center of the image, in the direction of the constellation Sagittarius. Although the disk appears wider in that direction, the center is not visible through the dust. The Large and Small Magellanic Clouds can be seen as bright swirls of light below the plane to the right of center. (This map of the Milky Way was hand-drawn from many photographs by Martin and Tatjana Keskula under the direction of Knut Lundmark; courtesy of the Lund Observatory, Sweden.)

Herschel's efforts culminated in his 1.2-m (48-inch) mirror, weighing almost 1 ton (1,000 kilograms) and housed within a 40-foot-long metal tube. The 40-foot reflector, as it was called, soon gained a reputation as the "Eighth Wonder of the World." Because England's King George III had financed the telescope, he was present at the dedication and told his guest, the Archbishop of Canterbury, that it would show him the way to heaven. Franz Joseph Haydn's visit to the colossal telescope apparently was the direct inspiration for "The Heavens Are Telling" passage in his great oratorio, *The Creation*.

Although Herschel concluded that the Sun is in the center of a flattened disk of stars with a disk diameter five times its thickness, he had no way to determine its size. Early in the twentieth century, the Dutch astronomer Jacobus C. Kapteyn and his colleagues resumed the star counts, arriving at a similarly flattened, Sun-centered distribution of stars with the greatest extent in the Milky Way. Measurements of the distance of some of these stars, using their parallax, provided a scale to Kapteyn's universe, placing its edge at about 5,000 light-years perpendicular to the Milky Way and 40,000 light-years within it.

However, astronomers have never succeeded in deciphering the true extent of the Milky Way by observing its stars, even when looking much farther into it using larger telescopes and photographic or electronic techniques that permit long exposures. This is because the most powerful telescopes can discern only the visible parts of our stellar system, not its most distant, invisible parts that lie behind an opaque veil of interstellar dust.

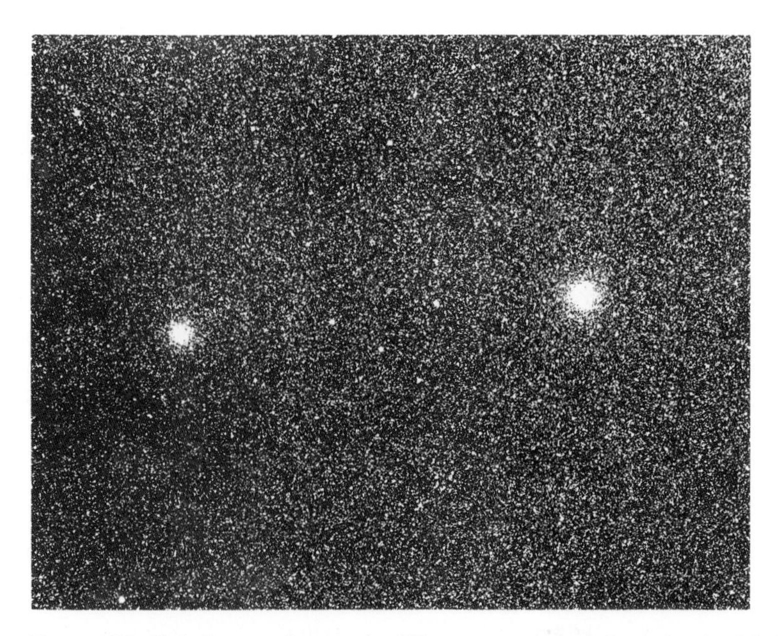

Figure 12.2. Globular star clusters. A million stars are crowded together in globular star clusters, like the two shown here. Many are located in a great spherical halo that encloses our Milky Way. A relatively few, like these, are concentrated toward the central nucleus. These globular clusters are designated NGC 6522 and NGC 6528. (Courtesy of KPNO/AURA.)

This dust blocks our view when we look deep into the plane of the Milky Way. The total amount of dust through which we are looking builds up with distance and eventually makes an impenetrable barrier, becoming so thick and dense that it blocks the light of distant stars. We can see only that far; more distant objects are hidden from view. New perspectives were required to look outside and eventually beyond this barrier to the heavens.

The Sun Is Not at the Center of Our Stellar System

The true enormity of our stellar system was discovered by using Cepheid variable stars to gauge the distances of globular star clusters (Fig. 12.2) located outside the plane of the Milky Way. These yellow supergiant Cepheids periodically brighten and dim in a period that increases with a star's luminosity. Measurements of this period and, therefore, the stellar luminosity can be combined with observations of a star's brightness to determine its distance (Focus 12.1). Because the Cepheid variable stars are very luminous, they are conspicuous and can be seen to exceptional distances, where conventional parallax methods of determining stellar distance do not work. At large distances, the parallax angles are too small to be reliably measured, even from space.

It was the American astronomer Harlow Shapley who observed Cepheids in globular star clusters outside the plane of the Milky Way. He showed that they are distributed within a roughly spherical system, which is centered far from the Sun in the direction of the constellation Sagittarius (Fig. 12.3). The Sun is located at a distance of about 27,700 light-years from this center, which is about 1.7 billion times the distance between the Earth and the Sun.

Focus 12.1 Cepheid Variable Stars

The luminosity of some stars does not remain constant but instead fluctuates over regular periods. These stars do not only turn on and off, like a switched house light, but instead gradually vary from dimmer to brighter and then back to dimmer again – like ocean waves slowly rising and falling – with periods ranging from a few days to a few months.

The very luminous variable stars are know as Cepheid variables, from their prototype star, Delta Cephei, whose variability was first noted by the deaf English astronomer John Goodricke. They have luminosities up to 100,000 times that of the Sun and masses of 4 to 20 times the solar mass. Because they are so luminous, these stars can be seen over a wide range of distances, from Delta Cephei, located only 890 light-years from the Earth, to galaxies 100 million light-years away.

The more luminous a Cepheid variable star is, the more slowly it varies and the longer the period of its luminosity change. This period–luminosity relationship was first discovered from observations of variable stars in the Large and Small Magellanic Clouds, which are nearby satellites of the Milky Way. These stellar systems, visible from the Earth's southern hemisphere, are named for the Portuguese nobleman Ferdinand Magellan, who observed them when his ships were completing the first circumnavigation of the world.

At the end of the nineteenth century, Harvard College established an observatory at Arequipa, Peru, with a 0.6-m (24-inch) refractor that was used in a photographic survey of the southern sky, including the Magellanic Clouds. Because of their proximity, the clouds could be resolved into stars. From these photographs, Henrietta Leavitt, a researcher in Cambridge, Massachusetts, found an extraordinary total of 1,777 variable stars. She reported that the brighter stars tended to have the longer cycles of variation. Because the extent of the Magellanic Clouds is small compared to their distance, the relationship of period to apparent brightness also implied a real connection with luminosity. Four years later, Leavitt had obtained precise apparent brightness and period data for 25 variable stars in the Small Magellanic Cloud, thereby establishing the important period–luminosity relationship for the Cepheid variables.

Once this relationship is calibrated suitably, by the measurement of a precise distance to one Cepheid variable star using independent methods, observation of the variation period leads to determination of a star's luminosity. Then, using the observed brightness, the distance of a star can be calculated. This technique of measuring distances with Cepheid variable stars has been used to demonstrate the vast extent of the Milky Way – as well as the Sun's place within it – and, subsequently, to discover the extragalactic nature of spiral nebulae.

Leavitt had no idea why the luminous output of these stars varies but, within a few years, the inquisitive English astronomer Arthur Eddington showed how. The stars are pulsating with a regular beat, expanding out and contracting in, rising and falling back, breathing in and out like a giant heart.

The Russian astronomer Sergei A. Zhevakin eventually showed that an outer region of doubly ionized helium acts as a valve for the heat engine that drives the pulsations of Cepheid variable stars. It absorbs the outward flow of energy from the star's center during stellar contraction and repeatedly returns it during expansion.

Because the pulsation depends on a critical stage of ionization, a star can maintain them only for a specific combination of size and temperature, which defines a strip of instability in the Hertzsprung–Russell diagram.

We now know that most of the stars and interstellar gas in the Milky Way are located within a flattened, plate-shaped, rotating disk and that the Sun is located within its outer precincts. This disk has a radius of about 50,000 light-years and a thickness of about 3,000 light-years.

How many stars are in the Milky Way? The distribution of stars mapped by the *HIPPARCOS* mission indicates that there are some 2.2 million Sun-like stars in a cube of nearby space measuring 1,000 light-years across. Assuming a uniform distribution of stars in the disk and multiplying this star density by the disk volume, we obtain a lower limit of at least 50 billion stars with a mass equal to that of the Sun. Because the stars are more concentrated toward the central regions, there is at least twice this amount; therefore, the Milky Way contains roughly 100 billion stars like the Sun. We can see only about 5,000 of them with the unaided eye.

Relatively young stars are found in the disk; some are seen even in the earliest stages of formation. These stars are designated Population I stars. In addition to their cosmic youth, they contain a relatively high abundance of the heavier elements, commonly called the metals. The Sun is a Population I star; the open star clusters found in the Milky Way contain Population I stars.

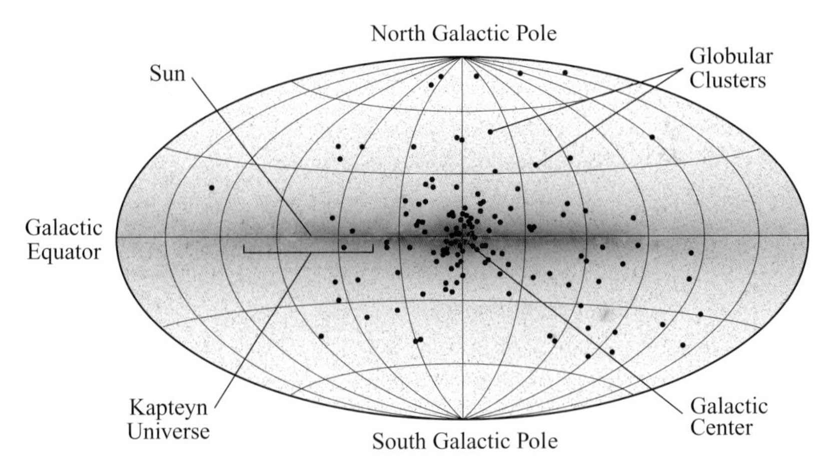

Figure 12.3. Edge-on view of the Milky Way. As shown in 1918 by the American astronomer Harlow Shapley (1885–1972), the globular star clusters are distributed in a roughly spherical system whose center coincides with the core of our Milky Way. The Sun is located in the disk, about 27,700 light-years away from the center. The disk and central bulge are shown edge-on in a negative print of an infrared image taken from the InfraRed Astronomical Satellite. The infrared observations can penetrate the obscuring veil of interstellar dust that hides the distant Milky Way from observation at optically visible wavelengths. It is this dust that limited astronomers' view of stars to a much smaller Kapteyn Universe, centered on the Sun.

Table 12.1. *Physical properties of the Milky Way disk*

R_{disk} = radius of disk = 50,000 light-years ≈ 15,000 pc ≈ 4.6 × 10^{20} m
L_{disk} = thickness of disk = 3,000 light-years ≈ 1,000 pc ≈ 3.0857 × 10^{19} m
$D_\odot = R_0$ = Sun's distance from the center = 27,700 light-years = 8.5 kpc = 2.6 × 10^{20} m
$V_\odot = V_0$ = Sun's orbital velocity about center = 220 km s^{-1}
$P_\odot = P_0$ = Sun's orbital period about center = 7.6 × 10^{15} s = 2.4 × 10^8 years
M_{disk} = mass of disk = 10^{11} $M_\odot$ ≈ 2 × 10^{41} kg
N_{disk} = number of stars in Milky Way = 100 billion, or 10^{11}, stars like the Sun
L_{Bdisk} = luminosity of blue band = 1.9 × 10^{10} $L_{B\odot}$ ≈ 7.2 × 10^{36} J s^{-1}
ρ_{disk} = mass density of disk near the Sun ≈ 0.515 10^{-20} kg m^{-3} = 0.076 $M_\odot$ pc^{-3} = 0.0022 $M_\odot$ (light-year)$^{-3}$
N_{disk} = number density of stars in disk near the Sun = 2.59 × 10^{-51} m^{-3}
S_{disk} = separation of adjacent stars in disk ≈ 6.5 light-years ≈ 2 pc ≈ 6.2 × 10^{16} m
Age = oldest disk stars = (6 to 13.5) × 10^9 yr = 6 to 13.5 Gyr; young stars are still forming
Oort's constants A = (+ 14.82 ± 0.84) km s^{-1} kpc^{-1} and B = (−12.7 ± 0.64) km s^{-1} kpc^{-1}
Center of the Milky Way: right ascension α(2,000) = 17^h 45^m 40.04^s,
declination δ(2,000) = −29° 00′ 28.1″

The Population II stars are found mainly outside the Milky Way, in globular star clusters. They include the oldest known stars and have a relatively low abundance of elements heavier than hydrogen or helium. A spherical aggregation of Population II, metal-poor stars also is found near the center of the Milky Way.

The Rotating Galactic Disk

As suggested by the German philosopher Immanuel Kant near the end of the eighteenth century, the flattened shape of the Milky Way can be attributed to its formation from a large, collapsing, rotating nebula, much like the origin of our solar system from a considerably smaller nebula. Observations of the motions of nearby stars and interstellar gas in the disk indicate that the entire stellar system indeed is rotating around a remote axis that pierces the center of the Milky Way. The enormous mass at this central hub steers stars into circular orbits, with an orbital speed that decreases with increasing distance from the center – all in accordance with Kepler's third law.

Stars in orbits inside the solar orbit travel faster than the Sun, thereby forging ahead of it, whereas the stars moving in orbits outside the Sun's orbit are falling behind. When viewed from the Earth, nearby stars that are a little closer than the Sun to the center therefore seem to move in one direction, whereas those a little farther away appear to move in the opposite direction, in two star streams.

We can measure the radial, or line-of-sight, component of a star's motion by observing the Doppler shift in the wavelength of its spectral lines. Moreover, radio astronomers can use the same Doppler effect with the spectral line of interstellar hydrogen atoms, emitted at a wavelength of 21 cm, to trace out the motions of interstellar gas. Both techniques indicate that the Sun and nearby gas and stars are revolving about the center at a speed of 220 km s^{-1}, with an orbital period of about 240 million years at their distance from the center of 27,700 light-years. Stars that are nearer the center of the Milky Way revolve about it at faster speeds than more distant stars and take less time to circle the center.

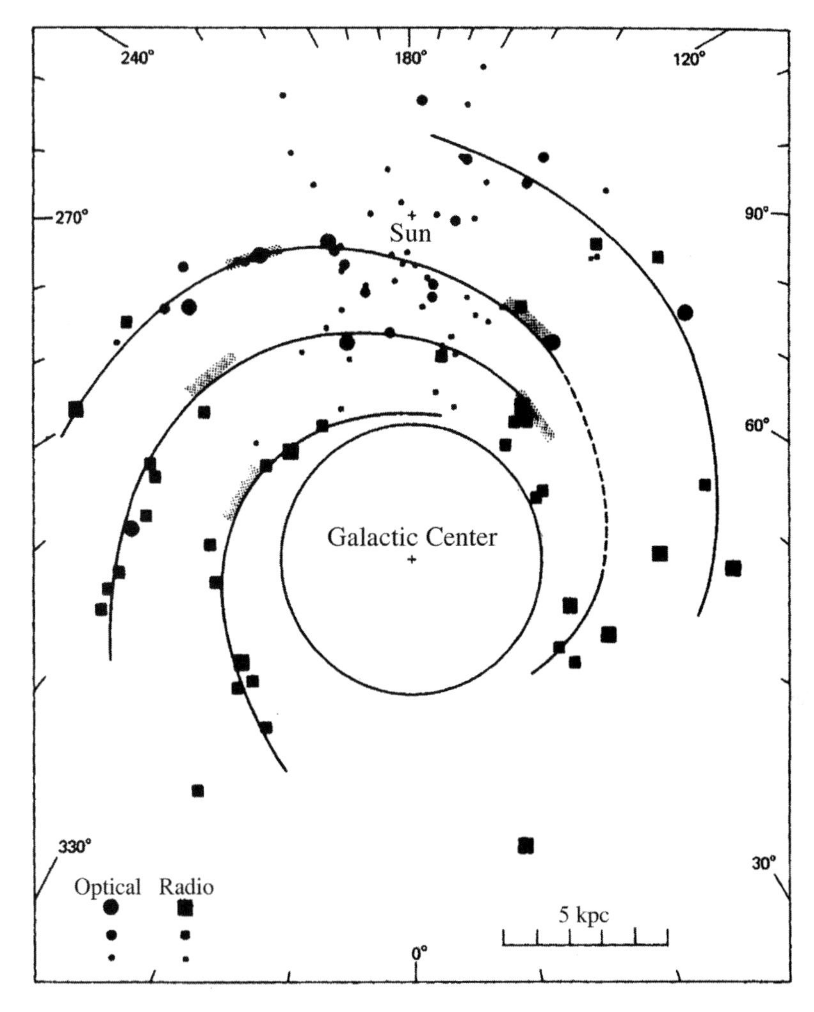

Figure 12.4. Spiral arms of the Milky Way. Luminous emission nebulae, known as ionized hydrogen (or H II) regions, act like beacons that mark out the inner spiral structure of our Milky Way. The Sun and solar system are located at the upper center, in a local spur from the nearest arm. The figure is centered at the center of our Milky Way. The scale, shown in the lower right, extends across 16,300 light-years.

We can use Kepler's third law with measurements of the Sun's distance from and velocity about the remote center of the Milky Way to infer the mass that gravitationally controls solar motion. It is an enormous mass equivalent to about 100 billion, or 10^{11}, stars like the Sun. This and other physical parameters of the Milky Way disk are listed in Table 12.1.

Whirling Coils of the Milky Way

The stars of the Milky Way do not reside in a uniform whirling disk. They instead are concentrated into arms that coil out from the center of the Milky Way, giving our stellar system a spiral shape. These features are delineated by relatively young, very luminous, and massive stars, which light up the nearby arms like city streetlights seen from an airplane (Fig. 12.4). They coincide with the well-known emission nebulae, or H II regions.

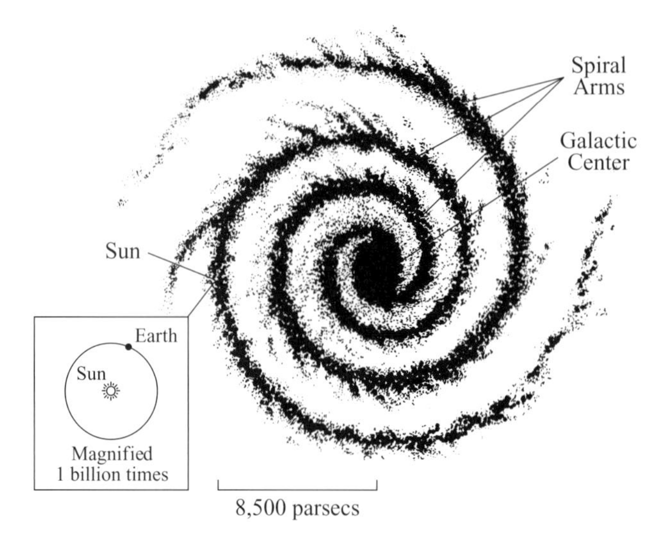

Figure 12.5. Structure of our stellar system. This drawing depicts our Milky Way as viewed from above its plane. The stars and interstellar material are concentrated within spiral arms. The Sun lies within one of these spiral arms at a distance of 27,700 light-years from the center, designated here as 8,500 parsecs, or 8.5 kpc. This distance is 1.75 billion times the distance between the Earth and the Sun.

Because the Sun is embedded in one of these arms, astronomers must look through that arm to see the rest of the Milky Way. This obscures their distant vision, hiding most of our stellar system from view in optically visible light. However, radio waves pass unimpeded through the obscuring material, permitting detection of most of the Milky Way. This is because the long radio waves are not absorbed by the relatively small particles of interstellar dust.

By observing the radio emission of interstellar hydrogen atoms at a wavelength of 21 cm, radio astronomers constructed a face-on view of the Milky Way, which we might see if we were transported into distant space and looked down on the plane of the Milky Way from above. They delineated extensive, arm-like concentrations that extend out from the short segments defined by young massive stars in the vicinity of the Sun (Fig. 12.5).

The Sun has circled the center of the Milky Way more than 19 times during the Sun's 4.6-billion-year lifetime. So, the spiral arms should have been wrapped around the massive center many times during the lifetime of the Sun, and a persistent dilemma has been understanding why they have not wound up to form a featureless ball of stars and interstellar gas and dust. The explanation seems to be density waves that control the concentrations of stellar and interstellar material, similar to highways that direct the motions of vehicles.

The wave pattern orbits the galactic center at a steady rate and does not wind up; it moves independently of the motions of individual stars, which follow their own orbit around the center. The spiral arms are places where the interstellar material and stars linger – like traffic at a stoplight – and they mark the locations where new stars tend to form and hot, massive, luminous, young stars are found.

A Central Super-Massive Black Hole

Radio astronomers have looked right through interstellar dust and detected the center of the Milky Way, where an exceptionally powerful and compact radio source has been found. It is in the direction of the constellation Sagittarius and therefore is named Sagittarius A* (pronounced "A star" and abbreviated Sgr A*).

Radio interferometer measurements with very long baselines show that Sgr A* is considerably smaller than our solar system. The compact radio-emitting center has a radius of about half the distance between the Earth and the Sun. It seems likely that an exceptionally massive black hole is energizing the extremely bright and small radio source.

Because such a black hole is very massive, dense, and compact on a cosmic scale, its formidable gravity can dominate a star's motion if it is close enough. The *super-massive black hole* guides nearby stars into a rapid orbital dance, betraying its presence. These stars can be seen at infrared wavelengths that also penetrate interstellar dust. By watching the motions of infrared stars that are near the center of the Milky Way and orbit it, a central super-massive black hole has been found.

After an unprecedented study lasting more than a decade, researchers were able to track the full revolution of one infrared star around the invisible black hole (Fig. 12.6). This star moves within 17 light-hours of the unseen center, with speeds of up to 5,000 km s^{-1}. These orbital parameters imply that the mass of the central black hole is a colossal 4 million times the mass of the Sun – that is, 4 million invisible solar masses not shining but rather gravitationally confining the observed stellar orbit.

Where did this super-massive black hole come from and why is it located at the center of the Milky Way? Its formation probably coincided with the origin of our stellar system, by the collapse of a huge rotating mass with a nucleus at the center and the flattened Milky Way spinning around it. The globular star clusters probably date back to the beginning of the collapse, about 14 billion years ago.

The super-massive black hole may have originated at the central nucleus, perhaps as the result of the gravitational collapse of an exceptionally massive cloud of gas located there. Or, it may have grown by the coalescence of smaller stellar black holes, each formed at the end of the lifetime of the first massive stars. The central black hole would have continued to gather in nearby, smaller black holes and surrounding stars and gas, with an active youth and a more sedate old age.

Dark Matter Envelops the Milky Way

As the result of differential rotation, stars and gas near the Sun should revolve about the galactic center at faster speeds than those at greater distances from the center. Yet, the stars and gas observed way out near the apparent edges of the Milky Way rotate at speeds that do not decrease with distance. This means that the Milky Way does not end where the light does, and that it extends farther than the eye can see directly, even with the aid of a telescope. There are appreciable amounts of dark unseen matter well outside the boundary of the visible Milky Way, and that dark invisible matter keeps the fast-spinning visible material connected to our stellar system.

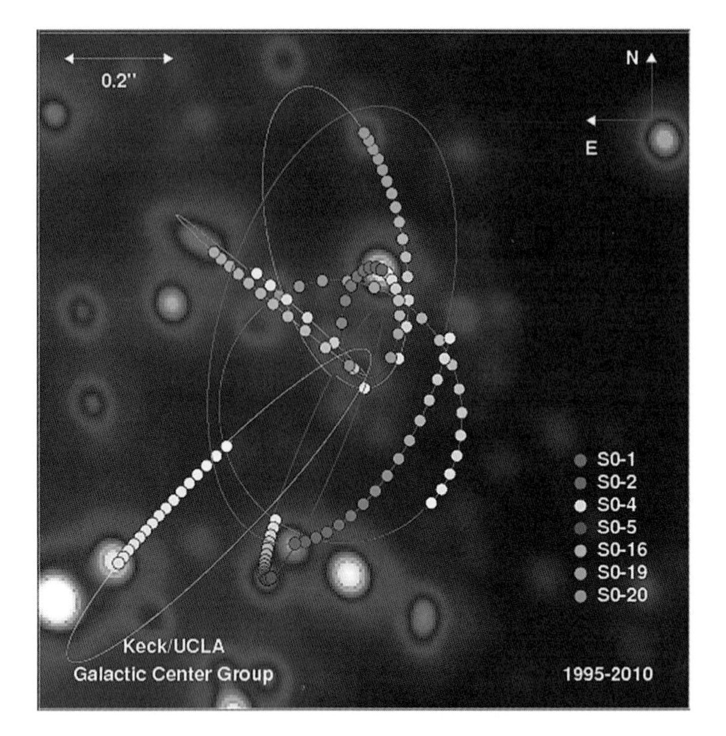

Figure 12.6. Super-massive black hole at the center of the Milky Way. Stars slowly revolve about an unseen center whose gravity controls their motions. This diagram portrays the orbits of infrared stars near the center of our Milky Way. The annual positions of seven stars were determined during a 15-year period (*colored dots*), determining their curved trajectories and inferring the mass of the invisible center. These orbits indicate that a super-massive black hole is located at the center of the Milky Way; it has a mass of 4.1 million times the mass of the Sun. The display covers the central 1.0 × 1.0 seconds of arc, which at a distance of about 27,700 light-years corresponds to a width of about 0.13 light-year or 1,140 light-hours. (Courtesy of Andrea M. Ghez/UCLA galactic center group.) (See color plate.)

The mass of the Milky Way within the Sun's orbit around the center of the Milky Way is roughly 100 billion, or 10^{11}, Suns. However, the rapid motions of dwarf satellite collections of stars, which revolve about the Milky Way at distances of up to 1 million light-years, indicate that a great reservoir of unseen matter envelops the observed disk of stars. A mass of roughly 1 trillion, 10^{12}, times the mass of the Sun, or about 10 times the mass of its visible stars, is required to hold onto these dwarf systems. This invisible, massive, outer region is known as the *dark halo*. It surrounds the Milky Way and outweighs it by a factor of about 10.

Thus, our Milky Way is held together by the gravity of *dark matter*, which is beyond the range of our vision. A similar darkness pervades and controls much of the universe, giving off neither light nor any other radiation to let us know it is there. The dark matter is studied by its gravitational influence on the motions of the stars that we can see.

Moreover, even the observable universe, the part we can see directly, is not limited to our stellar system, the Milky Way, but instead is populated by more than 100 billion galaxies, each composed of about 100 billion stars and perhaps containing 10 times as much mass in unseen dark matter. These galaxies stretch as far as the largest telescope can see – and perhaps beyond.

12.2 Out Beyond the Stars

Bigger Telescopes See Farther

When William Parsons, third Earl of Rosse, built the largest telescope of the time at Birr Castle in Northern Ireland, he thought that it might resolve an ongoing controversy about the diffuse patches in the night sky known as *nebulae*, the Latin word for "clouds." The famous English astronomer, William Herschel, had attributed their cloud-like, nebulous forms to some sort of dispersed, shining, fluid-like material; other astronomers thought the nebulae were composed of stars that are too far away to be detected and resolved by small telescopes.

The unsurpassed light-gathering power of Lord Rosse's 1.8-m (72-inch) metallic mirror only compounded the problem of the nebulae. It revealed an entirely new class of objects, called the *spiral nebulae*. Rosse worked from a catalogue of more than 100 prominent nebulae assembled by the French comet hunter, Charles Messier, to avoid misinterpreting them as comets. Within weeks of first pointing his telescope at the sky in 1845, Rosse found that the 51st nebula in Messier's list, designated M 51, exhibits a spiral pattern, which he attributed to its rotation (Fig. 12.7). This spiral nebula now is commonly known as "The Whirlpool Nebula," suggesting a whirling vortex of light.

The devastating potato famine in Ireland interrupted Rosse's astronomical work, which he abandoned to devote himself to his sick and dying tenants. However, he returned to his telescope 3 years later and eventually found 14 nebulae that exhibit a spiral pattern.

Rosse's drawing of M 51 was reproduced widely, appearing in *Les Etoiles*, a popular book published in 1882 by the French astronomer Camille Flammarion. It is possible that the Dutch artist Vincent Van Gogh saw the shape there, reproducing it in his second portrayal of *The Starry Night* executed from Saint-Rémy. Van Gogh's rendition of the active, swirling form of a spiral nebula plays a dominant role in the painting, and it bears a striking resemblance to Flammarion's description of M 51, which "flows like a celestial river" and "seems to tear into shreds like a fleece combed by the winds of the sky."[25]

Near the end of the nineteenth century, the American astronomer James E. Keeler began systematic photography of the spiral nebulae with the 0.91-m (36-inch) Crossley reflector at the Lick Observatory; he discovered thousands. He estimated that 120,000 spirals could be seen with the telescope, always avoiding the plane of the Milky Way. Only one of the spiral nebulae could be seen by the unaided eye – the Andromeda Nebula, or M 31 – as a faint, fuzzy glow, and telescopes were used to portray its detailed beauty (Fig. 12.8).

Large telescopes revealed an entirely new, otherwise unseen universe of more than 100,000 spiral nebulae. We still marvel at their beauty, revealed by the most sophisticated telescopes of our time (Fig. 12.9).

No one realized the possibilities of large telescopes more than George Ellery Hale, probably the greatest scientific entrepreneur of the twentieth century. While he was a young associate professor at the University of Chicago, Hale convinced the wealthy Charles T. Yerkes that he should endow the Yerkes Observatory, funding the construction of the world's largest refractor there, with a lens 1-m (40-inch) in diameter. Early in the twentieth century, Hale moved to Pasadena, California, near Los Angeles, and founded a solar observatory on nearby Mount Wilson, which rose above the dense smog below. A few years later, Hale inaugurated

Figure 12.7. Spiral shape of Messier 51. Lord Rosse (1800–1867) discovered the spiral structure of Messier 51, abbreviated M 51, using his 1.8-m (72-inch) telescope in the spring of 1845, and he subsequently found at least a dozen other nebulae with a spiral shape. In his description of this drawing of M 51, published in 1850, Rosse attributed the spiral pattern to rotation of the nebula. Camille Flammarion (1842–1925) included this sketch in his popular books about astronomy, leading to a growing awareness of spiral nebulae, and it may have inspired the swirls of starlight found in Vincent Van Gogh's (1853–1890) painting of the *Starry Night*. We now know that M 51, also designated NGC 5194 and called the Whirlpool Galaxy, is a magnificent, rotating spiral galaxy located 35 million light-years away, with a small, irregular companion NGC 5195 separated from the center of M 51 by about 10 million light-years. (Reproduced from The Earl of Rosse, Observations of the Nebulae, *Philosophical Transactions of the Royal Society*, pages 110–124, plate 35 [1850].)

a new telescope on Mount Wilson to look at the entire cosmos, using a 1.5-m (60-inch) mirror provided by Hale's father.

Hale then persuaded John D. Hooker, a Los Angeles businessman, to finance a 2.5-m (100-inch) mirror weighing 4.5 tons. It was installed on Mount Wilson using funding from the Carnegie Institution of Washington, DC. The Carnegie Institution is a private foundation supported largely from endowments by Andrew Carnegie, former owner of the Carnegie Steel Company, once the richest man in the world and a philanthropist who built more than 2,500 libraries in the English-speaking world. The Carnegie Institution also funded the Mount Wilson Observatory, a sort of "gentlemen's club" that ran the 100-inch Hooker telescope for three decades to discover the galaxies and the expanding universe.

Figure 12.8. The Andromeda Nebula. The nearest spiral galaxy, the Andromeda Nebula, also known as M 31 and NGC 224, is located at a distance of about 2.6 million light-years, so its light takes about 2.6 million years to reach us. Both the Andromeda Nebula and our Galaxy are spiral galaxies with total masses of about 1 million million, or 10^{12}, solar masses, and roughly 100 billion, or 10^{11}, optically visible stars. The several distinct stars surrounding the diffuse light from Andromeda are stars within our own Galaxy; these stars lie well in front of Andromeda. Two smaller galaxies also are shown in this image: M 32, also designated NGC 221, shown at the edge of the Andromeda Nebula; and NGC 205, located somewhat farther away. These are elliptical systems at about the same distance as M 31 but with only about 1/100th of its mass. (Courtesy of Karl-Schwarzschild Observatorium, Tautenburg.) (See color plate.)

Hale, the indefatigable fundraiser, did not stop with the 100-inch telescope; he wrote a stirring call for yet a larger telescope, published in 1928 in *Harper's Magazine*. Following is an excerpt from his article:

> Like buried treasures, the outposts of the Universe have beckoned to the adventurous from immemorial times. Princes and potentates, political or industrial, equally with men of science, have felt the lure of the uncharted seas of space, and through their provision of instrumental means the sphere of exploration has rapidly widened.... Each expedition into remoter space has made new discoveries and brought back permanent additions to our knowledge of the heavens.[26]

Hale asked the editor of *Harper's* to send a copy of his article to the Rockefeller Foundation, where it had a catalytic effect in bringing forth 6 million dollars for building a 5.0-m (200-inch) telescope, which was finished in 1948 on Palomar Mountain. This instrument reigned as the unrivaled leader in astronomy at optically visible wavelengths for almost a half-century, peering deeper into the universe than ever before and playing an important role in contemporary discoveries of the violent universe.

Hale also founded the California Institute of Technology (Caltech) and the Rockefellers preferred to give their money to the university. So, Caltech owned the 200-inch telescope and

Figure 12.9. Spiral galaxy. This natural-color Hubble Space Telescope image shows the spiral galaxy NGC 4911 in the Coma cluster of galaxies, which lies about 320 million light-years away from the Earth. Central clouds of interstellar gas and dust are silhouetted against glowing, young star clusters and clouds of hydrogen gas. (Courtesy of NASA/ESA/Hubble Heritage Team/STScI/AURA.) (See color plate.)

the Carnegie Institution owned the 100-inch telescope. The two institutions ran both telescopes together as the Mount Wilson and Palomar Observatories until the 1960s, when they were renamed the Hale Observatories. It was the 100-inch Hooker telescope, however, that resulted in the most profound discoveries, at about the same time that Hale was dreaming of bigger telescopes to come.

Spiral Nebulae Are Extragalactic and They Are Not Nebulae

As one of the first astronomers to use the new 2.5-m (100-inch) Hooker telescope, Edwin Hubble settled an ongoing controversy about the nature of spiral nebulae and showed that the Milky Way does not contain everything there is. A brilliant, colorful, and polished young man, Hubble already had turned down the chance to become a professional boxer, served as one of the first Rhodes Scholars at Oxford, obtained a doctorate at the University of Chicago, and served briefly in the Army. In 1919, at the age of 30, he moved to California and took a position at the Mount Wilson Observatory, where he showed that the spiral nebulae do not belong to the Milky Way.

The issue over the nature of spiral nebulae was included in the now-famous Shapley–Curtis debate over "The Scale of the Universe" during a meeting in 1920 of the National Academy of Sciences at the Smithsonian Institution in Washington, DC. Harlow Shapley, of the Harvard College Observatory, defended his novel conception of a much larger Milky Way than previously had been supposed, with a distant center and the Sun in its remote outer parts, but he supposed that the spiral nebulae are embedded in the cozy confines of the Milky Way. In contrast, Heber D. Curtis, of the Lick Observatory, attempted to defend a smaller Sun-centered stellar system but provided cogent arguments that the spiral nebulae are distant stellar systems located far beyond the Milky Way. Shapley was correct about the shape and size of the Milky Way, and Curtis was correct in supposing that the spiral nebulae are distant "island universes" composed of numerous stars.

The argument over the location of the spiral nebulae was finally and definitely resolved when Hubble used the Hooker telescope to photograph the spiral nebula Andromeda, or M 31, night after night, comparing hundreds of photographs to find Cepheid variable stars whose brightness waxed and waned like clockwork in a period of several days. On New Year's Day 1925, Hubble's results were read *in absentia* at a meeting of the American Astronomical Society in Washington, DC and caused an overwhelming sensation. The landmark paper, titled *Cepheids in Spiral Nebulae*, combined the known period–luminosity relationship of Cepheids with observations of the variable stars in M 31 and M 33, another spiral nebula, to derive a distance of about 900,000 light-years for the two spiral nebulae. Their size or linear extent, determined from this distance and their angular extents, was roughly comparable to that of our Milky Way.

These two spirals were way out there, far beyond any other known stars, and well beyond the Milky Way. This meant that the spiral nebulae were not nebulae but were instead remote island universes ablaze with stars – galaxies in their own right and separated from the Milky Way by wide gulfs of apparently empty space. It is as if the spiral galaxies had broken free of any local constraints, somewhat like the Zen death poem written more than eight centuries ago:

> Four and fifty years
> I've hung the sky with stars.
> Now I leap through –
> What shattering![27]

Like the poet, Hubble broke through the stars and moved the outer boundaries of the universe far out into space, enlarging our horizons in the process.

It took so long to establish the true nature of the spiral nebulae because the method of establishing the distances of the remote, luminous Cepheid variable stars needed to be developed, and a large powerful telescope was needed to detect the stars in spiral nebulae and collect enough of their faint starlight for a reliable measurement of their periodic brightness variation and, therefore, distance.

Once the observed periodic variations of these stars was correctly calibrated, the distance to Andromeda turned out to be 2.54 million light-years, about three times farther away than Hubble initially supposed – but his dramatic conclusion remained unchanged. The universe was no longer limited to the objects on which our unaided eyes can focus, and our stellar system had become just one of myriad galaxies located far beyond the Milky Way – which became our Galaxy, written with an uppercase G to show that it is special. All of the other galaxies were shown to be extragalactic, or outside of our Galaxy.

Hubble called the spirals extragalactic nebulae, but they are not large gaseous nebulae. The spectral characteristics of the light from the spirals are similar to that of our Sun, indicating stellar temperatures of thousands of K. If a spiral were filled with gas at this temperature throughout its enormous dimensions, the nebula would be more than 1 billion billion times more luminous than the Sun, rather than the much smaller luminosity of Andromeda at about 100 billion times the solar luminosity. This means that light is not coming from the entire surface of a spiral nebula but instead from individual stars, like the Cepheids that Hubble observed, separated by vast spaces without any stars.

Therefore, we now have dropped the nebula designation and use the term *spiral* or *elliptical galaxy*, depending on their shape. Each galaxy contains about 100 billion stars, just like our Galaxy. The designation *nebulae* now is being reserved for cloudy, gaseous material enveloping bright stars.

When looking up at the night sky, we see only stars and the black spaces between them. The galaxies are out there, but we cannot see them without a telescope. They are so far away that their brightness is below the detection threshold of the human eye.

The light we receive from the most distant galaxies was emitted before the Earth and the Sun were formed. Even more fantastic, they are all in flight, rushing away from us at speeds that increase with their distance.

12.3 The Universe Is Expanding

At the time of their discovery in enormous numbers, most astronomers thought that the spiral nebulae were nascent planetary systems, not galaxies. The bright center was supposed to be a newborn star, and the spiral arms surrounding it were thought to be developing planets, whirling and rotating around the central star just as the Earth revolves around the Sun.

The wealthy Bostonian, Percival Lowell, had built an observatory in Tucson, Arizona, primarily to detect canals on Mars supposedly built by parched, industrious Martians. He also believed that the spiral nebulae resemble our solar system in its early formative stages; he therefore instructed his staff astronomer, Vesto M. Slipher, to measure their rotations, hoping to gather insight about our own planetary system.

As often happens in astronomy, Slipher's measurements resulted in an entirely unexpected discovery. When using the 0.6-m (24-inch) refractor at the Lowell Observatory to record the spectra of bright spiral nebulae, he found that they almost unanimously are moving away from us at high velocities. They were also rotating, and a few were approaching – but at modest speeds in comparison to the outward motion of most spirals. The Andromeda Nebula, for example, was moving toward the Earth at an apparent velocity of 300 km s^{-1}. However, the other bright spirals were moving in the opposite direction, usually with higher velocities of up to 1,100 km s^{-1}, much faster than any star in the Milky Way.

By 1917, Slipher had accumulated spectra of 25 spiral nebulae, using the Doppler effect to measure their radial velocities, and he showed that none of them are at rest. All but three were rushing away from us and from each other, dispersing, moving apart, and occupying an ever-increasing volume, like a puff of smoke dispersing into the air.

According to Slipher, the observed motions of the majority of the spirals indicated a general fleeing from the Milky Way or us. It certainly was difficult to believe that objects with such enormous speeds could long remain a part of our stellar system. The combined gravitational pull of the entire 100 billion stars in the Milky Way is not enough to retain any spiral nebula moving at speeds in excess of 1,000 km s^{-1}.

Today, with larger telescopes and electronic detectors readily available, it is difficult to comprehend how much time Slipher required to record the spectra of even the brightest spirals. They were faint and the light that reached the Earth was reduced further in intensity to near-invisibility when dispersed to obtain their spectra. Heroic exposure times of 20, 40, and even 80 hours were needed to obtain the elusive spectra and to infer radial velocities from the Doppler shifts of the spectral lines recorded in them. Subsequently, bigger telescopes collected more light, thereby reducing the necessary observation time for the brighter spirals. The larger telescopes also could see farther into space, where the fainter spirals are located, and they could resolve individual components of the nearby spirals, which eventually led to measurements of their distances.

In 1929, Hubble showed that the measured distances of spirals, which he had established by using the superb light-gathering power of the 100-inch Hooker telescope, were roughly correlated with Slipher's velocities. The comparison indicated that the farther a spiral is, the faster it is moving away from us. This relationship now is attributed to the expanding universe, which no one had anticipated at the time Slipher made his measurements.

All of the other galaxies are not rushing away from our Galaxy in particular but rather from one another. They apparently are moving apart because space is getting bigger. In this interpretation, the galaxies are not moving through space but instead are imbedded within expanding space, like dots painted on a balloon that is being inflated.

In his publication of these results, titled *A Relation between Distance and Radial Velocity Among Extra-Galactic Nebulae*, Hubble drew a straight line through a plot of the observed data. However, there was a wide dispersion between the plotted points and only a mild tendency for velocity to increase with distance (Fig. 12.10). Nevertheless, his conclusion subsequently was confirmed by more comprehensive observations of a much greater number of galaxies.

In just two years, for example, Hubble extended the velocity–distance relationship to substantially greater distances, with the help of Milton Humason, who made the velocity measurements. Humason was a tobacco-chewing gambler and reputed "ladies' man" whose formal education ended in the eighth grade. He began his career on Mount Wilson as a mule driver and janitor. After filling in as a substitute night assistant, however, he became an astronomer more skilled at observing than Hubble when it came to obtaining the spectra needed to infer a radial velocity from the Doppler effect. Even with the 100-inch telescope, the velocity measurements of those remote galaxies required photographic exposures of up to 50 hours.

Because of the impossibility of measuring distances to such faint objects, Hubble and Humason simply assumed that all galaxies have the same intrinsic luminosity. They inferred a velocity–distance relationship by comparing the observed velocities of the galaxies to their apparent brightness. Thus, they reformulated Hubble's law and showed that a linear relationship between distance and recession velocity is valid, within the observational uncertainties, to distances as far as 300 million light-years and radial velocities of nearly 20,000 km s^{-1}.

The connection between velocity and distance is known now as the *Hubble law*, and the ratio of the velocity of recession of any galaxy and its distance from us now is called the

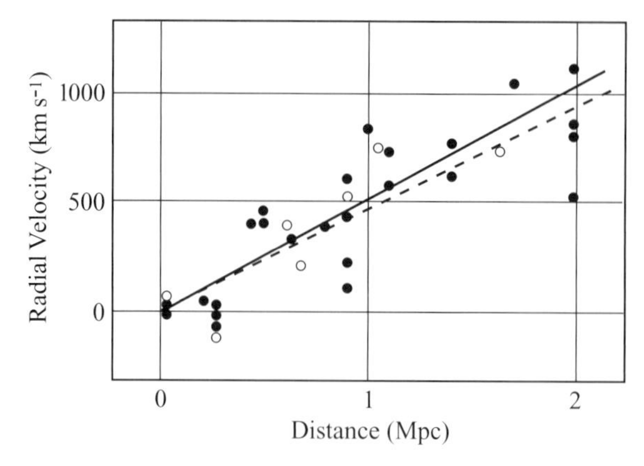

Figure 12.10. Discovery diagram of the expanding universe. A plot of the distance of extragalactic nebulae, or galaxies, versus the radial velocity at which each galaxy is receding from the Earth, published in 1929 by the American astronomer Edwin Hubble (1889–1953). The linear relationship between the distance and radial velocity indicates that the universe is expanding. Vesto M. Slipher (1875–1969) determined most of these velocities more than a decade before this diagram was drawn. Here, the velocity is in units of kilometers per second, abbreviated km s^{-1}, and the distance is in units of millions of parsecs, or Mpc, where 1 Mpc is equivalent to 3.26 million light-years. Hubble underestimated the distances of the spiral nebulae; therefore, the distance scale for modern versions of this diagram is about seven times larger. The filled circles and solid line represent the solution for individual nebulae; the open circles and dashed line are for groups of them.

Hubble constant, a fundamental measure of the universe. It is designated by the symbol H_0, in which the H is for Hubble and the zero subscript denotes its current value.

The units of Hubble's constant are given in kilometers per second per Megaparsec, abbreviated km s^{-1} Mpc^{-1}, and Hubble's initial estimate was pegged at 530 in these units. The radial velocity is in units of kilometers per second, abbreviated km s^{-1}, and the distance is in Megaparsecs, abbreviated Mpc, in which 1 Mpc is equivalent to 3.26 million light-years. Galaxies typically are separated by a few Mpc, or about 10 million light-years, which is about 100 galaxy diameters. So, the universe is largely empty space relative to galaxies.

There is an ongoing controversy about the exact value of this important constant, with estimates ranging between 50 and 100 km s^{-1} Mpc^{-1} and a favored value of about 75 (Fig. 12.11). It sets the physical scale of the universe, with the distance to any galaxy given by the measured Doppler effect, which is known as the *redshift*. The largest observed redshift, denoted by the lowercase letter z, corresponds to the greatest distance and looks the farthest back in time. Astronomers currently are detecting galaxies out to redshifts as great as $z = 8.6$, the radiation of which was emitted about 13.1 billion years ago.

Altogether, there are roughly 1 billion galaxies in the volume of space that modern telescopes can detect, and there is no end in sight. As Hubble realized, astronomers see only as far as their telescopes permit, eventually reaching a limit – a dim boundary to the observable universe where they measure the shadows. Even now, there is no telescope powerful enough to detect where the galaxies might end. There is no edge to the observable universe and it has no detectable center, like the darkness of night.

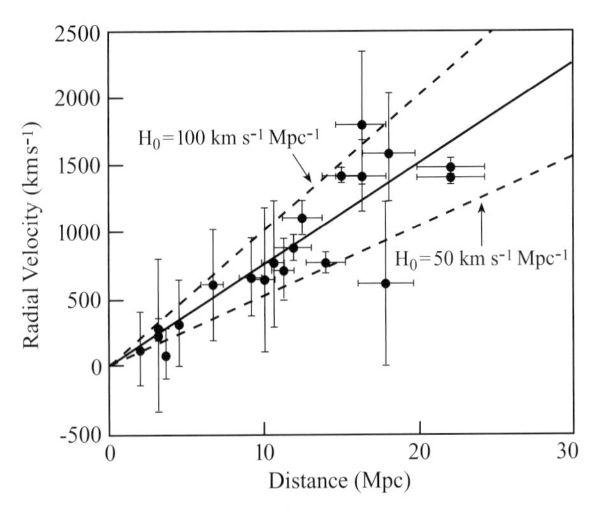

Figure 12.11. Hubble diagram for Cepheid variable stars. This plot of galaxy distance versus recession velocity is analogous to that obtained by Edwin Hubble (1889–1953) in his 1929 discovery of the expansion of the universe (see Fig. 12.10). The slope of the linear fit (*solid line*) to the data (*dots*) measures the expansion rate of the universe, a quantity called the *Hubble constant*, designated H_0. The data shown here summarize 11 years of effort to measure this constant by using the Hubble Space Telescope to determine the distances and velocities of Cepheid variable stars in nearby galaxies. The distance is in units of 1 million parsecs, or Mpc, where 1 Mpc is equivalent to 3.26 million light-years; the radial velocity is given in units of kilometers per second, denoted as km s^{-1}. The fit to these data indicate that $H_0 = 75 \pm 10$ km s^{-1} Mpc^{-1} and that this constant lies well within the limits of 50 and 100 in the same units (*dashed lines*). (Adapted from Wendy L. Freedman et al., "Final Results from the Hubble Space Telescope Key Project to Measure the Hubble Constant," *Astrophysical Journal* 553, 47–72 [2001].)

12.4 The Cosmic Web

Shape, Content, and Form

Spiral galaxies are not the only denizens of the outer reaches of the cosmos; there also are the elliptical and irregular galaxies, distinguished by their shapes (Table 12.2). The spiral galaxies are disk-shaped when viewed edge-on and contain dusty, curving arms of spiral shape when viewed face-on. The elliptical galaxies have a smooth, rounded elliptical shape, regardless of viewing angle. They are nearly devoid of the interstellar gas and dust found in the arms of spiral galaxies. As therefore might be suspected, elliptical galaxies contain relatively few newly formed stars and open star clusters. The ellipticals contain older, more evolved stars, resembling those found in the globular star clusters that surround the Milky Way.

There is much more to all of the galaxies than meets the eye. As much as 90 percent of the mass of all three types of galaxies consists of invisible dark matter. The galaxies are not placed randomly throughout expanding space, and they are not uniformly strewn here and there like dust scattered in the wind. The galaxies are not isolated from one another but instead knot together in great clusters that are millions of light-years across (Fig. 12.12); they also contain large additional quantities of unseen matter. A rich cluster of galaxies typically

Table 12.2. *Physical properties of galaxies*[a]

R_g = radius of a spiral or elliptical galaxy $\approx$ 33 light-years to 326 light-years = 10 kpc to 100 kpc $\approx (3 \text{ to } 30) \times 10^{20}$ m

M_S = mass of a spiral galaxy $\approx (10^{11} \text{ to } 10^{12})\ M_\odot \approx (2 \text{ to } 20) \times 10^{41}$ kg

M_E = mass of an elliptical galaxy $\approx (10^{12} \text{ to } 10^{13})\ M_\odot \approx (2 \text{ to } 20) \times 10^{42}$ kg

M_S/L_B = mass to light ratio of spiral galaxy within radius R $\approx 60\ h\ (R/0.1 \text{ Mpc}) \approx 42\ (R/0.1 \text{ Mpc})$

M_E/L_B = mass to light ratio of elliptical galaxy within radius R $\approx 200\ h\ (R/0.1 \text{ Mpc}) \approx 140\ (R/0.1 \text{ Mpc})$

L_B = mean galaxy luminosity density in blue band $= 1.93 \times 10^8\ h\ L_\odot \text{ Mpc}^{-1} \approx 5.2 \times 10^{34} \text{ J s}^{-1} \text{ Mpc}^{-1}$

L_g = mean galaxy luminosity density $= (2 \text{ to } 3) \times 10^8\ h\ L_\odot \text{ Mpc}^{-1} \approx (5 \text{ to } 8) \times 10^{34} \text{ J s}^{-1} \text{ Mpc}^{-1}$

L_x = x-ray luminosity of spiral or elliptical galaxy $= 10^{31}$ to 10^{35} J s^{-1}

M_{Sgas} = mass of cool gas of hydrogen atoms and molecules in spiral galaxy $\approx 10^9$ to $10^{10}\ M_\odot \approx (2 \text{ to } 20) \times 10^{39}$ kg

M_{Egas} = mass of hot gas in elliptical galaxy $\approx 10^9$ to $10^{10}\ M_\odot \approx (2 \text{ to } 20) \times 10^{30}$ kg

N_g = average volume density of galaxies $= 5.52 \times 10^{-2}\ h^3 \text{ Mpc}^{-3} \approx 1.89 \times 10^{-2} \text{ Mpc}^{-3}$

ρ_g = mass density of galaxies $\leq 0.4 \times 10^{-27}\ h^2 \text{ kg m}^{-3} \approx 2 \times 10^{-28} \text{ kg m}^{-3}$ (for visible stars and dark matter with a galaxy mass of about 10^{12} solar masses)

[a] For Hubble constant, $H_0 = 100\ h \text{ km s}^{-1} \text{ Mpc}^{-1} \approx 75 \text{ km s}^{-1} \text{ Mpc}^{-1}$. The symbols $M_\odot$ and $L_\odot$ denote the Sun's mass and luminosity, respectively.

spans 10 to 20 million light-years and contains hundreds and even thousands of individual galaxies (Table 12.3).

In 1958, George O. Abell, then a graduate student at Caltech, used a newly completed photographic survey to identify clusters of galaxies over a large fraction of the sky. He published a famous catalogue of 2,712 rich clusters, and even today, these dense concentrations of galaxies are referred to simply as "Abell clusters" – designated by the word "Abell" or the letter "A" followed by the number in his catalogue.

In addition, as Abell noticed, the clusters gather and congregate together into larger superclusters, which he called *second-order clustering*. Our Galaxy lies in the outskirts of one, known as the Local Supercluster, which is oriented perpendicular to the Milky Way and extends all the way to the rich Virgo cluster of galaxies.

Dark Matter in Clusters of Galaxies

Clusters of galaxies are bound together by gravity, even though the expansion of the universe is pulling the galaxies away from one another. We might think that the combined gravitational pull of the numerous galaxies would be sufficient to hold the clusters together, but, in 1937, Fritz Zwicky showed that there must be significant quantities of unseen material to keep the galaxies from dispersing. In his extraordinarily prescient paper, titled *On the Masses of Nebulae and of Clusters of Nebulae*, he concluded that there must be substantial amounts of unseen matter in clusters of galaxies or else they would be unstable dynamically. In a German-language article discussing many of the same topics four years earlier, Zwicky introduced

Figure 12.12. Inside the Coma cluster of galaxies. More than 1,000 identified galaxies are located within the Coma cluster, also designated Abell 1656, which has a mean distance of 321 million light-years and is more than 20 million light-years in diameter. Each of these galaxies contains hundreds of billions of stars. Most of the galaxies that inhabit the central portion of this cluster are giant elliptical galaxies. Several spiral galaxies are found farther out from the center, such as the one shown in the upper left of this mosaic of images taken from the Hubble Space Telescope. Nearly every object in this picture is a galaxy. It is a section of the cluster that is several million light-years across, and it is located about one third of the way out from the center. (Courtesy of NASA/ESA/Hubble Heritage/STScI/AURA.) (See color plate.)

Table 12.3. *Physical properties of rich clusters of galaxies*[a]

N_T = total number of galaxies in a rich cluster of galaxies = 30 to 300
R_C = central radius of galaxy cluster ≈ 3 million light-years ≈ (1 to 2) h^{-1} Mpc
≈ (4.4 to 8.8) × 10^{22} m
N_{cl} = volume density of galaxies in galaxy cluster ≈ 100 Mpc^{-3}
σ_V = velocity dispersion of galaxy motions ≈ 100 to 1,400 km s^{-1} = 10^5 to 1.4 × 10^6 m s^{-1}
M_C = virial mass of galaxy cluster = $\sigma_v{}^2 R_{cl} / G$
≈ (10^{14} to 2 × 10^{15}) h^{-1} $M_\odot$ ≈ (2.8 to 57) × 10^{44} kg
T_C = cluster crossing time ≈ $2 R_{cl}/\sigma_v$ ≈ 6 × 10^{16} s ≈ 2 × 10^9 yr
L_B = luminosity of galaxy cluster in blue band
= (6 × 10^{11} to 6 × 10^{12}) h^{-2} $L_\odot$ ≈ (4.6 to 46) × 10^{38} J s^{-1}
M_C/L_B = mass to light ratio of galaxy cluster ≈ 300 h $M_\odot/L_\odot$ ≈ 210 $M_\odot/L_\odot$
L_X = x-ray luminosity of galaxy cluster = ($10^{35.5}$ to 10^{38}) h^{-2} J s^{-1} ≈ 2.0 ($10^{35.5}$ to 10^{38}) J s^{-1}
n_{cl} = cluster number density ≈ (10^{-5} to 10^{-6}) h^3 Mpc^{-3} ≈ 0.34 (10^{-5} to 10^{-6}) Mpc^{-3}

[a] For Hubble constant, H_0 = 100 h km s^{-1} Mpc^{-1} ≈ 75 km s^{-1} Mpc^{-1}. The symbols $M_\odot$ and $L_\odot$ denote the Sun's mass and luminosity, respectively.

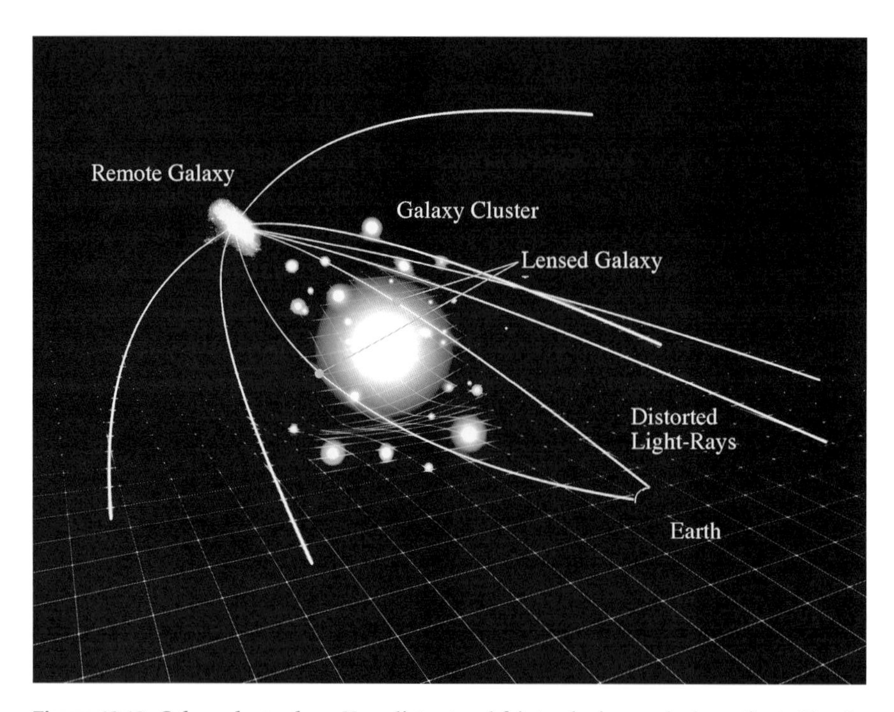

Figure 12.13. Galaxy cluster lens. Very distant and faint galaxies can be investigated by observing them through a cluster of galaxies. The powerful gravitation of the cluster acts like a lens, bending, focusing, and magnifying the light of more distant galaxies that lie behind it (see Fig. 12.14). The gravitational lens action can distort the light from the background galaxies into faint arcs or produce magnified images of individual galaxies that otherwise would remain invisible. (Courtesy of NASA/JPL-Caltech.)

the term *dunkle materie*, or "dark matter," for the invisible stuff, and he concluded that *dark matter* might have a greater density than luminous matter.

Zwicky measured the amount of mass required to keep the Coma cluster of galaxies stable, assuming that the motions of its constituent visible galaxies are balanced by the gravitational pull of their combined mass. He found that the total mass of the Coma cluster must be about 10 times the sum of the masses of the individual galaxies it contains. If an unseen mass of about 10 times that of the visible galaxies were not there, the Coma cluster would be flying apart.

Zwicky also proposed that the formidable gravity of dark matter in clusters of galaxies would act as a powerful lens, diverting and focusing the light of more distant galaxies. However, it was not until relatively recent times that astronomers began to observe the effect.

Images of rich clusters of galaxies taken from the Hubble Space Telescope (HST) reveal the highly stretched, distorted and magnified images of faint galaxies lying far behind them (Figs. 12.13 and 12.14). The images contain arcs that are not centered or anchored on any particular visible galaxy in the cluster. Dark matter, in which the galaxies are embedded, spreads out the arcs. As Zwicky proposed, the gravitational-lens effect provides information on both the visible and unseen matter. Moreover, the dark matter can act like a zoom lens, magnifying distant galaxies too faint to be seen and bringing them into view.

Figure 12.14. Cluster of galaxies and gravitational lens. A Hubble Space Telescope image of a rich cluster of galaxies designated Abell 2218. It is about 3 billion light-years away from the Earth. A typical rich cluster contains hundreds and even thousands of galaxies, each composed of hundreds of billions of stars and possibly up to 10 times more mass within invisible dark matter. The galaxy cluster Abell 2218 is so massive and so compact that its gravity bends and focuses the light from galaxies that lie behind it. Multiple images of these background galaxies are distorted into long faint arcs. Magnified or ring images of individual background galaxies also can be observed. (Courtesy of NASA/STScI/Andrew Fruchter/the ERO team [Sylvia Baggett/STScI, Richard Hook/ST-ECF/Zoltan Levay, STScI].)

The presence of two galaxies along the same line of sight, one more distant than the other, was suggested from the spectroscopic database of the Sloan Digital Sky Survey, and these gravitational-lens candidates were confirmed using the high-resolution imaging capability of the HST. Hundreds of these cosmic gravitational lenses have been found. Some exhibit partial or complete "Einstein" rings, which indicate near-perfect alignment of the foreground and background galaxies. Even a double ring arising from the light of three aligned galaxies has been found; the galaxies are located at distances of 3 billion, 6 billion, and 11 billion light-years (Fig. 12.15).

Cosmic Streams

The galaxies are not simply flying outward with the expansion of the universe, in a smooth and regular manner. Individual galaxies dart here and there, like mosquitoes in a swarm, and entire groups are streaming in concert over vast distances, like powerful currents awash in the cosmic sea.

These so-called *peculiar motions* are caused by the gravitational pull of huge assemblages of galaxies, and they are unrelated to the uniform expansion of the galaxies, known as the *Hubble flow*.

For instance, the nearest large galaxy, Andromeda (M 31), is moving toward us. Due to their mutual gravitational attraction, the Milky Way and Andromeda are set on an irrevocable collision course. In a few billion years, they will meet in a possibly destructive encounter.

Figure 12.15. Einstein ring. When a background and foreground galaxy are aligned perfectly, the closer galaxy acts as a gravitational lens, bending and magnifying the light of the more distant galaxy and forming a glowing "Einstein" ring. A double ring is captured in this Hubble Space Telescope image, indicating an exceptionally rare alignment of a massive foreground galaxy with two background galaxies; the distances of the three galaxies are estimated at 3 billion, 6 billion, and 11 billion light-years from the Earth. (Courtesy of NASA/ESA/Raphael Gavazzi and Tommaso Trea, University of California Santa Barbara and the SLACS team.)

Occasionally, we observe other galaxies colliding, merging, or passing through each other (Fig. 12.16), when they should be moving farther apart. This mingling also is due to the gravitational attraction of neighboring galaxies, drawing them together and producing local eddies within an outward Hubble flow.

In addition, an entire swarm of galaxies can set off on a trajectory that is independent of the expansion. Because the galaxies are moving together over vast distances of hundreds of millions of light-years, their collective behavior is known as a *large-scale streaming motion*, which is large in space but not so big in velocity – generally no more than 1,000 km s^{-1}.

All of the galaxies in our region of space, within a volume of 100 million light-years across, are not just along for the ride in the expanding universe. They are rushing en masse, like a high-speed celestial convoy, toward the same remote point in space. Alan Dressler, of the Carnegie Observatories, and his colleagues showed that all of these galaxies are being pulled through space, forced into mass migration by the gravitational pull of "The Great Attractor." They pinpointed its distance at about 150 million light-years away and estimated that its

Figure 12.16. Colliding galaxies. Gravitational interaction of the antennae galaxies, catalogued as NGC 4038 and NGC 4039, produces long arms of young stars in their wake. The colliding galaxies are located about 62 million light-years from the Earth and have been merging for the past 800 million years. As the two galaxies continue to churn together, clouds of interstellar gas and dust are shocked and compressed, triggering the birth of new stars. This composite image is from the Chandra X-Ray Observatory (*blue*), the Hubble Space Telescope (*gold* and *brown*), and the Spitzer Space Telescope (*red*). The blue x-rays show huge clouds of hot interstellar gas, the red data show infrared radiation from warm dust clouds that have been heated by newborn stars, and the gold and brown data reveal both star-forming regions and older stars. (Courtesy of NASA/ESA/SAO/CXC/JPL-Caltech/STScI.) (See color plate.)

mass is equivalent to about 50 million billion (5×10^{16}) stars like the Sun and at least 500,000 galaxies like the Milky Way. The Great Attractor most likely is a rich and massive cluster of galaxies, part of an even larger supercluster.

Vast rivers of galaxies may be flowing through more distant regions, as great concentrations of mass pull the galaxies here and there in a cosmic "tug of war." It makes us wonder where we are going, like riding in a taxi with the meter running in a city that we have never before visited.

It is the gravitational interaction of galaxies with one another that distorts the smooth cosmic expansion, producing the peculiar motions superposed on the expanding universe. However, the uniform Hubble flow gathers speed with distance, like a great river flowing downhill. Because the localized streaming motions are limited in velocity, they are relatively slow when compared with the expansion speed of remote galaxies. The very existence of the large-scale streaming motions nevertheless indicates a decidedly uneven and lumpy distribution of galaxies, which eventually was mapped across billions of light-years.

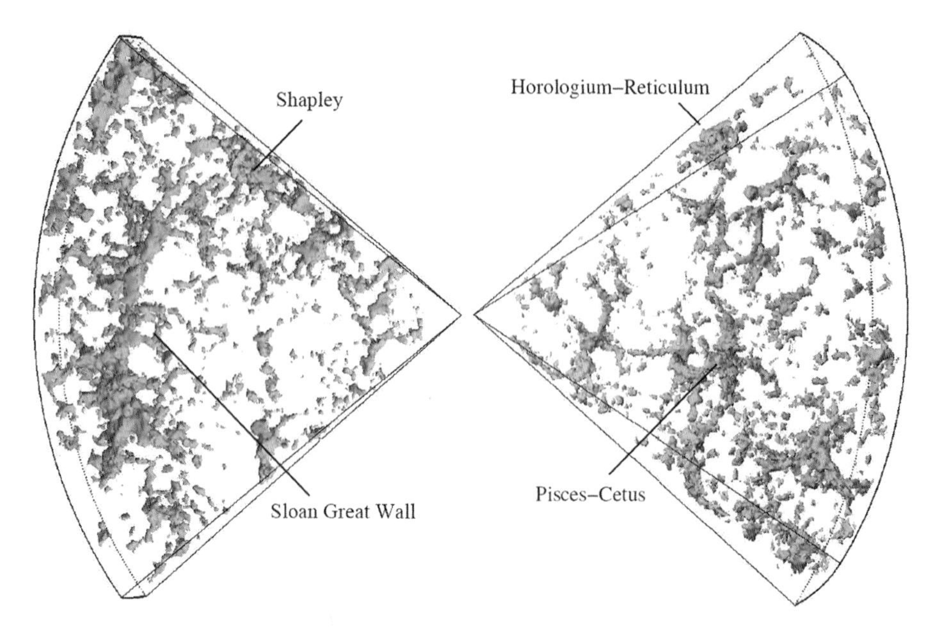

Figure 12.17. Great walls and empty places. By measuring the recession velocity, or redshift, of galaxies, astronomers determined their distance and combined it with their location in the sky to obtain the three-dimensional distribution of galaxies. The map shown here is for galaxies within 1 billion light-years (*far left* or *far right*) from the Earth (*center*). Because galaxies started to form about 12 billion years ago, this is a relatively nearby part of the universe. It includes recession velocities of up to 30,000 km s^{-1}, at a redshift of 0.1. The galaxies are concentrated in long, narrow sheet-like walls encircling large empty places known as *voids*, about 100 million light-years across. The Sloan Great Wall (*left*) spans about 1.4 billion light-years. It may be gravitationally unbound, perhaps beginning to fall apart, but it includes superclusters of galaxies that may stay bound together by their mutual gravitational pull. The Sloan Great Wall was discovered using data from the Sloan Digital Sky Survey in 2004. Other superclusters, or clusters of galaxy clusters, are labeled in the diagram, which is from the Two Degree Field Galaxy Survey. (Courtesy of Willem Schaap, Kapteyn Institute, University of Groningen et al., 2dF Galaxy Redshift Survey.)

Cosmic Connections

Astronomers map the distribution of galaxies, thereby determining our place in space and also establishing the framework and structure of the observable universe. Like the initial exploration of the Earth, mapping the universe has proceeded gradually. First the nearby places were examined and then the more distant horizons were explored. Eventually, the very fabric of the universe was specified, with maps that reveal its texture. These maps show that all of the galaxies are tied and bound to one another, interconnected by cosmic threads resembling a spider's web.

Two-dimensional maps, derived from the catalogued positions of galaxies on the celestial sphere, provide a blurred picture of the galaxy distribution in space. They portray galaxies near and far, piled on one another at any given location in the sky. Chance superposition of distant and nearby groups of galaxies therefore could be mistaken for actual physical structures. To overcome this problem, astronomers used the radial velocities, or redshifts, of galaxies to gauge their approximate distances, which provided the crucial third dimension.

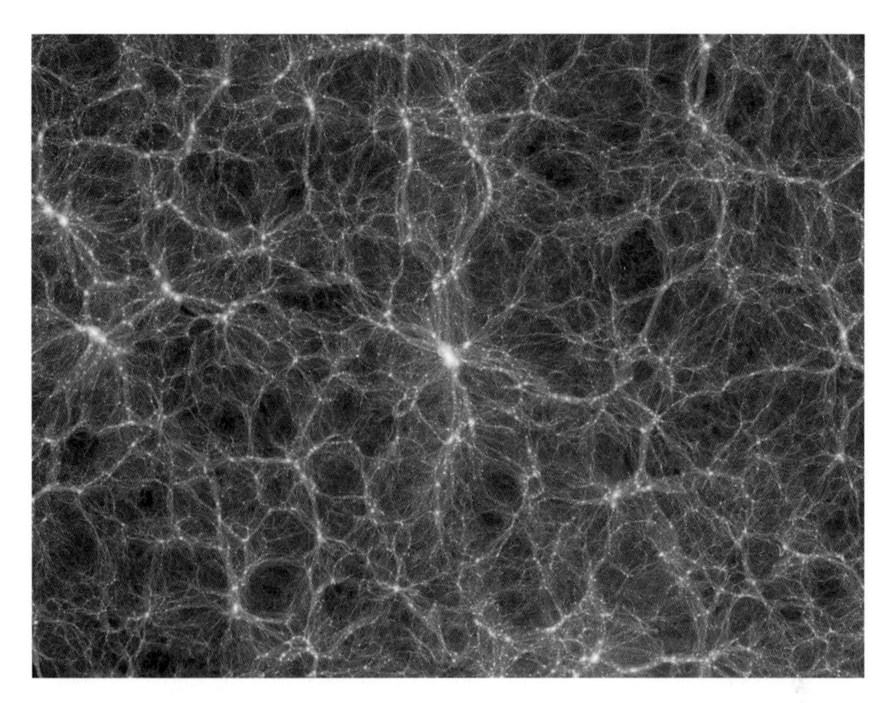

Figure 12.18. Cosmic web. One moment in the ever-changing distribution of galaxies studied using a supercomputer to trace out their formation, evolution, and clustering. The width of this image is about 10 million light-years. (Courtesy of Volker Springel, the Millennium Simulation Project/Max Planck Institute for Astrophysics, Garching, Germany.)

By determining the concentrations of galaxies in different directions and at various redshifts, or depths, astronomers located places where the collective force of gravity pulled galaxies together and locally reversed the uniform expanding motion. The three-dimensional maps reveal fascinating lace-like patterns that connect the galaxies and curving filaments that enclose dark, seemingly vacant places. This web of interconnected galaxies resembles the tangled ribbons of light we see when looking out from the bottom of a calm ocean bay or a swimming pool.

The galaxies apparently are distributed along the peripheries of gigantic hollow bubbles that nest together in some sort of cosmic foam, like soapsuds in a kitchen sink. Thin elongated strands of galaxies, clusters of galaxies, and superclusters swirl across the astronomers' maps in curved structures that resemble the border of wetness left by waves on a beach or the flowing lines of a Matisse painting. The relatively compact galaxy clusters appear as bright, concentrated knots and clumps in those strands, whose tangled texture resembles *spaghetti alle vongole* (which means spaghetti with small clams); the longer strands delineate the boundaries of huge voids, each seemingly as empty as an immense vacuous bubble.

The early redshift surveys also delineated an enormous sheet of galaxies, dubbed "The Great Wall," at distances ranging from 350 million to 500 million light-years away. Subsequent three-dimensional maps were obtained using electronic technology that permits the simultaneous measurement of hundreds of galaxy redshifts in a single exposure at a large telescope. The Sloan Digital Sky Survey, for example, revealed the longest sheet of galaxies yet seen. Dubbed the "Sloan Great Wall," it measures 1.37 billion light-years across (Fig. 12.17), and it is the largest observed structure in the universe – at least so far.

Thus, everywhere they look, in whatever direction and near or far, modern telescopes are finding a complex and richly textured universe, filled with luminous concentrations of matter. Even when looking across 10 percent of the observable universe, astronomers continue to find galaxy structures crossing their maps from edge to edge, as well as smaller bubbles, walls, and voids that are nestled together.

All parts of the observable universe are bound within this all-encompassing fabric, glued together by the invisible forces of gravity, suspended in space by motion, and linked by radiation. They combine, interconnect, entangle, interpenetrate, and overlap one another. This all-embracing cosmic web extends throughout the observable universe (Fig. 12.18).

13

Birth, Life, and Death of the Universe

13.1 Hotter Than Anything Else

Regardless of the direction in which we look out into space, almost all of the distant galaxies are flying apart, dispersing and moving away at speeds that increase with their distance, as if they had been ejected by a cosmic bomb. Astronomers call it the "big bang." We are participating in its explosion, watching it blow up before our eyes.

We can envision this early state by putting the observed expansion of the galaxies in reverse, pushing the galaxies back closer together until a time – about 13.7 billion years ago – when the universe was incredibly small and all of its mass was compressed to a very high density.

Because gases become hotter when they are compressed and cool when they expand, the observed universe must have been incredibly hot in its earliest, most compact state. As we look back in time, the universe becomes increasingly hotter, eventually becoming so exceptionally hot that radiation was the most powerful force, dominating the expansion of the universe.

In the earliest moments of the big bang, there were no stars or galaxies, only intense radiation and subatomic particles from which the material universe subsequently grew. This condition seems to have been described in *Genesis*, the first book of the Bible, in which God said "Let there be light " or, as in the popular spiritual, "I got a home in glory land that outshines the Sun."

In the early stages, the radiation controlled the expansion of the universe, essentially because it was incredibly hot. Just 1 second after the big bang, the radiation had a temperature of about 10 billion, or 10^{10}, K. As the universe continued to expand, the radiation steadily cooled and began its long descent into darkness, when matter eventually took over the expansion. However, the big bang was so intense and so hot that we are still immersed within the radiation.

13.2 Three Degrees above Absolute Zero

An Unexpected Source of Noise

The discovery of the faint afterglow of the big bang was a serendipitous event, involving a horn-reflector antenna that had been used at the Bell Telephone Laboratories in the first tests of a communication satellite. Arno Penzias and Robert Wilson were identifying noises in the horn-antenna system so they could use it to make accurate measurements of the intensity of several extragalactic radio sources. A persistent, ubiquitous, and unvarying noise source was detected at a microwave wavelength of 0.0735 m, contributing an antenna temperature of only 3 degrees above absolute zero, or about 3 K. It was equally strong in all directions, wherever the antenna was pointed, independent of the time of day and year and with no dependence on the location of any known cosmic radio source.

Penzias and Wilson did not know what they had found and avoided any mention of the cosmological implications in the report of their discovery, published in 1965 with the modest title, *A Measurement of Excess Antenna Temperature at 4090 MHz*. However, a group at Princeton University, which was attempting to make a similar measurement at the time, drew attention to the implications in a companion paper. The unexpected source of noise was the faint, cooled relic of the hot big bang, now known as the *3-degree cosmic microwave background radiation*. It is called the background radiation because it originated before the stars and galaxies were formed and it lies behind them.

This particular discovery was not entirely unanticipated. In the late 1940s and early 1950s, George Gamow, Ralph A. Alpher, and Robert C. Herman had speculated that the 1-billion-degree, or 10^9 K, radiation of the early universe would have cooled to about 5 K during the past billions of years of expansion, but no one had attempted to observe the relic radiation. Penzias and Wilson were unaware of this previous calculation until after their discovery. They received the 1976 Nobel Prize in Physics for their discovery of the cosmic microwave background radiation.

Blackbody Spectrum

At the high temperatures prevailing in the early history of the expanding universe, the radiation and subatomic particles frequently interacted, achieving a thermal equilibrium characterized by a single temperature. Later, when the universe thinned out and cooled by expanding into a greater volume, the matter and radiation quit interacting, going their separate ways. However, the radiation would have retained its thermal nature as it cooled and the temperature slowly decreased.

A perfect thermal radiator is known as a *blackbody*, which absorbs all thermal radiation falling on it and reflects none – hence, the term *black*. The distribution of the radiation emitted by the blackbody, its spectrum, peaks at a wavelength that is inversely proportional to the temperature, dropping precipitously at shorter wavelengths and falling off gradually at longer ones.

The expansion of the universe preserves the blackbody spectrum of the radiation for all time. No process can destroy its shape, but the location of maximum intensity will stretch to increasingly longer wavelengths as time goes on and the radiation gets colder. In the present epoch, with a temperature of only about 3 K, the blackbody-radiation intensity peaks at a wavelength of about 0.001 m. Unfortunately, the Earth's atmosphere absorbs cosmic radiation at this short wavelength where the most intense radiation occurs.

The definitive spectral measurements had to be made from above the atmosphere using NASA's COsmic Background Explorer (COBE), launched in November 1989. Less than two months after COBE went into orbit, but a quarter-century after the discovery of the

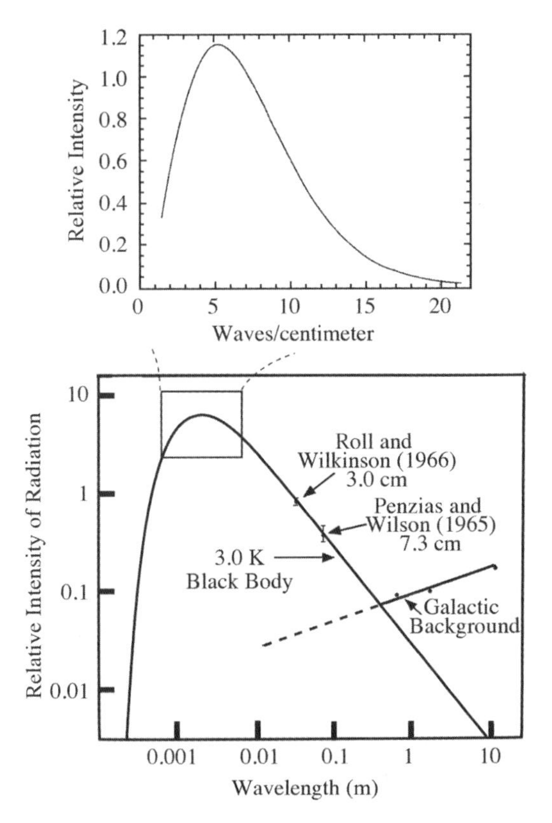

Figure 13.1. Spectrum of the cosmic microwave background radiation. The intensity of the cosmic microwave background radiation plotted as a function of wavelength. This thermal radiation was formed about 390,000 years after the big bang, which occurred about 14 billion years ago. The observed radiation has a nearly perfect blackbody spectrum. Pioneering measurements by Arno A. Penzias (1933–) and Robert W. Wilson (1936–) in 1965 and Peter G. Roll (1935–) and David T. Wilkinson (1935–2002) in 1966, at 7.35 and 3.0 cm wavelength, respectively, are compared to the expected spectrum of a 3-degree blackbody and radiation from our Galaxy (*bottom*). The full spectrum at millimeter wavelengths (*top*) was obtained from instruments aboard the COsmic Background Explorer (COBE) in late 1989. These data are so accurate that the error bars of the individual points all lie within the width of the plot curve. This solid line, which matches the shape and peak location of the observed data, corresponds to a thermal radiator, or blackbody, with a temperature of 2.725 K.

cosmic radiation, John C. Mather reported the combined results of millions of COBE spectral measurements at an American Astronomical Society meeting near Washington, DC. The spectrum fit the Planck blackbody curve with a precision of 1 part in 10,000 (Fig. 13.1), establishing a temperature of precisely 2.725 K, with an uncertainty of 0.002 K. The presentation caused the audience to break into a standing ovation.

Such a thermal spectrum could not have happened in the universe as it is now. Matter currently has a very different temperature than the background radiation. In other words, the observed spectrum is proof that the observable universe must have expanded from a very hot, dense state in the past, when matter and radiation were in thermal equilibrium and at the same temperature.

Every "nook and cranny" of space now is filled with background radiation. Any cubic meter of space in every part of the observable universe now contains about 500 million photons of the cosmic radiation. These are tiny bundles of radiation energy that originated about 13.7 billion years ago.

As Smooth as Silk

What alerted astronomers to the importance of the background radiation was its equal brightness wherever one looked, indicating that it uniformly fills all of space. This spatial isotropy satisfied one of the basic tenets of modern cosmology, *the cosmological principle*, which asserts that except for local irregularities, the universe presents the same aspect from every point. The COBE instruments, for example, could detect no regions brighter than others to 0.0003 K, or 1 part in 10,000.

Somewhere in all of that smoothness, there should be departures from silky uniformity. If the background radiation were completely uniform, it would remain that way as the universe expanded, and nothing would draw together to make stars or galaxies. Concentrations in the radiation intensity were needed to act as seeds for the subsequent formation of the material universe. They must have acted as a template or blueprint, encoding the information required to explain the subsequent formation of stars and galaxies – the astronomical equivalent of the human genome.

Cosmic Ripples

In 1992, George Smoot and his colleagues announced measurements of the temperature fluctuations in the cosmic microwave background radiation using four years of data gathered by COBE. After subtracting the known microwave emission of the Milky Way and using mathematical averaging techniques on about 100 million observations, the COBE team found that the temperature varies ever so slightly over large angular sizes. Mather and Smoot were awarded the 2006 Nobel Prize in Physics for their discovery of the blackbody form and anisotropy of the cosmic microwave background radiation.

After the COBE results, experiments from the ground and balloons brought the temperature fluctuations into sharper focus for localized regions of the background radiation. However, another satellite experiment was needed that would scan the entire sky without the confusion of microwave radiation from the atmosphere and ground. This time, the spacecraft would not only detect the cosmic ripples; it instead would determine their distribution and characteristic sizes, filling in the gaps between the large features seen with *COBE* and the smaller features detected by other instruments.

David T. Wilkinson, of Princeton University, joined Charles L. "Chuck" Bennett of the Goddard Space Flight Center to create a small team of experts and design a spacecraft that could accomplish the goal. The resultant Microwave Anisotropy Probe (MAP) was launched in June 2001. The name was changed in early 2003 to Wilkinson Microwave Anisotropy Probe (WMAP) to honor Wilkinson after his death. Instruments aboard WMAP provided definitive measurements of the rippling departures from uniformity, with temperature fluctuations of 1 part in 100,000, or at about 0.00003 K (Fig. 13.2). This anisotropy is given with other physical properties of the background radiation in Table 13.1.

Table 13.1. *Physical properties of the cosmic microwave background radiation*

Parameter	Name	Value
$T_0 = T_{CMB}$	Temperature	2.725 ± 0.002 K
N_{CMB}	Photon density	$(410.4 \pm 0.9) \times 10^6$ m^{-3}
ρ_{CMB}	Mass-energy density of photons	4.648×10^{-31} kg m^{-3}
$\Delta T/T_0$	Anisotropy	$(1.1 \pm 0.1) \times 10^{-5}$

The WMAP instruments also showed that temperature variations are concentrated within certain angular sizes that are displayed in an angular power spectrum – a plot of the relative strength of the hot and cold spots against their angular sizes (Fig. 13.3). This spectrum is not flat but rather varies in power. Gravity explains the fluctuations of the power spectrum, the relative amplitudes of which can be used to infer the gravitational pull that caused them.

The ratio of the heights of the first and second peak of the angular power spectrum was used to determine the amount of "ordinary" matter with which we are familiar, the baryonic type that comprises atoms. The neutrons and protons found in the nuclei of all atoms are *baryons*.

When the height of the third peak is compared to the other two, the amount of dark, nonbaryonic matter is estimated. It is about five times more abundant than ordinary baryonic matter, and the combined gravitational pull of both types of matter is not enough to stop the future expansion of the universe.

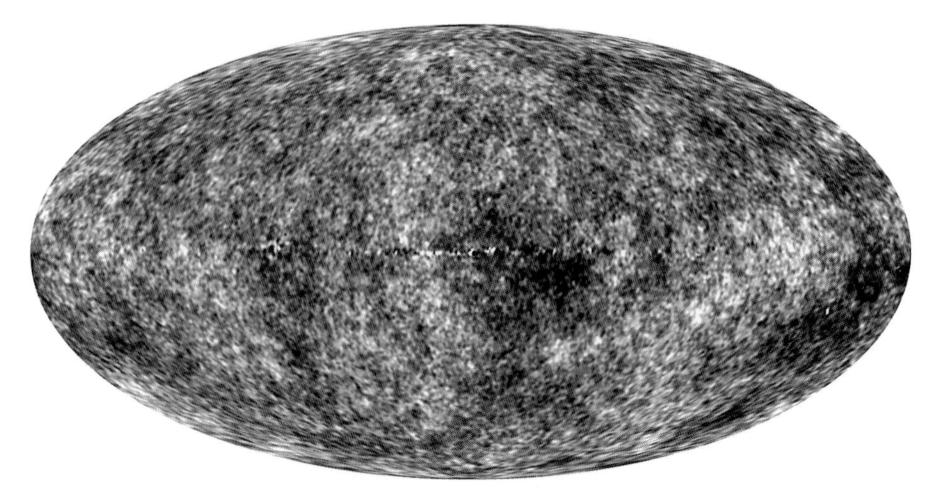

Figure 13.2. Maps of the infant universe. An all-sky view of the 3-degree cosmic microwave background radiation emitted from the universe in its infancy, just 390,000 years after the big bang that occurred 13.7 billion years ago. The data, taken in 2003 from the Wilkinson Microwave Anisotropy Probe (WMAP), are shown here after seven years of data analysis. The temperature fluctuations range up to 0.0002 K above and below the average value. Darker regions are cooler and lighter regions are hotter. These temperature fluctuations provided the seeds from which galaxies subsequently grew. (Courtesy of the NASA/*COBE* and NASA/*WMAP* Science Teams.)

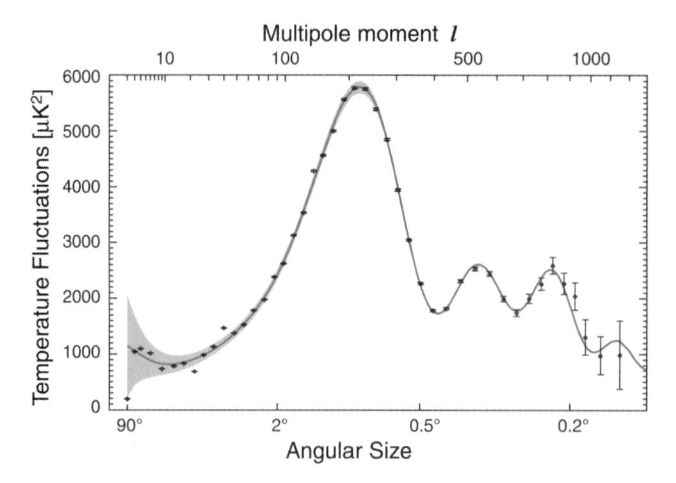

Figure 13.3. Ripple data. The angular fluctuation strength, or power, of the cosmic microwave background radiation in which temperature fluctuations, in units of square micro kelvin (10^{-6} K and designated μK), are displayed as a function of their angular extent in degrees, denoted o. This plot shows the relative brightness for the all-sky map observed from the Wilkinson Microwave Anisotropy Probe (WMAP) (see Fig. 13.2) at various sizes. The solid line is the model that best fits the observed data (*solid dots*); the gray band represents uncertainties in the model. Anisotropy data obtained by previous experiments are denoted by dots with error bars. The observed power spectrum has been compared to other astronomical observations and different theoretical models, providing estimates for the amount of dark matter and dark energy in the universe. (Courtesy of the NASA/*WMAP* Science Team.)

The COBE and WMAP results have carried cosmology beyond the esoteric realms of theoretical speculation and into precision scientific tests when combined with other astronomical observations (Table 13.2). Definitive new observational descriptions of the background radiation, with refined cosmological consequences, are expected from the *Planck* mission, launched in May 2009. In the early years, the *Planck* scientists identified all of the foreground microwave emission from our Galaxy, which had to be removed to reveal an all-sky image of the background radiation. In the process, cold radiation was found to be enveloping the center of our Galaxy in a murky haze.

13.3 The Beginning of the Observable Universe

The First 3 Minutes

Soon after the end of World War II (1939–1945), physicists began to apply their wartime knowledge of atomic bombs to investigations of the origin of the material constituents of the universe. They knew how elements were produced during nuclear chain reactions, and they had observed the creation of new chemical elements in the early tests of nuclear explosions.

George Gamow looked back to the dawn of time, before the stars were born, and proposed that the first elements were formed during the most energetic explosion of them all: the big bang that propelled the universe into expansion. He was partly correct; the lightest atomic

Table 13.2. *Cosmological parameters*[a]

Parameter	Name	Value
H_0	Hubble constant	73.8 ± 2.4 km s^{-1} Mpc^{-1}
t_0	Age of expanding universe	$(13.73 \pm 0.12) \times 10^9$ years (after big bang)
t_{eq}	Equality of matter and radiation	$76{,}000 \pm 5{,}000$ years
t_r	Recombination time (time radiation originated)	$(3.79 \pm 0.08) \times 10^5$ years (After the big bang)
Ω_{tot}	Matter and energy density	1.01 ± 0.02
Ω_Λ	Dark energy density	0.728 ± 0.015
Ω_m	Total matter density	0.273 ± 0.010 (for $h = 0.75$)
Ω_B	Baryonic matter[a]	0.040 ± 0.0010 (for $h = 0.75$)

[a] Adapted from Charles L. Bennett et al., *Astrophysical Journal Supplement* 148, 1 (2003); and N. Jarosik et al., *Astrophysical Journal Supplement* 192, 14 (2011). The parameter Ω is the ratio of the specified quantity to the critical amount required to keep the expansion of the universe on the brink of closure. The total density parameter Ω_{tot} is the sum of the contributions from visible matter, dark matter, and dark energy; $\Omega_{tot} = 1.00$ is consistent with inflation and a universe that is described by Euclidean geometry without space curvature. The Hubble constant $H_0 = 100\ h$ km s^{-1} Mpc^{-1} and $h = 0.75$ have been assumed in the estimates of the matter densities.

nuclei were synthesized before any atoms were created, during the first 3 minutes following the big bang; the nuclei of less abundant, heavier atoms were manufactured at a later time, within the stars (see Section 8.4).

During the first moments of the expansion, some of the incredibly energetic big-bang radiation was being transformed into electrons and their anti-matter counterparts – the positrons, or positive electrons – and an electron would collide with a positron just as often to make radiation again. Neutrons and protons also were around and these subatomic particles also would turn back and forth into one another.

As time went on, the radiation cooled as a result of the expansion of the universe into a greater volume, until the temperature was no longer hot enough to create anti-matter. The leftover positrons then were consumed by interactions with electrons, and an equilibrium was established in which the relative amounts of neutrons and protons were governed by their mass difference and the temperature.

Modern computations by David N. Schramm and others conclusively demonstrated that all of the deuterium nuclei and most of the helium nuclei that are found in the universe today were synthesized from the remaining neutrons and protons. Most of the protons did not contribute to forming these nuclei, and abundant leftover protons eventually became the nuclei of hydrogen atoms. Within just 3 minutes, production of light atomic nuclei was over, with vastly more hydrogen than anything else.

The nuclei of the hydrogen and deuterium atoms and most of the nuclei of helium atoms that now are present in the universe were synthesized in the first 3 minutes of the expansion, in the immediate aftermath of the big bang about 13.7 billion years ago. All of the hydrogen found in stars, within interstellar space, in the Earth's water, and in our bodies was produced by this big-bang nucleosynthesis. Deuterium is destroyed inside stars and, although helium is synthesized in main-sequence stars, the amount of helium formed inside stars over the lifetime of the expanding universe is no more than 10 percent of what is now observed in cosmic objects.

The agreement of observed light-element abundances and predictions from big-bang nucleosynthesis works only if the density of ordinary baryonic matter in the universe – in both visible and invisible forms – is substantially less than the critical mass density required to eventually stop expansion of the universe. That is, the results set an upper limit to the density of baryons, in which the number of baryons is equal to the number of neutrons and protons in matter in either seen or unseen forms. When measurements of the Hubble constant are considered, the upper limit becomes a definite value, which indicates that the baryon density is now 0.05, or 5 percent, of the critical mass density required to ever halt the current expansion of the universe in the future. A similar and completely independent estimate of the baryon density is obtained from the power spectrum of anisotropies in the cosmic microwave background radiation.

Formation of the First Atoms

With the further passage of time, the universe grew larger and the radiation energy density decreased more rapidly than the mass density. Eventually, at about 76,000 years after the big bang, the mass-energy density of the radiation had become equal to that of the matter; and thereafter, it was mass that dominated the expansion of the universe.

Whole atoms were not formed until the expanding universe cooled enough for electrons to combine with protons and helium nuclei to form long-lived hydrogen and helium atoms. This recombination occurred about 400,000 years after the big bang, when the temperature had fallen to about 3,000 K. The rate of recombination was then higher than the rate of ionization by the intense radiation. By the end of recombination, all of the nuclei and electrons had been bound up in atoms, and the universe became transparent to the radiation that then could travel through space without scattering on free, unattached electrons. The cosmic microwave background radiation that we observe in the present was released about 13.3 billion years ago.

Because there is no stable nucleus of atomic mass 5 or 8, elements heavier than helium (of mass 4) could not be synthesized by successive collisions with protons (of mass 1). Big-bang nucleosynthesis therefore stopped at helium 4. Heavier elements needed to be synthesized inside stars where the densities are high enough for triple collisions of helium to form carbon, rather than the big bang in which the density had become too low by the time helium nuclei were formed for triple collisions to become significant at the prevailing temperature (see Section 8.4).

The *recombination time* also marks the beginning of the *dark ages* of the expanding universe, for there were no sources of radiation other than the gradually cooling and darkening cosmic background radiation until stars and galaxies began to form, between about 100 million and 1,000 million years after the big bang. They provided beacons of bright light that

Table 13.3. *Crucial times during the expansion of the universe*

Time (after the big bang)	Redshift, z (K)	Temperature	Key events
10^{-14}	10^{27}	10^{27}	Inflation ends, $\Omega_m + \Omega_\Lambda = 1$
10 seconds	4×10^9	10^{10}	Neutron and positron production stops
3 minutes	4×10^8	10^9	Big-bang nucleosynthesis ends; light elements H, D, He formed
76,000 years	3,196	10^4	Radiation domination equals matter domination
400,000 years	1,100	3,000	Hydrogen atoms recombine, universe becomes transparent to background fluctuations, dark ages begin
100 million years	6 to 20	20 to 60	First stars and galaxies form, to 1 billion years universe re-ionized by their radiation, and dark ages end
6 billion years	1	5	Dark energy begins acceleration of universe expansion
13.7 billion years	0	2.725	Today, present epoch

could ionize surrounding matter. By then, the universe had thinned out enough that the low-density ionized hydrogen remained transparent, ending the dark ages. These milestones in the history of the expanding universe are summarized in Table 13.3.

Looking Back into Time

Because light travels at a finite speed, to look far into space is to look back into time; when we look farther out into space, we travel back more in time. Large telescopes that detect the faint light of distant objects therefore can be used as time machines to see objects as they were in the past, when the light detected at the telescope was emitted, as they were then and not as they are now. In effect, astronomers watch cosmic history race toward us at the speed of light, 299,792 km s^{-1}. The look-back time is simply the amount of time it takes for light to travel from the object to us at that speed.

Moving at the speed of light, it takes 2.3 million years for light to travel from the nearest spiral galaxy, Andromeda, to the Earth. Astronomers have observed radiation from distant galaxies whose light was emitted 13 billion years ago, before the Sun was formed about 4.6 billion years ago. Therefore, the look-back times for galaxies range from millions to billions of years, spanning an enormous period in which we can watch them evolve.

Some of the most distant galaxies may no longer exist, but they were embryonic galaxies when the light now reaching the Earth began its journey. These galaxies may have perished over time, but their light can survive unchanged, helping us trace out the history of the observable universe from the big bang – about 13.7 billion years ago – to now. That beginning time of the expanding universe was not always known with such precision, and its story describes interesting diversions along the bumpy road of cosmic knowledge (Focus 13.1).

13.4 When Galaxies Formed and the First Stars Began

Pulling Primordial Material Together

Immediately after the big bang, the universe was distributed smoothly, with almost no structure at all; because the subatomic neutrons and protons were then in thermal equilibrium with the radiation, the material universe must have had a smooth beginning.

Figure 13.4. Merging galaxies. Two spiral galaxies (*left*), each represented by M 81, can merge to form an elliptical galaxy. A pair of colliding galaxies (*right*), designated NGC 6240, illustrates such a merger just before becoming a single larger galaxy. The prolonged violent collision drastically altered the shape of both galaxies and created large amounts of heat and infrared radiation. All of these images were taken by the infrared camera aboard the Spitzer Space Telescope, with the inclusion of optically visible data for M 81 taken from the Hubble Space Telescope. (Courtesy of NASA/JPL-Caltech/University of Arizona/CfA/NOAO/AURA/NSF [*left*] and NASA/JPL-Caltech/STScI-ESA [*right*].) (See color plate.)

The First Stars and Galaxies

With the help of cold dark matter, the first stars and galaxies appeared more than 10 billion years ago. We can observe these embryonic galaxies when they were cosmic infants; the light now reaching us began its journey long before the Sun came into existence and life began on the Earth.

Each galaxy may have formed through the gravitational collapse of a larger, protogalactic cloud, which would become a rotating disk like the Milky Way. These flattened, spinning galaxies often show spiral structure, with arms of gas and dust in which new stars are forming.

Not all galaxies have a disk or spiral shape, and the most massive are the giant, rounded, featureless, elliptical galaxies. They may have resulted from the collision and subsequent merger of two spiral galaxies (Fig. 13.4). During the encounter, the ordered motions of the stars in the spiral galaxies would be transformed by tidal forces, which would tear their disks and arms apart and randomize the orbits of their stars. When the merger is complete, a single elliptical galaxy remains, composed of old stars with little or no gas and dust left to form new stars.

When astronomers use infrared telescopes, aboard the Herschel and Spitzer spacecraft, they can peer behind veils of local interstellar dust to see infant stars in distant galaxies. Some starburst galaxies are very powerful infrared emitters, with an infrared output that is 100 times their visible-light emission. An exceptional amount of interstellar dust in these galaxies absorbs the intense ultraviolet radiation produced by enhanced star formation, and the dust reradiates in the infrared part of the electromagnetic spectrum. Their intense

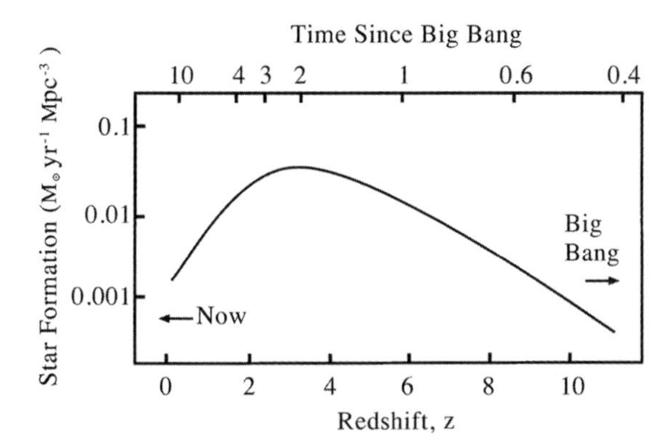

Figure 13.5. Star formation rates. The star-formation rate, in solar masses per year per cubic Megaparsec, or $M_{\odot}$ yr^{-1} Mpc^{-3}, plotted as a function of redshift, z (*bottom axis*) and time since the beginning of the expanding universe (*top axis*), in units of 10^9 years, or 1 billion years and a Gyr. The rate of star formation peaked at a redshift of about 3, or roughly 2 billion years after the expansion began, and this rate subsequently has decreased as gravitation pulls more material into stars.

infrared emission and implied dust suggests that these galaxies are forming stars more vigorously than our present-day Milky Way (Fig. 13.5).

About 2 billion years after the big bang and roughly 12 billion years ago, some starburst galaxies had an exceptionally high rate of star formation exceeding 100 stars per year for hundreds of millions of years – much greater than the rate in most galaxies and currently in the Milky Way, at about one new star every year. After the bursts of star formation, there may have been less material available for forming new stars because some of that material already had been used up in creating other stars.

The bursts of star formation might be feeding off gas stirred up as a result of collisions or close encounters between galaxies, or they could be associated with a voracious "feeding frenzy" of hydrogen gas, which was observed in greater abundance back then compared to more recent times. A steady supply of gas may have streamed in from filaments of dark matter.

When the first stars formed out of collapsing clouds of gas and ignited the nuclear reactions that make them shine, the early universe consisted of nothing more than the light elements, hydrogen and helium. These young stars must have been clean and uncontaminated by heavier elements. Some of these "infant" stars most likely were very massive, perhaps with about 100 times as much mass as the Sun; therefore, they would have a relatively short lifetime on the cosmic time-scale. The first massive stars would have exploded as supernovae, spewing out ashes of dust made of heavy elements synthesized within them and spawning the next generation of stars.

The interstellar medium would have become steadily enriched in heavy elements as subsequent generations of massive stars were born, lived, and died explosively. They seeded their surroundings with elements such as carbon, oxygen, and iron, which were needed for the formation of Earth-like planets and life.

The oldest stars in our Milky Way Galaxy, which were formed when the universe was only 1 billion years old, are deficient in heavy elements when compared to stars that are now

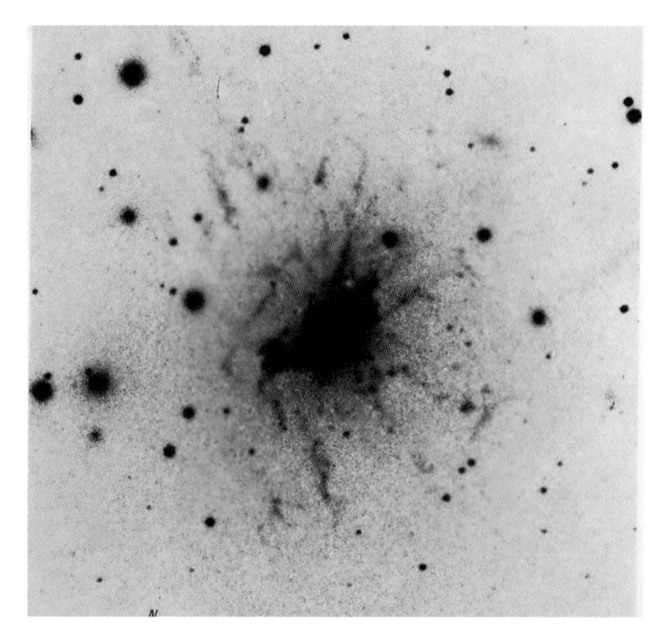

Figure 13.6. Seyfert galaxy. A negative print of the optically visible hydrogen emission from the Seyfert galaxy NGC 1275. (Courtesy of KPNO.)

forming there. No one has yet found a completely pure star that formed out of uncontaminated hydrogen. Perhaps such stars now are inaccessible to direct observation, awaiting the next space telescope that can peer deeper into the remote, shadowy past.

An Active Youth and Calmer Older Age

Because no significant change in the equilibrium of galaxies can be produced without a substantial change in the distribution of mass and angular momentum, it was long believed that no significant departures from a stable equilibrium in their shape, form, mass, or luminosity would be produced during most of their lifetime. Nevertheless, it now is known that the centers of galaxies are locations of pronounced activity that disrupts the expected stability and that, like people, the galaxies tend to be more active in their youth, calming down into a more dignified old age.

The American astronomer Carl K. Seyfert provided early observational evidence that the central regions of some galaxies are not in equilibrium when he examined the intense blue-colored centers of certain spiral nebulae – a type subsequently named *Seyfert galaxies*. Although most spirals exhibit spectral lines in absorption, similar to the absorption spectra of stars, the central regions of Seyfert galaxies exhibit intense emission lines of the type produced by ionized emission nebulae.

The emission lines of Seyfert galaxies are unexpectedly wide due to Doppler broadening by high-velocity motions of the emitting ions at a speed of up to 8,500 km s^{-1}. These motions exceed the expected escape velocity of an entire galaxy, and they provide the first evidence for violent explosive events in the nuclei of galaxies. Their matter could be flowing out into intergalactic space; some of the Seyfert galaxies exhibit bright filaments that suggest the ejection of gas (Fig. 13.6).

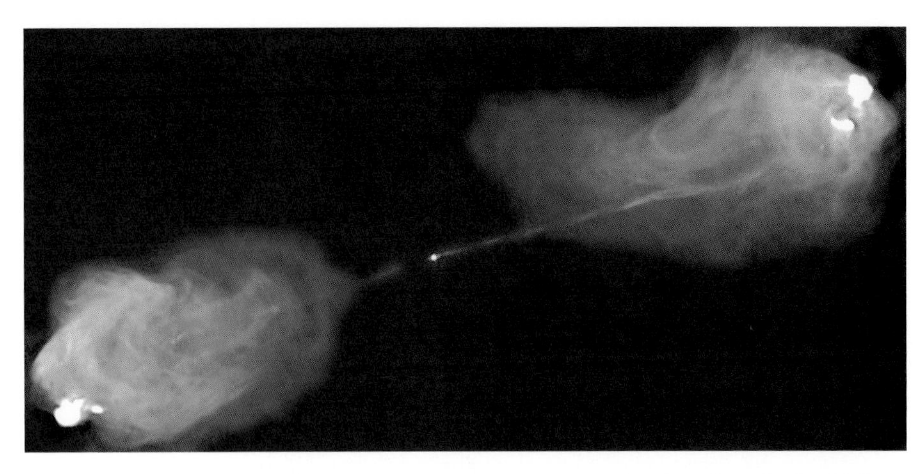

Figure 13.7. Radio galaxy Cygnus A. The radio galaxy Cygnus A, listed as 3C 405 in the third Cambridge catalogue of bright radio sources, which has a radio output 1 million times more powerful than the radio emission of a normal galaxy like the Milky Way. This radio image, taken with the Very Large Array at a wavelength of 6 cm with a field of view of 0.038×0.022 degrees, shows two narrow, straight radio-emitting jets of particles that protrude in opposite directions from a giant elliptical galaxy at the center. The redshift of the optically visible elliptical indicates a distance of about 780 million light-years, and a linear extent for the radio galaxy of about 1 million light-years from end to end. The radio jets probably were ejected along the rotation axis of a super-massive black hole located within a central elliptical galaxy. It had to be active for millions of years to produce the two radio lobes. (Courtesy of NRAO/AUI/NSF.)

Powerful cosmic radio sources provide additional evidence for intense activity in the central regions of young galaxies. The bright *radio galaxy* named Cygnus A – also numbered 3C 405 in the third Cambridge catalogue of radio sources – for example, is associated with a giant elliptical galaxy that emits the strong "forbidden" emission lines found in the nuclei of Seyfert galaxies, with comparable widths corresponding to velocities of a least 1,000 km s^{-1}. Moreover, Cygnus A is emitting as much power at radio wavelengths as the optical visible luminosity of 1 million million, or 10^{12}, stars like the Sun, which has relatively dim radio emission.

When the optical image of the elliptical galaxy associated with Cygnus A is combined with maps of the radio signals, it is found that the radio emission is not confined to its optically visible counterpart. Instead, it is concentrated in two radio lobes that are separated from the central visible galaxy by hundreds of thousands of light-years. It is as if the radio-emitting clouds were expelled from the central elliptical galaxy, which is detectable only at optically visible wavelengths. The astonishing radio power is attributed to the nonthermal synchrotron radiation of high-speed electrons supplied from the visible center along two oppositely directed jets that feed the radio lobes (Fig. 13.7). These dual jets remain extraordinarily straight and surprisingly stable, energizing the radio lobes and pushing them farther and farther apart. If the radio galaxy has been sending out radio power at the present rate at an estimated million-year lifetime, then it has emitted radio energy equivalent to the complete annihilation of about 100,000 stars.

Even more dramatic sources of energy were found still deeper in space and generated a longer time ago in the quasars. The discovery of the first quasar, numbered 3C 273, resulted from the accurate location of a bright radio source, which was determined when the Moon happened to pass in front of it. The observed occultation indicated that 3C 273 is a double radio source, one component of which apparently coincided with a blue object whose spectral lines indicated an exceptionally high recession velocity of 0.16 percent of the speed of light. When this velocity is used to infer a distance using the Hubble law, it is found that 3C 273 is located at a distance of billions of light-years and that it is shining with the visible blue light of 10 million million, or 10^{13}, Sun-like stars. Because the bright optically visible object appeared star-like in color and of small angular size, its radio counterpart became known as a *quasistellar radio source*, a term that soon was shortened to *quasars*.

Astronomers gradually came to realize that quasars are brilliant, tiny cores, sometimes smaller than the solar system and embedded in much larger, extremely active galaxies, whose outer parts are difficult to detect in the intense quasar glare. Looking at them is like driving directly into the bright headlights of an oncoming vehicle, blinding us from seeing anything else. From its vantage point in space, the Hubble Space Telescope resolved the core quasar light and removed it from the computerized images to detect the faint, fuzzy halo of a host galaxy that is as large as the elliptical galaxies found at the centers of many intense radio sources.

Quasars are believed to be very luminous versions of the same blue nuclei that Seyfert observed in the center of nearby spiral galaxies. The visible-light emission of quasars exhibits the same emission lines as both the Seyfert galaxies and the central elliptical galaxies of radio galaxies. They all belong to a common class, known collectively as *active galactic nuclei*.

Super-massive black holes can account for the prodigious energy output, violent activity, and rapid variations of active galactic nuclei, as well as jets of material that move out of them with extremely high, relativistic velocities approaching the speed of light.

Super-Massive Black Holes

The tremendous luminosity of radio galaxies and quasars most likely is supplied by a *super-massive black hole*, which emits luminous radiation as its powerful gravity pulls in surrounding stars and gas. The gravitational pull of a mass equivalent to 100 million Suns is needed to balance the visible quasar luminosity; otherwise, its radiation pressure would blow away the quasar. Such a super-massive black hole would be sufficiently small and powerful enough to explain the tiny sizes and the colossal brightness of quasars. The super-massive black hole's rotational energy is used to accelerate charged particles and spew them out in diametrically opposite directions along its rotation axis at about the speed of light, continuously feeding the two radio lobes commonly found symmetrically placed from the center of radio galaxies and quasars.

The classic example is M 87, a giant elliptical galaxy whose central spinning disk of hot gas and stars indicates that a super-massive black hole resides at its center. M 87 is close enough to measure the motions of stars, and their increasing velocities toward the center indicate that billions of solar masses must to be crammed within a very small, unseen central volume to keep the high-velocity stars from flying into space.

A one-sided jet of gas emerges from the center of M 87 and stretches out into one of the two lobes of the bright radio galaxy Virgo A (numbered 3C 274 in the Cambridge survey)

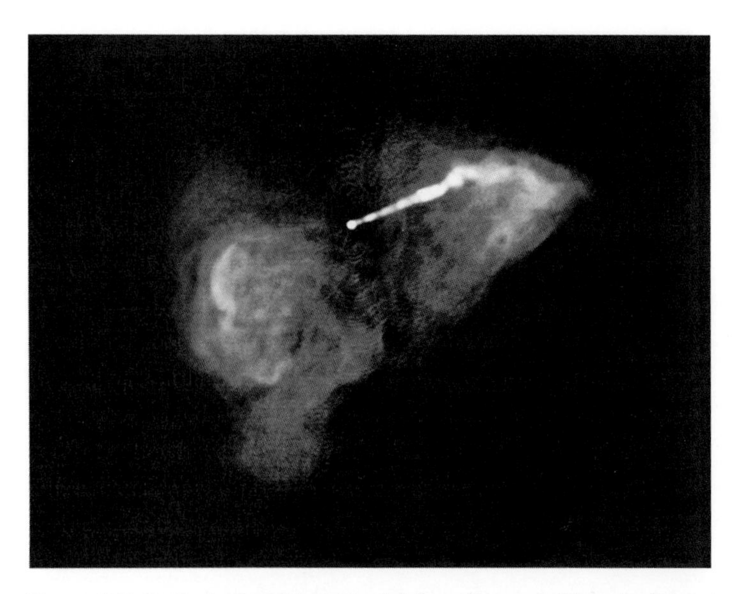

Figure 13.8. Radio jet from M 87. The bright radio source Virgo A, also designated 3C 274, coincides with M 87, a giant elliptical galaxy located at a distance of about 56.8 million light-years. M 87 is the largest and brightest galaxy within the Virgo cluster of galaxies. The core of M 87 contains a super-massive black hole of about 6.3 billion solar masses, which is 1,500 times more massive than the black hole at the center of our Milky Way Galaxy. This radio map, made with the Very Large Array, shows two elongated lobes, one on either side of the center of M 87, apparently fed by the super-massive black hole. The most intense radio emission comes from a jet that emerges from the core of the galaxy and extends about 5,000 light-years into one of the two lobes. The observed high-speed motion of bright knots in the jet implies that its radio-emitting electrons are traveling at nearly the speed of light. (Courtesy of NRAO/AUI/NSF.)

(Fig. 13.8). The motions of bright knots in the jet indicate that they are traveling outward at about half the speed of light. Interferometric observations with widely separated radio telescopes reveal that the M 87 jet emerges from a region at most 6 light-years across, most likely harboring the super-massive black hole that produces the jet.

Monstrous, super-massive black holes seem to inhabit the centers of all galaxies (Figs. 13.9 and 13.10). They are massive, scaled-up versions of stellar black holes, with millions if not billions of times the mass of the Sun packed into a region only a few light-years across. Like stellar black holes, the super-massive black holes cannot be observed directly. Their presence is inferred from the orbital motion of nearby visible stars, the trajectories of which are guided by the otherwise invisible black holes.

The faster the stars are moving, the more gravity – and therefore mass – is needed to hold the stars in their orbits. By measuring the sharp rise in orbital velocity at close distances from galaxy centers, astronomers have weighed unseen super-massive black holes in nearby giant elliptical galaxies, which are the brightest galaxies in clusters of galaxies. The central black-hole powerhouse in relatively nearby galaxies, designated M 87, NGC 3842, and NGC 4889, tips the scales at 6.3 billion, 9.7 billion, and 21 billion solar masses respectively. Without a gravitational pull equivalent to about 10 billion Sun-like stars, the close, fast-moving stars would fly away from the galaxies. Such central, super-massive black holes most likely reside in more distant galaxies that are too far away to resolve central stars and measure their motions.

Figure 13.9. Black hole jets from NGC 5128 and Centaurus A. Composite image of the optically visible galaxy NGC 5128 and the bright radio galaxy Centaurus A. X-ray jets and radio lobes emanate from the active galaxy's central, super-massive black hole. Microwave observations (*orange*), at a wavelength of 0.0087 m with the APEX array, show the radio lobes; the jets are seen in an x-ray image (*blue*) from the Chandra X-Ray Observatory, and the visible light image is from a 2.2-m (87-inch) telescope of the European Southern Observatory. The central black hole, thought to have a mass of about 55 million solar masses, apparently ejects infalling matter in opposing particle jets at about half the speed of light. The jets inflate the two radio-emitting lobes of Centaurus A, one of the biggest and brightest objects in the radio sky and nearly 20 times the angular extent of the full Moon. The galaxy NGC 5128 has unusual dust lanes and is about 12 million light-years from the Earth. The radio lobes are about 1 million light-years in extent. (Courtesy of ESO/WFI [*visible light*], MPIfR/ESO/APEX/A. Weiss et al. [*microwave*], NASA/CXC/CFA/R. Kraft et al. [*x-ray*].)

Because the super-massive black holes are many billions of times more massive than any dead star or any stellar black hole, we wonder how they grew so big. Their rapid increase in weight probably occurred in the galaxy's youth, in the early stages of the observable universe when embryonic black holes fed on smaller ones or gorged themselves on plentiful surrounding gas.

Quasars and active galactic nuclei become increasingly numerous as we look deeper into space. The number density of quasars peaks at about 10 billion years ago, shortly after the first galaxies were born. The spurt of activity apparently became worn out and used up as the galaxies grew older.

To power the youthful activity of a quasar, there must be about 1 solar mass per year of gas flowing into the super-massive black hole. Therefore, billions of stars or the equivalent amount of gas must be consumed as its active nucleus evolves during the course of billions

Figure 13.10. Supermassive black hole. This barred spiral galaxy, designated NGC 1097, is located about 44 million light-years from the Earth. It is a Seyfert galaxy and a moderate example of an active galactic nucleus (AGN) with jets shooting out of its core. A super-massive black hole, with about 100 million times the mass of the Sun, is located at the center of the galaxy. As shown in this infrared view of NGC 1097, taken from the Spitzer Space Telescope, a star-forming ring surrounds its center. As gas and dust spiral into the central black hole, they cause the ring to light up with hundreds of new stars. The galaxy's spiral arms and the swirling spokes between them show dust heated by other newborn stars as well as older stars. (Courtesy of NASA/JPL-Caltech.) (See color plate.)

of years. The supply dwindles away over time and the activity dies down, but the black hole does not disappear.

Most galaxies probably contain super-massive black holes at their center. Those in the older, nearby galaxies are the starving remains of former quasars, with a dwindling supply of material that once fed a higher rate of activity. They are found in ordinary nearby galaxies, such as Andromeda, whose cores are surviving fossils of former quasars. Our Galaxy, the Milky Way, is almost as old as the expanding universe, and it contains a central super-massive black hole. However, its mass is equivalent to only about 1 million stars like the Sun, rather than the billions in some super-massive black holes.

13.5 How Did It All Begin?

What We Know and Do Not Know

Despite recent advances in astronomy and the other sciences, there are fundamental questions that remain unanswered. How did the universe come into being? Where did everything come from? Why is the universe the way it is? How did life begin? Why are we here and what is the point of living? No one knows, with certainty, the answer to any of these questions.

Nevertheless, we can begin an inquiry by taking an inventory of what we do know. That understanding depends on our education and experience, which establish an individual perception. This singular point of view is always incomplete, only a small part of the totality of things. In other words, no one sees it all and no one is ever completely aware. Astronomers, for example, observe, measure, and quantify the known constituents of the universe. They often look at planets, stars, or galaxies in new ways, using novel technology and telescopes, and this can lead to the discovery of unanticipated and previously unknown ingredients of the cosmos. Like every other way of perceiving the world, the astronomer's view is always limited, only a small part of a much vaster, concealed universe that remains to be found. After all, most of the universe consists of dark matter and dark energy, of which we know almost nothing. There are other ways of seeing, including art, music, and religion; they all make connections with different states of awareness and help us interpret and understand the world in different ways.

This does not mean that science is not important. Science provides unique insights to the wonders of the universe, showing that nature has regularities that can be described by laws – even though we do not have the foggiest notion of where those laws came from or why they exist.

Equations Can Be Useful but Limited

Physicists use mathematical equations to describe underlying patterns in the observed universe. The equations can be used to extrapolate beyond the observed regularities into the future, making predictions of what may happen next. When these predictions are tested successfully, the equations may become laws, like the conservation of energy or momentum, which are valid within a restricted range of conditions.

These laws are extremely useful for they describe how inanimate objects will behave, but they also provide only one, limited perspective. The equations may tell us how the stars will move in the future, subject to the gravity of a nearby mass, but this does not mean that we know why stars move. The laws must be extended and revised in proximity to very massive objects, such as black holes.

Unfortunately, some theoretical cosmologists and physicists concluded that equations rule and that they can explain all that is. The equations sometimes are disconnected from the observable world, and testable predictions are no longer used. Or, the theories contain so many variable parameters that adjusting them or adding a new one can explain any contradictory observation. Moreover, the theories are created from complex mathematics that almost no one can understand; and the experts sometimes lack humility. One prominent MIT theoretician even proposed that his equations demonstrated that there is no civilization in the visible universe more advanced than ours; he then changed his speculation by stating that the parameters he used may not be correct.[28]

Some modern cosmologists suppose that tiny, invisible, vibrating *strings* occupy unseen places and extra dimensions, springing out of nothing and accounting for everything. Others speculate that our big bang may not have been the only one; our universe may be just one small part of a multitude of possible universes, dubbed *multiverses*, produced in eternal inflations – not one world without end but rather an infinite eternity of worlds that might have bubbled up out of vacuous nothing. They say that nothingness has zero total energy, with negative gravitational energy balancing all the other sorts – therefore, anything is possible, and no one really knows how to describe a nothing with no energy, no matter, no space, and no time.

There simply are no observational tests of the hypothetical strings or disconnected multiverses that are permanently invisible. Persistent appeals to the unobservable and unverifiable essentially are unscientific.

Equations Do Not Explain the Beginning

What happened before the big bang that propelled the expanding universe into existence? No one knows! The theories do not work for they fall apart at the first crucial instant, at the beginning of the big bang. The equations cannot be extended to anything that occurred before, and there is no observational evidence for prior events.

Mathematicians use the term *singularity* to describe the conditions when an equation blows up, as in dividing a nonzero number by zero. Cosmologists say that a singularity is an isolated point where conditions are undefined. It is certainly singular, happening only in exceptional, highly compressed situations – such as the big bang or at the center of black holes – and it appears in a particular set of equations, Einstein's *General Theory of Relativity*.

These equations cannot be extended to anything that occurred before the big-bang singularity, so we have just pushed the mystery of the ultimate origin of the universe back about 13.7 billion years, to a point that science cannot penetrate.

An inflation theory describes what may have happened in the first fraction of a second of the big bang, when a repulsive gravity – unlike the attracting type we are used to – blew up the universe, enlarging it by an enormous factor. Due to its inherent instability, the burst of inflation soon decayed and came to an end, in a time far less than 1 second. It released its remaining energy into material particles, creating the heat of the big bang, the primeval fireball.

This accelerated expansion in the first miniscule moments of the big bang – this inflation – supposedly obliterated evidence of previous space, time, energy, and matter, erasing previous history. That cosmic forgetfulness closes the door to the very beginning, conveniently avoiding the question of ultimate origins – the original genesis – and removing it from any observational consequences.

In other words, according to this theory, the big bang or its immediate consequences destroyed all evidence of what came before. It may resemble the shock of human birth, which seems to obliterate any memory of previous events; or the big bang may have initiated time, on a day without a yesterday. So, there is no before – or perhaps the explanation lies outside space and time or another way of understanding things.

We still do not know how the universe came into being and equations cannot explain it. For this fundamental question, the equations are no more than primitive markings that describe a paper world, detached from the observable universe, the beginning of which remains a captivating mystery.

What Is The Point Of It All?

As Albert Einstein proclaimed, "What I see in nature is a grand design that we can understand imperfectly, one with which a responsible person must look at with humility. This is a genuine religious feeling and has nothing to do with mysticism."[29] He also knew that "all our science, measured against reality, is primitive and childlike"[30] and that there are underlying patterns in the universe that exist independent of humans. For Einstein, we stand before Nature, awaiting discovery and understanding of a great hidden mystery. To him, anyone who cannot experience that mystery, who does not know it and can no longer wonder, no longer feel amazement, "is as good as dead, a snuffed-out candle."[31]

The sum total of all things, the complete picture show, lies shrouded behind a cosmic veil; curiosity, joy, and wonder help to penetrate it. They can sweep us into the boundless celestial realms, keeping the world forever young, which describes what the Chilean poet Pablo Neruda felt when he wrote his first line of poetry:

> And I, tiny being,
> drunk with the great starry
> void,
> likeness, image of
> mystery,
> felt myself a pure part of the abyss.
> I wheeled with the stars,
> my heart broke lose on the wind.[32]

This brings us to the opposite end of the pendulum swing and the ultimate fate of the observable universe.

13.6 When Stars Cease to Shine

An Accelerating Expansion

According to the Hubble law, the velocity of a galaxy steadily increases with its distance; however, this relationship becomes nonlinear at exceptionally large distances and in an opposite way from that expected. Astronomers used observations of exploding stars known as Type Ia supernovae to measure how fast the expansion of the universe was slowing down, due to the gravitational pull of its combined matter. However, light from the distant supernovae was fainter than predicted (Figs. 13.11 and 13.12), which meant that the galaxies are speeding up, expanding at a quickening pace and accelerating instead of slowing down. In other words, the distant galaxies were not where they were supposed to be, and the space they are in seems to be expanding at a faster rate as time goes on.

The 2011 Nobel Prize in Physics was awarded to the American astronomer Saul Perlmutter, the Australian Brian P. Schmidt, and another American, Adam G. Riess, for the discovery of this accelerating expansion of the universe through observations of distant supernovae.

The distant galaxies are being accelerated by the antigravity push of a mysterious *dark energy* that pushes matter apart. It is not the same as dark matter, which encourages attraction. The fate of the universe is no longer supposed to depend on its mass but rather on

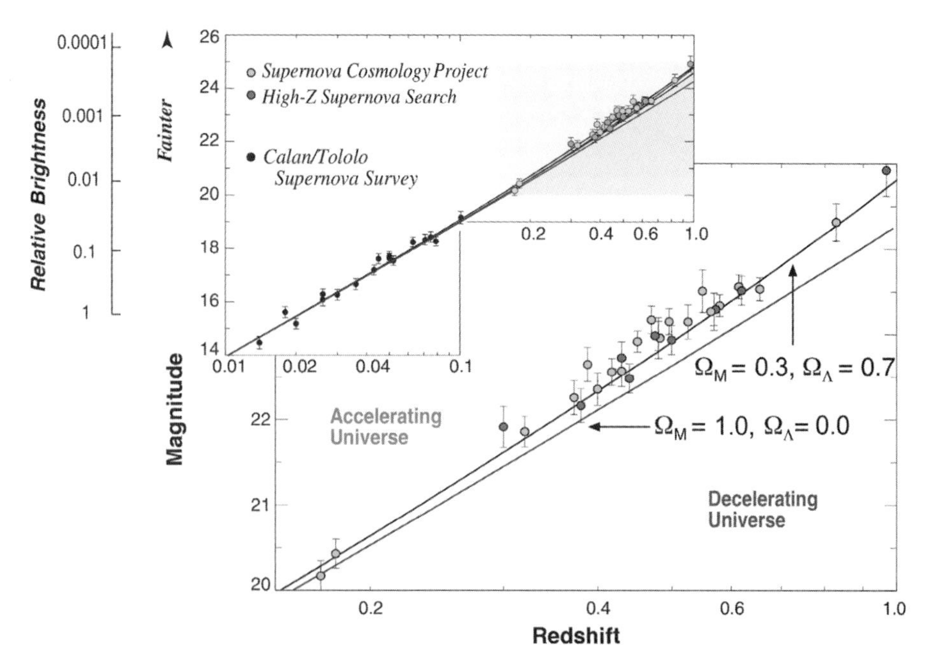

Figure 13.11. The accelerating expansion. The Hubble diagram plot of the apparent magnitude of Type Ia supernovae plotted as a function of their redshift. At a redshift below about $z = 0.1$, there is a linear fit to the data, but at larger redshifts, the observations begin to diverge from a straight line. The curved departures for distant supernovae at high redshift indicate an acceleration in which the speed of expansion is increasing. The observed data can be compared to cosmological models with different values of the omega parameter, Ω. It is the ratio of the inferred density to the critical mass density needed to stop the expansion of the universe in the future. The subscript Λ denotes the cosmological constant, a possible form of dark energy, and the subscript m denotes matter. (Adapted from Saul Perlmutter, *Physics Today,* April 2003.)

its energy. If dark energy retains its vigor, the universe is doomed to expand forever, perhaps keeping the universe at the brink of closure within "flat" space, which is described by Euclidean geometry.

The trouble is, no one understands the "mysterious something," this dark energy, which permeates space and eventually overwhelms the gravitational self-attraction of the entire material universe. However, an old idea, termed the *cosmological constant*, has been revived to give dark energy another name and couch it in mathematical terms. This is the antigravity "fudge factor" that Einstein introduced to stabilize a nonmoving universe against collapse.

Cosmic Destiny

Ever since the discovery of the expansion of the universe, we have known that it is slowly and inexorably approaching an end. There has never been any known force that can prevent it from steadily moving into darkness. As Georges Lemaître so eloquently stated, "The evolution of the world can be compared to a display of fireworks that has just ended: some few red wisps, ashes and smoke. Standing on a well-chilled cinder, we see the slow fading of the suns, and we try to recall the vanished brilliance of the origin of the worlds."[33]

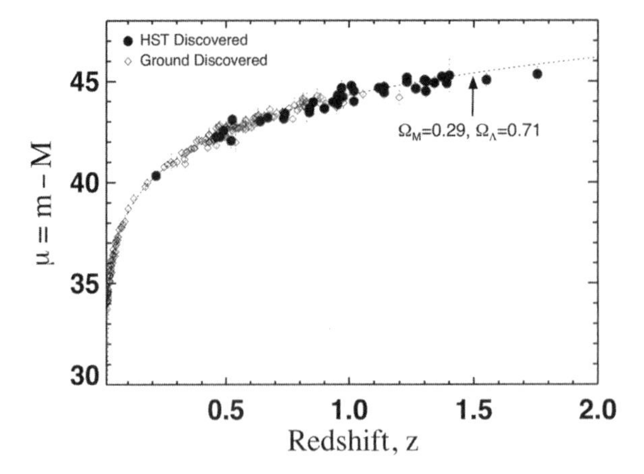

Figure 13.12. Hubble diagram of Type Ia supernovae. The distance modulus $\mu = m - M$, or the difference between apparent magnitude m and absolute magnitude M, for Type Ia supernovae is plotted against their redshift, z. The omega parameter, Ω, is the ratio of the inferred density to the critical mass density needed to stop the expansion of the universe in the future. A comparison between the observations and models suggests a mass density of at most $\Omega_m = 0.3$, significant dark energy with $\Omega_\Lambda = 0.7$, and an inflationary universe without spatial curvature with $\Omega_m + \Omega_\Lambda = 1.0$. (Courtesy of Adam G. Riess.)

This should not be surprising; everything has an end. Each of us will eventually perish, and every star that we see at night inevitably will expend its nuclear energy and vanish into the darkness. Sooner or later, everything in nature must come to an end. We can think ahead to a distant time when everything we now see in the universe either will no longer exist or will be altered beyond recognition.

Dark energy has been found but that discovery may not help matters. It gives an extra outward push to the expansion and further reduces the power of mass to stop the expanding universe in the future. If the acceleration caused by dark energy continues unabated at the current rate, all of the galaxies will be moving apart so quickly that they cannot communicate with one another in about 150 billion years, disappearing over the cosmic horizon. Eventually, in about 100 trillion years, all of the interstellar gas and dust from which new stars condense finally will be used up, and new stars will cease to be born in any galaxy; all of the stars eventually will wink out. The poet T. S. Eliot foretold of this type of cosmic destiny with the following verse:

> This is the way the world ends
> This is the way the world ends
> This is the way the world ends
> Not with a bang but a whimper.[34]

Time may be running out for the entire observable universe; the apocalypse may be imminent. However, there are other equally plausible worldviews. There are, for example, several cycles of world birth and death in Hinduism and in Buddhism. For Buddhists, the world will come and disappear and then come and disappear. Eventually, the world we live in will become desert and even the oceans will dry up; but then another new world is reborn – it is an endless cycle.

If dark energy weakens as time goes on, expending its strength, then gravity and mass may take over and eventually pull back on the outward moving galaxies, ultimately reversing the expansion and dragging the universe back, melting it down and remaking the big bang.

The possible forecasts now belong to the mysterious unknown, which operates in the dark, beyond the range of direct perception. All possibilities are open and any destiny is possible for the universe. The poet Robert Frost put the choices in human terms with the following:

> Some say the world will end in fire,
> Some say in ice.
> From what I've tasted of desire
> I would hold with those who favor fire.[35]

Quotation References

[1] Harrison, George (1943–2001): "Here Comes the Sun," from the 1969 Beatles album, *Abbey Road*. Words and music by George Harrison © 1969 Harrisongs Ltd.

[2] Francis William Bourdillon (1852–1921): "Among the Flowers" (1878). In *The Oxford Dictionary of Quotations*, Fourth Edition (Angela Partington, ed.), Oxford University Press, New York, 1992, p. 138. Also in John Bartlett, *Familiar Quotations*, Sixteenth Edition (Justin Kaplan, ed.), Little, Brown and Company, Boston, 1992, p. 563.

[3] Johannes Kepler (1571–1630): In the prayer near the end of his *Harmonice Mundi*, translated by Owen Gingerich in "Kepler's Anguish and Hawking's Query," *The Great Ideas Today 1992, Encyclopedia Britannica*, Chicago, 1992. The prayer begins with "I give Thee thanks, O Lord Creator, because I have delighted in thy handiwork." Kepler was a deeply religious thinker, convinced that his Christian God had created a rational, ordered universe, whose ultimate secrets could be discovered. He hoped to show that "the heavenly machine is a kind of clockwork, insofar as nearly all the manifold motions are caused by a most simple, magnetic, and material force . . . given numerical and geometrical expression," as written in a letter dated February 10, 1605 to Herwart von Hohenberg, the Bavarian Chancellor and Kepler's patron, reproduced in Max Casper and Walther von Dyck (eds.), *Kepler: Gesammelte Werke, Volume XV*, C. H. Beck, Munich, 1938, p. 145; and translated by Arthur Koestler, *The Sleepwalkers: A History of Man's Changing Vision of the Universe*, Arkana, Penguin Books, New York, 1959, p. 345.

[4] Letter written by Robert William Bunsen (1811–1899) to the English chemist Henry Enfield Roscoe (1833–1915) in November 1859: Quoted by Roscoe in *The Life and Experiences of Sir Henry Enfield Roscoe*, London, 1906, p. 71. It is reproduced by A. J. Meadows in his article, "The Origins of Astrophysics," *The General History of Astronomy, Vol. 4, Astrophysics and Twentieth-Century Astronomy to 1950, Part A* (Owen Gingerich, ed.), Cambridge University Press, New York, 1984, p. 5. Also see Gustav Robert Kirchhoff (1824–1877): "On the Chemical Analysis of the Solar Atmosphere," *Philosophical Magazine and Journal of Science* 21, 185–188 (1861); reproduced by A. J. Meadows in *Early Solar Physics*, Pergamon Press, Oxford, 1970, pp. 103–106; Gustav Kirchhoff and Robert Bunsen, "Chemical Analysis of Spectrum Observations," *Philosophical Magazine and Journal of Science* 20, 89–109 (1860); 22, 329–349, 498–510 (1861).

[5] Ronald W. Gurney (1898–1953) and Edward U. Condon (1902–1974): "Wave Mechanics and Radioactive Disintegration," *Nature* 122, 493 (1928).

[6] Wolfgang Pauli (1900–1958): Remarks at the Seventh Solvay Conference, October 1933. Reproduced in the original French in *Collected Scientific Papers of Wolfgang Pauli, Vol. 2* (Ralph Kronig and Victor F. Weisskopf, eds.), Wiley Interscience, New York, 1964, p. 1319. Quoted in English by Christine Sutton in *Spaceship Neutrino*, Cambridge University Press, New York, 1992, p. 19.

[7] Frederick Reines (1918–1998) and Clyde L. Cowan (1919–1974): Telegram to Pauli dated 14 June 1956. Quoted in "Proceedings of the International Colloquium on the History of Particle Physics," *Journal de*

Physique 43, Supplement C8, 237 (1982). Also quoted by Christine Sutton in *Spaceship Neutrino*, Cambridge University Press, New York, 1992, p. 44.

[8] *Bhagavad-Gita, Book 11, Sections 12, 32*, English translation by Winthrop Sargeant, State University of New York Press, Albany, New York, 1984, pp. 464 and 484.

[9] Jules Robert Oppenheimer (1904–1967): "Physics in the Contemporary World: Enhancement of Science, with Knowledge Imparted for Man's Benefit," Second Arthur D. Little Lecture, *Technology Review*, February 1948, p. 203, reproduced in *Time Magazine*, February 23, 1948, p. 94.

[10] William Thomson (1824–1907): "On the Age of the Sun's Heat," *Macmillan's Magazine*, March 5, 288–293 (1862). *Popular Lectures* I, 349–368. William Thomson is better known today as Lord Kelvin. Also see J. D. Burchfield, *Lord Kelvin and the Age of the Earth*, Science History Publications, New York, 1975.

[11] Arthur Stanley Eddington (1882–1944): "The Internal Constitution of the Stars," *Nature* 106, 14–20 (1920); *Observatory* 43, 341–358 (1920). Reproduced by Kenneth R. Lang and Owen Gingerich in *A Source Book in Astronomy and Astrophysics, 1900–1975*. Harvard University Press, Cambridge, MA, 1979, pp. 281–290.

[12] Arthur Stanley Eddington (1882–1944): ibid., Reference 11.

[13] Hans A. Bethe (1906–2005): "My Life in Astrophysics," *Annual Review of Astronomy and Astrophysics* 41, 6–7 (2003).

[14] John Updike (1932–2009): Cosmic Gall, *The New Yorker*, 17 December 1960, p. 36. Reproduced by Updike in *Telephone Poles and Other Poems*, Alfred A. Knopf, New York, 1979, p. 5, and in *Collected Poems 1953–1993*, Alfred A. Knopf, New York, 1993, p. 315.

[15] Dante Alighieri (1265–1321): *La Divina Commedia, Cantica I, L'Inferno, Canto XXXIV, line* 129. Dorothy L. Sayers (translator). *The Divine Comedy, Volume I. Hell, Canto XXXIV, line* 139, Basic Books, New York, 1963, p. 289.

[16] "Star Light, Star Bright" is an English-language nursery rhyme, which first began to be recorded in the late nineteenth century.

[17] Dante Alighieri (1265–1381): *La Divina Commedia, Cantica III, Il Paradiso, Canto XXXIII, line* 145. Dorothy L. Sayers (translator), *The Divine Comedy, Volume III, Paradise, Canto XXXIII, line* 145, Basic Books, New York, 1963, p. 347.

[18] Arthur Stanley Eddington (1882–1944): "The Internal Constitution of the Stars," *Nature* 106, 14–20 (1920); *Observatory* 43, 341–358 (1920). Reproduced by Kenneth R. Lang and Owen Gingerich in *A Source Book in Astronomy and Astrophysics, 1900–1975*. Harvard University Press, Cambridge, MA, 1979, pp. 281–290.

[19] Percy Shelley (1792–1822): *Hellas, A Lyrical Drama*, Chorus lines 196–199. London, Charles and James Ollier, 1822.

[20] Henry Norris Russell (1877–1957) recalled this discovery of the anomalous spectrum of 40 Eridani B in a colloquium at Princeton University Observatory in 1954, A. G. Davis Philip and D. H. de Vorkin (eds.), "In Memory of Henry Norris Russell," *Dudley Observatory Report No. 13* (1977).

[21] Arthur Stanley Eddington (1882–1944): *Stars and Atoms*, Oxford Clarendon Press, 1927, p. 50.

[22] Dylan Thomas (1914–1953): *Do Not Go Gentle into That Good Night*, written for the Welsh poet's dying father and first published in the journal *Botteghe Oscure* in 1951.

[23] Walter Baade (1893–1960) and Fritz Zwicky (1898–1974): "Cosmic Rays from Super-Novae," *Proceedings of the National Academy of Sciences* 20, No. 5, 259–263 (1934).

[24] John Michell (1724–1793): "On the means of discovering the distance, magnitude, etc., of the fixed stars, in consequence of the diminution of their light, in case such a diminution should be found to take place in any of them, and such other data should be procured from observations, as would be further necessary for that purpose," *Philosophical Transactions of the Royal Society (London)* 74, 35 (1784).

[25] Camille Flammarion (1842–1925): *Les Etoiles*, Paris, 1882, p. 181. English quotation by Charles A. Whitney, "The Skies of Vincent Van Gogh," *Art History* 9, 358 (1986).

[26] George Ellery Hale (1868–1938): *Harper's Magazine* 156, 639–646 (1928).

[27] Eihei Dogen (1200–1253): *Death Poem.*

[28] Alan H. Guth (1947–): "Eternal Inflation and Its Implications," *Journal of Physics* A40, 6811–6826 (2007).

[29] Albert Einstein (1879–1955). Quoted in Helen Dukas and Baresh Hoffmann, *Albert Einstein, the Human Side*, Princeton University Press, Princeton, NJ, 1979, p. 132. Quotation reproduced in *The Expanded Quotable Einstein*, collected and edited by Alice Calaprice, Princeton University Press, Princeton, NJ, 2000, p. 298.

[30] Albert Einstein (1879–1955): Quoted in Banesh Hoffmann: *Albert Einstein: Creator and Rebel*, New York, Viking, 1972. Also see Alice Calaprice (ed.), *The Expanded Quotable Einstein*, Princeton University Press, Princeton, NJ, 2000, page 261.

[31] Albert Einstein (1879–1955): "What I Believe," *Forum and Century* 84, 193–194 (1930). Also see Alice Calaprice (ed.), *The Expanded Quotable Einstein*, Princeton University Press, Princeton, NJ, 2000, p. 295.

[32] Pablo Neruda (1904–1973): *Poetry*, translated by Alastair Reed (1926–).

[33] Georges Lemaître (1894–1966): "L'expansion de l'espace," *La Revue des Questions Scientifiques, 4e Série*. November 1931. English translation in *The Primeval Atom: An Essay on Cosmogony*, by Georges Lemaître, D. Van Nostrand Co., New York, 1950, p. 78.

[34] T. S. Eliot (1881–965): *The Hollow Men.*

[35] Robert Frost (1874–1963): *Fire and Ice.*

Author Index

Subject Index